DIGITAL IMAGE
PROCESSING

DIGITAL IMAGE

PROCESSING

KENNETH R. CASTLEMAN

California Institute of Technology
Jet Propulsion Laboratory

PRENTICE-HALL, INC., Englewood Cliffs, New Jersey 07632

Library of Congress Cataloging in Publication Data

CASTLEMAN, KENNETH R
  Digital image processing.

  (Signal processing series)
  Includes bibliographical references and index.
  1. Image processing. I. Title. II. Series.
TA1632.C37   621.3815′42   78-27578
ISBN 0-13-212365-7

Editorial/production supervision and interior design by Virginia Rubens
Cover design by Mark Binn
Manufacturing buyer: Gordon Osbourne

Printed in the United States of America

10   9   8   7   6   5   4

PRENTICE-HALL INTERNATIONAL, INC., *London*
PRENTICE-HALL OF AUSTRALIA PTY. LIMITED, *Sydney*
PRENTICE-HALL OF CANADA, LTD., *Toronto*
PRENTICE-HALL OF INDIA PRIVATE LIMITED, *New Delhi*
PRENTICE-HALL OF JAPAN, INC., *Tokyo*
PRENTICE-HALL OF SOUTHEAST ASIA PTE. LTD., *Singapore*
WHITEHALL BOOKS LIMITED, *Wellington, New Zealand*

To
"Uncle Bob" Nathan, pioneering pixel pusher
and
"Cruel Jewel" Haynes, professor of poetry and prose
because they started it

CONTENTS

Preface   xiii

Acknowledgements   xv

PART I   BASIC TECHNIQUES

1   Images and Digital Processing   3

Introduction   *3*
The Elements of Digital Image Processing   *4*
Some Philosophical Considerations   *9*
Digital Image Processing in Practice   *11*

2   Digitizing Images   14

Introduction   *14*
Characteristics of an Image Digitizer   *15*
Types of Image Digitizers   *16*
Image Digitizing Components   *17*
Electronic Image Tubes   *26*
Other Digitizing Systems   *30*
Film Scanning   *31*
Bibliography   *38*

3   Digital Image Display   39

Introduction   *39*
Display Characteristics   *40*
Display Technologies   *49*
Bibliography   *51*

4   Image Processing Software   52

Introduction   *52*
Software Organization   *53*
The Processing Sequence   *57*
References   *67*

5   The Gray Level Histogram   68

Introduction   *68*
Uses of the Histogram   *73*
Relationship Between Histogram and Image   *78*
Summary of Important Points   *83*
References   *83*

6   Point Operations   84

Introduction   *84*
Uses for Point Operations   *85*
Linear Point Operations   *86*
Point Operations and the Histogram   *86*
Applications of Point Operations   *90*
Summary of Important Points   *94*

7 **Algebraic Operations 96**

Introduction *96*
Algebraic Operations and the Histogram *97*
Applications of Algebraic Operations *101*
Summary of Important Points *108*

8 **Geometric Operations 110**

Introduction *110*
Gray Level Interpolation *112*
The Spatial Transformation *115*
Applications of Geometric Operations *119*
Summary of Important Points *134*
References *135*

PART II   LINEAR FILTERING

9 **Linear System Theory 139**

Introduction *139*
Harmonic Signals and Complex Signal Analysis *141*
The Convolution Operation *145*
Applications of Digital Filtering *150*
Some Useful Functions *150*
Convolution Filtering *156*
Conclusion *159*
Summary of Important Points *159*
References *160*

10 **The Fourier Transform 161**

Introduction *161*
Properties of the Fourier Transform *166*
Linear Systems and the Fourier Transform *173*
The Fourier Transform in Two Dimensions *180*
Correlation and the Power Spectrum *186*
Summary of Fourier Transform Properties *187*
Summary of Important Points *187*
References *189*

11 **Filter Design 190**

Introduction *190*
Examples of Common Filters *190*
Optimal Filter Design *199*
Summary of Important Points *224*
References *225*

12 **Processing Sampled Data 226**

Introduction *226*
Sampling *227*
Computing Spectra *234*
Truncation *238*
The Effects of Digital Processing *240*
Digital Filtering *246*
Summary of Important Points *248*
References *249*

13 **Optics and System Analysis 250**

Introduction *250*
Optics and Imaging Systems *251*
Diffraction-limited Optical Systems *254*
Aberrations in an Imaging System *262*
The Analysis of Complete Systems *264*
Summary of Important Points *273*
References *274*

PART III APPLICATIONS

14 **Image Restoration 277**

Introduction *277*
Approaches and Models *278*
Superresolution *288*
System Identification *290*
OTF from the Degraded Image Spectrum *292*
Noise Modeling *293*
Summary of Important Points *295*
References *295*

15   **Image Segmentation   299**

Introduction   *299*
Image Segmentation by Thresholding   *303*
Optimal Threshold Selection   *305*
Gradient Based Methods   *311*
Region Growing Techniques   *313*
Segmented Image Structure   *314*
Summary of Important Points   *319*
References   *319*

16   **Measurement and Classification   321**

Introduction   *321*
Size Measurements   *323*
Shape Measurements   *324*
Feature Selection   *332*
Classification   *334*
Summary of Important Points   *344*
References   *345*

17   **Three-Dimensional Image Processing   347**

Introduction   *347*
Multispectral Analysis   *350*
Optical Sectioning   *351*
Computerized Axial Tomography   *360*
Stereometric Ranging   *364*
Stereoscopic Image Display   *368*
Shaded Surface Display   *371*
Summary of Important Points   *377*
References   *379*

APPENDICES

I   **A History of Digital Image Processing at JPL   383**

II   **VICAR Program Index   401**

III   **Fourier Transforms   412**

IV   **Function Tables   418**

Index   421

PREFACE

This book is intended as a text for a graduate level course in computer science or engineering. It can also serve as a reference book for professionals using digital image processing in their work. It is designed to acquaint the reader with the basic techniques of image processing by computer, the mathematical rationale behind each technique, and its applications and limitations.

The book should be useful to students in a variety of fields who find digital image processing applicable to their research. This includes not only work in inherently pictorial data but also other applications making use of two-dimensional digitized data. Professionals in related fields can use this book to add digital image processing to their battery of problem-solving skills.

In this book we treat digital image processing as a tool for solving practical problems rather than as a discipline unto itself. The techniques are presented in a straightforward manner with emphasis on applications, limitations, and performance. Computational optimization techniques, such as data compression, image coding, and software minimization, are, by and large, left to other authors. The aim of this book is to prepare the reader to apply digital image processing techniques intelligently to the solution of his problems, avoiding the pitfalls that trap the inexperienced user. This book had its beginning in an Information Sciences course which the author taught at Caltech for three years. That course was designed to equip students to use digital image processing in their research work.

Since the readership comes from diverse scientific and engineering fields, this book includes brief coverage of several disciplines that may fall within the background of some readers. For example, relevant topics of Fourier optics, linear system theory, random processes, and signal detection theory are reviewed even though they are common to the background of some students. Other readers, however, will find the reviews helpful.

Part I covers the basic techniques of digital image processing, those which can be discussed without the aid of advanced mathematics. This represents a collection of relatively simple but nonetheless important concepts. In Part II we use linear system theory and the Fourier transform to explain the effects of linear filtering, sampling, and optical imaging. Finally, Part III gives examples from several areas of image processing applications. Each part of the book can be covered in one college quarter. The format of the presentation is (1) a definition of a class of techniques, (2) development of the relevant mathematical background, (3) interpretation of the mathematics with illustrative examples, and (4) demonstration of the usefulness of the technique with practical applications.

Since the book is a guide to practical applications of a tool rather than an esoteric study of a mathematical discipline, the level of mathematical rigor is reasonably relaxed. The underlying theory is interpreted in light of the practical constraints of available hardware systems.

Many of the illustrative examples in this book have been taken from work ongoing at the Jet Propulsion Laboratory (JPL). For that reason the unique philosophy which has developed around the image processing which supports the NASA unmanned planetary exploration missions, and a variety of other applications, is partially reflected herein. If certain sections of this book appear to imply that digital image processing was discovered, originated, or invented at JPL, such a claim is neither intended nor accurate. Numerous groups have made contributions to this field, some of which predate JPL digital image processing.

The multidisciplinary field of digital image processing has amassed a voluminous bibliography in its relatively short lifetime. We have made no attempt at a comprehensive review. Instead we list only selected references particularly useful to our discussion. The interested reader can make use of existing literature surveys in this field.

With few exceptions each of the techniques discussed herein is either conceptually straightforward, or firmly based upon well-established principles of mathematics and applied science and engineering, or both. This complicates attempts to trace a technique back to its origin. In view of this we make no attempt to sort out the exact parentage of each technique. We discharge our scholarly responsibilities with merely a caution against reading claims of discovery into the descriptive material presented herein.

Kenneth R. Castleman

|||||||||||||||||||||||||||||||||||||||||||||||||||||||||||||||||||||||||||||||||||||||||||||||||||||||||||||||||||||||||||||

ACKNOWLEDGMENTS

|||||||||||||||||||||||||||||||||||||||||||||||||||||||||||||||||||||||||||||||||||||||||||||||||||||||||||||||||||||||||||||

This book would not have been possible were it not for the efforts of the men and women, past and present, of the Image Processing Laboratory at JPL. Individually they developed many of the techniques described, contributed illustrative figures, and provided valuable comments on the manuscript. While their names and contributions are too numerous to mention here, their efforts are summarized in Appendix I. The author is particularly grateful for the close collaboration of Dr. Ray Wall, and for the valuable interaction of Dr. Michael Shantz, Dr. Henry Fuchs, Dr. Benjamin White, Jerre Pabst, Bruce Lane, and the author's students at Caltech.

Particularly noteworthy are the efforts of Cheryl Mills, who tirelessly transformed hours of tape recordings into elegantly executed manuscript, and kept track of endless details during the production of the book. With great patience Dorris Wallenbrock, JPL technical editor, brought correctness and consistency into the text. John Kempton and Robert Van Buren of the JPL Documentation Section also provided valuable production support. The author also wishes to thank Mr. Maurice Brundage of the Caltech Counsel's Office for his assistance.

Special thanks are due several individuals for special efforts to prepare illustrative examples of computer processed images from their own work. They are James Soha (Figure 6-8), Howard Frieden (Figure 8-2), Joel Mosher (Figure 8-15), Milan

Karspeck (Figure 15-11), R. J. Wall (Figure 16-5), and James F. Blinn (Figures 17-20 and 17-24). Sayuri Harami contributed to several more.

The author is particularly grateful to Robert Selzer and to Benn Martin, Dr. Douglas O'Handley, and Dr. William Spuck of the JPL Energy and Technology Applications Office for their role in obtaining contractual support for this undertaking.

The production of this manuscript was supported by the Technology Applications Division, Technology Utilization Office, NASA Headquarters as a part of an active program to make NASA-derived technology available to industry and the research community. The author is particularly grateful for the support of Jeff Hamilton, Bill Smith and Ray Whitten of NASA Headquarters.

DIGITAL IMAGE
PROCESSING

Part I

BASIC TECHNIQUES

IMAGES AND DIGITAL PROCESSING

IIIIIIIIIIIIIIIIIIIIIIIIIIIIIIIIIIIIIIIIIIIIIIIIIIIIIIIIIIIIIIIIIIIIIIIIIIIIIIIIIIIIIIIIIIIIIIIIIIIIIIIIIIIIIIIIIIIIIIIIIIIIIIIIIIIIIIIIIIIIIIIIII

INTRODUCTION

Digital image processing, the manipulation of images by computer, is a relatively recent development in terms of man's ancient fascination with visual stimuli. In its short history, it has been applied to practically every type of imagery, with varying degrees of success. The inherent subjective appeal of pictorial displays attracts perhaps a disproportionate amount of attention from the scientist and awe from the layman. Digital image processing, like other "glamour" fields, suffers from myths, misconceptions, misunderstandings, and misinformation. It is a vast umbrella under which fall diverse aspects of optics, electronics, mathematics, photography, and computer technology. It is a truly multidisciplinary endeavor plagued with imprecise jargon. This book attempts to collect the fundamental concepts into a self-consistent package for a relatively painless introduction to the field.

Several factors combine to indicate a lively future for digital image processing. A major factor is the declining cost of computer equipment. Both processing units and bulk storage devices are becoming less expensive year by year. A second factor is the increasing availability of equipment for image digitizing and display. There are indications that its cost will continue to decline. Several new technological trends promise to further promote digital image processing. These include parallel processing made practical by low-cost microprocessors, and the use of charge-coupled devices (CCDs) for digitizing, storage during processing and display, and large, low-cost image storage arrays.

Another impetus for development in this field stems from some exciting new applications on the horizon. Certain types of medical diagnosis, including differential blood cell counts and chromosome analysis, are nearing a state of practicality by digital techniques. The remote sensing programs are well suited for digital image processing techniques. Thus, with increasing availability of reasonably inexpensive hardware and some very important applications on the horizon, one can expect digital image processing to continue its phenomenal growth and to play an important role in the future.

THE ELEMENTS OF DIGITAL IMAGE PROCESSING

Digital image processing basically requires a computer upon which to process images. In addition, the system must have two pieces of special input/output equipment, an image digitizer and an image display device.

In the form in which they usually occur, images are not directly amenable to computer analysis. Since computers work with numerical rather than pictorial data, an image must be converted to numerical form before processing. This conversion process is called *digitization*, and a common form is illustrated in Figure 1-1. The image is divided into small regions called *picture elements*, or *pixels* for short. The most common subdivision scheme is the rectangular sampling grid shown in Figure 1-1. The image is divided into horizontal lines made up of adjacent pixels. At each pixel location, the image brightness is sampled and quantized. This step generates an integer at each pixel representing the brightness or darkness of the image at that point. When this has been done for all pixels, the image is represented by a rectangular array of integers. Each pixel has a location or address (line or row number and sample or

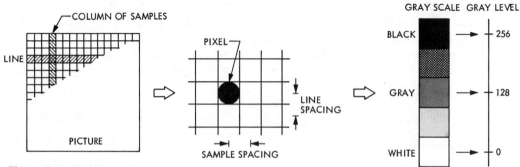

Figure 1-1 Digitizing an image

column number) and an integer value called the *gray level.* This array of digital data is now a candidate for computer processing.

Figure 1-2 shows a complete system for image processing. The digital image produced by the digitizer goes into temporary storage on a suitable device. In response to job control input, the computer calls up and executes image processing programs from a library. During execution, the input image is read into the computer line by line. Operating upon one or several lines, the computer generates the output image pixel by pixel. As the output image is created, it is written on the output data storage device line by line. During the processing, the pixels may be modified at the programmer's discretion in processing steps limited only by his imagination, patience, and computing budget. After processing, the final product is displayed by a process that is the reverse of digitization. The gray level of each pixel is used to determine the brightness (or darkness) of the corresponding point on a display screen. The processed image is thereby made visible and hence amenable to human interpretation.

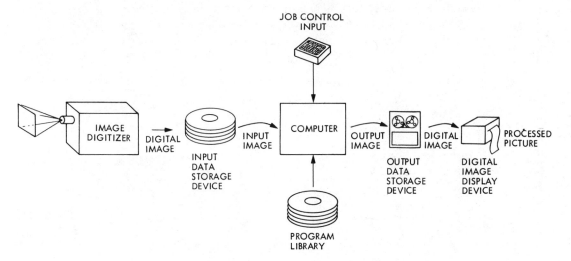

Figure 1-2 A digital image processing system

The Terminology of Digital Image Processing

Images occur in various forms, some visible and others not, some abstract and others real, some suitable for computer analysis and others not, and it is important to have an awareness of the different types of images. The lack of this awareness can lead to considerable confusion among people communicating ideas about images when they have differing concepts of what an image is. Since images form an overwhelming part of our experiences from birth, there is a tendency to take them for granted. This section is intended to establish a foundation upon which images of all forms can be discussed without confusion. Our definitions do not establish a standard for the field but are introduced to make this book self-consistent. The reader may wish to compare this section with the nomenclature of Ref. 1.

Before we can define digital image processing, we must agree upon a definition for the word *image*. While most people have a notion of what an image is, a precise definition is somewhat elusive. Among several definitions of the word in Webster's Dictionary are the following: "An image is a representation, likeness, or imitation of an object or thing, a vivid or graphic description, something introduced to represent something else." Thus, in a general sense, an image is a representation of something else. A photograph of Abraham Lincoln, for instance, is a representation of an American president as he once appeared before a camera. An image contains descriptive information about the object it represents. A photograph displays this information in a manner that allows the human eye and brain to visualize the subject itself. Notice that under this relatively broad definition of image fall many "representations" not perceivable by eye.

Images can be classified into several types based upon their form or method of generation. In this regard, it is instructive to employ a set theory approach. If we consider the set of all objects (Figure 1-3), the images form a subset. There is a correspondence between each image in the subset and the object that it is used to repre-

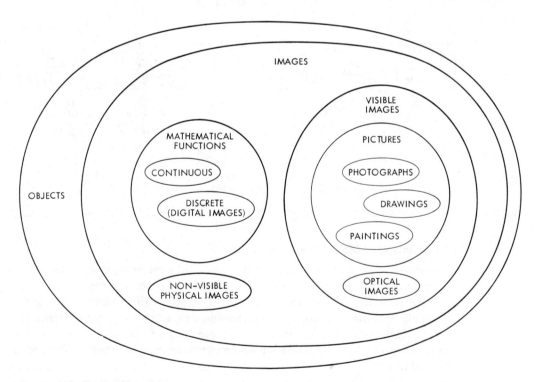

Figure 1-3 Types of images

sent. Within the set of images, there is a very important subset containing all the visible images, those that can be seen and perceived by eye. Within this set exist several subsets representing the various methods of generation. These include photographs, drawings, and paintings. Another subset contains the optical images, that is, those formed with lenses, gratings, and holograms.

The physical images are distributions of measurable physical properties. For example, optical images are spatial distributions of light intensity. These can be seen by the human eye and are visible images as well. Examples of nonvisible physical images are temperature, pressure, elevation and population density maps. A subset of the physical images is multispectral. These are images having more than one local property defined at each point. An example is the trispectral image as it is reproduced in color photography and color television. Whereas the black and white image has one value of brightness at each point, the color image has three values of brightness, one each in red, green, and blue. The three values represent intensity in different optical spectra which the eye perceives as different colors.

Another subset of images contains the abstract images of mathematics, the continuous functions and the discrete functions or digital images. Only the digital images can be processed by computer.

A *picture* is a restricted type of image. Webster defines a picture as "a representation made by painting, drawing, or photography. A vivid, graphic, accurate description of an object or thing so as to suggest a mental image or give an accurate idea of the thing itself." For our purposes we take the word picture to mean a distribution of matter that is visible when properly illuminated. In the vernacular of image processing, however, the word *picture* is sometimes used as equivalent to the word *image*.

According to Webster, the word *digital* relates to "calculation by numerical methods or by discrete units." We can now define a *digital image* to be a numerical representation of an object (which may itself be an image).

Processing is the act of subjecting something to a process. A *process* is a series of actions or operations leading to a desired result. Thus a series of actions or operations are performed upon an object to alter its form in a desired manner. An example is a car wash, where automobiles are processed to change them from dirty to clean.

Now we can define *digital image processing* as subjecting numerical representations of objects to a series of operations in order to obtain a desired result. In the case of pictures, the processing changes their form to make them more desirable or attractive, or to accomplish some other previously defined goal.

For purposes of discussion, it is convenient to restrict the general definition of a digital image. Unless otherwise stated, we shall use the restricted definition of a digital image, which is *a sampled, quantized function of two dimensions which has been generated by optical means, sampled in an equally spaced rectangular grid pattern, and quantized in equal intervals of gray level.* Thus a digital image is now a two-dimensional rectangular array of quantized sample values. In discussing images not so restricted, we shall make use of the following four generalizations: (1) nonoptical digital images generated from other than optical images; (2) higher-dimensional digital images defined on three or more dimensions (this includes multispectral images in which there is more than one gray level value at each point); (3) nonstandard sampling, in which the domain of the image is sampled by a scheme other than the equally spaced rectangular grid; (4) nonstandard quantization, where the quantizing levels are not equally spaced.

An image is usually a condensation or summary of the information in the object it represents. Ordinarily, an image contains less information than the original object. Therefore, an image is an incomplete, yet in some sense adequate, representation of the object.

Digital image processing starts with one image and produces a modified version of that image. It is therefore a process that takes an image into an image. *Digital image analysis* is taken to mean a process that takes a digital image into something other than a digital image, such as a set of measurement data or a decision. For example, if a digital image contains a number of objects, a program might analyze the image and extract measurements of the objects. The term *digital image processing*, however, is loosely used to cover both processing and analysis.

Digitizing is the process of converting an image from its original form into digital form. The term *conversion* is used in a nondestructive sense because the original image is not destroyed but is used to guide the generation of a digital image. The reverse operation is *display*, that is, the generation of a visible image from a digital image. Commonly used equivalents are the words "playback," "reconstruction," and "recording." The process again is nondestructive since displaying a digital image does not destroy the data. There are both *volatile* and *permanent* displays. The latter produce *hard-copy* output.

We take *scanning* to mean the selective addressing of specific locations within the domain of an image. Each of the small subregions addressed in the scanning process is called a *picture element*, which is abbreviated by the word *pixel*. When digitizing photographic images, *scanning* is the process of sequentially addressing small spots on the film. The term *scanning* is loosely taken as an equivalent to the term "digitizing." The rectangular grid scanning pattern is known as a *raster*.

Sampling is defined as measuring the gray level of an image at a pixel location. When digitizing images, it is frequently desirable to employ devices that convert one physical quantity to another. An example is the photomultiplier tube, which converts light energy into electrical energy. Devices of this type are called *transducers* and the process itself, *transduction*.

Quantization is the representation of a measured value by an integer. Since digital computers process numbers, it is necessary to reduce the continuous measurement values to discrete units and represent them by numbers.

The steps of scanning, sampling, transduction (if necessary), and quantizing are sufficient to generate a numerical representation of an image and therefore constitute the steps in digitization. We may reverse the process to display a digital image. With the ability to convert images into digital form and back into visible form, we are able to define and execute digital processing steps and observe the results.

When a process generates an output image from an input image, there must exist a correspondence between points in the two images. Each pixel in the output image must correspond to one pixel in the input image. Thus, when the operation is applied to one point or a neighborhood centered upon one point in the input image, the resulting gray level value is stored in the *corresponding point* in the output image.

The operations that can be performed on digital images fall into several classes. An operation is a *global operation* if it is applied equally throughout the entire digital image. A *point operation* is one in which the output pixel value depends only on the value of the corresponding input pixel. Point operations are sometimes called *contrast manipulation* or *stretching*. A *local operation* is one in which the output pixel value depends on the pixel values in a neighborhood of the corresponding input point.

The notion of *contrast* refers to the amplitude of gray level variations within an image. *Noise* is broadly defined as an additive (or multiplicative) contamination of an image. The *sampling density* of a digital image is the number of sample points per unit measure in the domain. *Gray scale resolution* is the number of gray levels per unit measure of image amplitude. *Magnification* refers to the size relationship between an image and the object or image it represents. It is defined only for linear geometrical relations where one can define the same metric in the domains of both images and where the size relationship is uniform over the entire image. Magnification is a meaningful relationship between input and output digital images in a processing step. However, the magnification from a physical image to a digital image is not a meaningful concept, and sampling density should be used.

SOME PHILOSOPHICAL CONSIDERATIONS

When one approaches a topic such as digital image processing, he cannot do so without bringing with him a set of notions and attitudes—in other words, a philosophy. In this section, we discuss three topics that are constructive in this regard.

The Continuous Versus Discrete Philosophy

There are two approaches one may take when considering image processing operations. We can think of the digital image as a set of discrete sample points (which it actually is), each having individual importance, or we may think in terms of the continuous function which the digital image represents. The theory underlying many of the processing operations is based on the analysis of continuous functions. Other operations are most easily thought of as logical operations performed on individual points. Thus it is important to be able to think of digital images in either way, but without confusion.

Since the digital image is fundamentally discrete, it is dangerous to overlook this characteristic. Frequently, when thinking in the continuous mode, one will be surprised by an unexpected characteristic of the processed image that has been brought about by the discrete sampling. When the processed image differs from that predicted by the analysis of continuous functions, this is usually termed a *sampling effect*.

The Finite Number of Pictures Exercise

It is interesting to consider the total number of different pictures that exist. Let us now consider as "different pictures" those that are perceived by the human eye to be different from each other. We are ignoring, for the time being, other uses for pictures. The human eye can resolve approximately 40 distinct shades of gray. If we consider an 8 × 10 inch photograph digitized to 1000 lines of 1000 samples each, that is, a million pixels, the pixels are in themselves indistinguishable when the picture is held at arm's length. This means that the sampling is fine enough to represent that image accurately insofar as the human eye is concerned. Thus an 8 × 10 photograph viewed at arm's length can be completely and uniquely represented by a million-point digital image having 40 levels of gray.

Let us now consider the question of how many different such photographs there are. One may be tempted to say that there are infinitely many different photographs, and in some sense there are. If all possible photographs can be represented by 1,000,000-point, 40-gray level digital images, however, then there are only finitely many. In fact, the number of different pictures of this type is $40^{1,000,000}$. While the exact value of this number depends on our assumptions, it does point out that there are only finitely many "pictures" in this restricted sense. Just as a finite change in a digital image is required to produce a different digital image, a finite change in a picture is also required before it will be perceived by the human eye as being different.

It is interesting to note that only one out of that large number of pictures is solid black. Indeed, there are only 40 pictures having the same shade throughout, one of which is solid white and one solid black. This large but finite set also contains many pictures of every person who ever lived. Digital image processing can be thought of as generating mappings between the elements of this finite set of images. It is likely that the vast majority of the images in this set would be unrecognizable if displayed. Perhaps we would be able to attach significance to only a very small percentage. If one exhaustively generated and cataloged all possible images, he would greatly assist his colleagues in the field. Such a task, however, poses severe practical problems.

The foregoing analysis assumes that the human eye is the ultimate consumer of the fruits of image processing. This is not always the case, and other applications may dictate other requirements for spatial and gray scale resolution. There are even cases in which processing images for visual interpretation does not require high resolution. It is important, however, to note that, practical considerations notwithstanding, one can digitize images with sufficient resolution that the sampling itself does not produce visible degradation.

Correspondence Between Images

In most digital image processing applications, we process the image of an object in order to derive information about the object itself. Since only digital images can be processed by computer, such images are merely "standing in" for the objects they represent. Thus we establish a correspondence between an object and the image that represents it. Since we cannot digitally process an object, or even a nondigital image, we are restricted to processing a corresponding digital image.

Figure 1-4 views an image processing sequence in terms of a chain of corresponding images. The camera forms an optical image that corresponds to the subject. The developed film has on it a negative image corresponding to the optical image. The film forms a corresponding optical image on the digitizer faceplate, and that produces an input digital image that gives rise to a series of six corresponding images, the last of which is the desired output picture. Even if the actual processing is a simple one-step operation, there is, in this case, a series of 10 corresponding images between the subject and the output picture. Although our casual parlance belies this fact, it is important to remember how many corresponding images are involved.

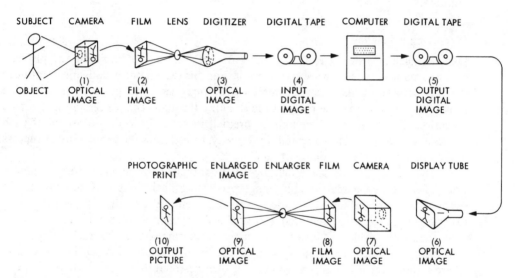

Figure 1-4 An image processing sequence

Each step in the process provides an opportunity for image degradation. To minimize this, the entire process should be well designed and properly controlled. One goal of this book is to develop means to analyze quantitatively the performance of each step and of the process as a whole.

DIGITAL IMAGE PROCESSING IN PRACTICE

Later sections of this book present in detail various techniques and applications of digital image processing. At this point, however, it is useful to discuss some general aspects of using this tool to produce results.

Problem Solving with Digital Image Processing

Digital image processing requires a varied background for its successful use. Practitioners in this field are called upon both for theoretical analysis and practical application. It requires a balanced knowledge of mathematical theory, optics, and computer technology, and the use of common sense.

Once the decision to use digital image processing has been made, the problem frequently guides the search for a solution. Processing and analysis techniques are drawn from a bag of tools consisting of formal theory, proved heuristics, and untested ideas. The solution often results from a combination of techniques. Whether or not the immediate problem is solved, the list of available techniques usually grows. Frequently, a heuristic is solidified in theory after its successful application, and new theoretical methods often suggest processing and analysis approaches. Both success and failure tend to increase the list of techniques, although failures are less frequently reported.

Processing Efficiency

Digital image processing can only be done with the use of a significant physical resource, namely the hardware system itself. For this reason one usually must optimize the productivity of his machine time by avoiding, insofar as possible, unprofitable runs. This can best be done by carefully thinking through each processing step before execution. The user should be able to predict, in general terms, the results of each processing step before it is executed. In this way, he increases the probability of success and avoids unprofitable computer runs.

The "prediction beforehand" approach contrasts with haphazard experimentation. There are so many possible image processing operations that the chances of selecting a successful one at random are quite remote. Certainly some intelligent experimentation may be required before the final result is obtained. This book is concerned with developing insight into image processing so that the reader may make intelligent and efficient use of his equipment. When one finds himself in a situation where system time is less expensive than his own personal time, then "artistic" experimentation may be justified. This, however, requires little instruction and is not treated here.

The "prediction beforehand" approach, of course, does not preclude surprises. It is quite likely that some factor unaccounted for in the planning will emerge to produce quite an unexpected image. Each time this happens, it offers the user an excellent chance to increase his knowledge of image processing, if he takes time to track it down.

Functional Requirements for Digital Image Processing

The following is a list of requirements a general-purpose image processing system should meet to be effective in its application:

1. The hardware must be adequate for the problems attempted. Inadequate

sampling in the spatial domain and inadequate gray scale quantization may not make success unattainable, but can render failure inconclusive. Processing algorithms usually assume that the image function is continuous. If the sampling and quantization used do not justify this assumption, performance may suffer considerably. Thus, the fear of large data quantities can be a threat to a successful solution. When system noise levels degrade the image, success is again in jeopardy. While image analysis requires a high-quality image digitizer, image processing requires a high-quality image display device as well.

2. For general-purpose work, the software system should allow simple library call-up and execution of processing and analysis programs. Convenient tape or disk storage of input and output digital images and library programs is a practical requirement.

3. The library programs should be maintained with an eye toward versatility. The power of the system is greatly enhanced if existing programs can be used to try out new approaches to old or new problems.

4. The program library should be easily expandable to include new programs as they are developed. In this way, the system experiences continual growth.

REFERENCE

1. R. M. HARALICK, "Glossary and Index to Remotely Sensed Image Pattern Recognition Concepts," *Pattern Recognition*, **5**, 391–403, 1973.

DIGITIZING IMAGES

IIIIIIIIIIIIIIIIIIIIIIIIIIIIIIIIIIIIIIIIIIIIIIIIIIIIIIIIIIIIIIIIIIIIIIIIIIIIIIIIIIIIIIIIIIIIIIIIIIIIIIIIIIIIIIIIIIIIIIIIIIIIIIIIIIIIIIIIIIIIIIIIIIIIIIIIIIIIIIIII

INTRODUCTION

Since computers can process only digital images, and nature affords images in other forms, a necessary precursor to digital image processing is the conversion of images into digital form. The specialized equipment for digitizing images is, by and large, what transforms an ordinary computer center into an image analysis laboratory. For image processing applications, display devices are also required. Image display, however, is obtainable from a line printer, although this process is rather cumbersome.

Until relatively recently, image digitizing equipment was so expensive and complex that only a few centers had such a capability. Advances in technology, however, are making image digitizers less expensive and their use more widespread. Widely diverse configurations of apparatus have been used to convert images into digital form. In this section, we discuss the elements of an image digitizer and examine several implementations. The aim is to develop an insight into the capabilities and limitations of the various approaches.

The Elements of a Digitizer

An image digitizer must be able to divide an image into picture elements and address each individually, measure the gray level of the image at each pixel, quantize that continuous measurement to produce an integer, and write out the set of integers on a data storage device. To accomplish this, a digitizer must have five elements. The first is a sampling aperture, which allows the digitizer to access picture elements individually while ignoring the remainder of the image. The second element of an image digitizer is a mechanism for scanning the image. This process consists of moving the sampling aperture over the image in a predetermined pattern. Scanning allows the sampling aperture to address pixels in order, one at a time.

The third element is a sensor, which can measure the brightness of the image at each pixel through the sampling aperture. The sensor is commonly a transducer that converts light intensity into an electrical voltage or current. The fourth element, a quantizer, converts the continuous output of the sensor into an integer value. Typically, the quantizer is an electronic circuit called an *analog-to-digital converter*. This unit produces a number that is proportional to the input voltage or current.

The fifth element of an image digitizer is the output medium. The gray level values produced by the quantizer must be stored in an appropriate format for subsequent computer processing. Technically, the output medium could be omitted if the image were being processed "on-line." Image digitizing is frequently done "off-line" from the main computer system, however, and the output medium is necessary. The output medium can be magnetic tape, magnetic disk, data cards, or even punched paper tape. Because of the size of typical digital images, the latter two formats are seldom practical.

CHARACTERISTICS OF AN IMAGE DIGITIZER

While image digitizers differ in the apparatus they use to perform their function, they may be equated on the basis of their relevant characteristics. Two important characteristics are the size of the sampling aperture and the spacing between adjacent pixels. If the digitizer has a lens system with variable magnification, the sample size and spacing at the input image are variable, and the range is of interest. Another important parameter is the image size capability of the instrument. In the case of a film scanner, the maximum input size might be 35-millimeter (mm) film, or perhaps 11- by 14-in. X rays. At the output, image size is specified by the maximum number of lines and of samples per line. Thus, a digitizer might be able to produce digital images having up to 1000 lines of 1000 samples each.

A third significant characteristic of an image digitizer is the physical parameter that it actually measures and quantizes. In the case of film scanners, for example, the instrument could measure and quantize either transmittance or optical density. Both are functions of the darkness or lightness of the film, but in certain applications one may be better than the other. The linearity of the digitization is also an important parameter. For instance, if the instrument digitizes light intensity, one should know to what accuracy the gray levels are proportional to the actual brightness of the image. The validity of subsequent processing may be jeopardized by a nonlinear digitizer. Of interest, too, is the number of gray levels to which the instrument can quantize the image. Early image digitizers had only two gray levels: black and white. In current practice, 8-bit (256-level) data is common, and considerably higher resolution is possible with recent instrumentation.

Finally, one of the most important characteristics of a digitizer is its noise level. If a uniformly gray image is presented to a digitizer, the noise inherent in the system will cause variations in the output even though the input is constant. Noise introduced by the digitizer is a source of image degradation and should be small relative to the contrast of the image.

These characteristics constitute a brief specification sheet for an image digitizer. They provide a basis upon which to compare different instruments or to decide whether a given instrument is adequate for a particular job. In some applications, digitizing images with relatively few lines, samples, and gray levels and with appreciable non-linearity and a high noise level may be adequate. However, many of the important applications of digital image processing require a high-quality image digitizer—one capable of digitizing large images to many gray levels with good linearity and a low noise level. In later sections, we discuss image digitizer requirements in light of processing applications.

TYPES OF IMAGE DIGITIZERS

An important and highly versatile type of image digitizer is the so-called digitizing camera, which has a lens system and can digitize an image of any object presented to it. An example is a television camera interfaced to a computer. Such a device can digitize not only physical objects but also other images such as photographic film. A restricted but nonetheless important type of image digitizer is the film scanner. This is an instrument made specifically for scanning images on film. Film scanners can digitize an image of an object only after it has been photographed initially by a film camera. Historically, film scanners have played a predominant role in image processing, but current practice favors direct digitizing cameras.

There are two important digitizing philosophies, "scan-in" digitizing and "scan-out" digitizing. In a scan-out system (Figure 2-1), the entire object or film image is illuminated continuously, and the sampling aperture allows the sensor to "see" only one pixel at a time. In a scan-in system (Figure 2-2), only one small spot of the object is illuminated, and all the transmitted light is collected for the sensor. In this case, the object is scanned with the illuminating beam, and the sensor is spatially nonspecific.

There is a third philosophy—a combination of the previous two. In a "scan-in/scan-out" system, the object is illuminated by a moving spot and sampled through a moving aperture that follows the spot. Such a system reduces the effects of glare and

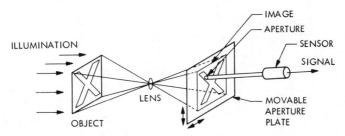

Figure 2-1 A scan-out digitizer

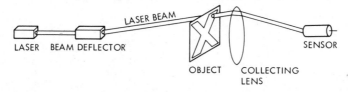

Figure 2-2 A scan-in digitizer

has found some application in digitizing microscope images. The complexity of such systems, particularly in tracking the illuminating spot with the sampling spot, has limited their application.

IMAGE DIGITIZING COMPONENTS

As discussed before, an image digitizer must have a light source, a light sensor, and a scanning system. Furthermore, either the light source or the light sensor, or both, must be behind a sampling aperture. In this section, we discuss various types of light sources, light sensors, and scanning systems; in the following section, we put them together to form complete image digitizers.

Light Sources

The most common man-made light source is the incandescent bulb. For scan-out systems, incandescent lighting is convenient for general illumination of the object or image being digitized. For scan-in work, the filament of a small bulb can be imaged with a lens to form a small bright spot.

Highly concentrated beams of light can be produced with a laser. The laser generates a narrow, intense, coherent beam of light by first raising the atoms of an active material (helium, neon, ruby, etc.) to a high-energy state and then stimulating a simultaneous transition back to the normal state. This transition gives rise to a high-intensity beam of coherent light that is easily focused and deflected. While the laser could be used for general illumination in a scan-out system, its principal advantage lies in producing small high-intensity spots for scan-in digitizers.

Certain phosphors emit light when irradiated with an electron beam. If an electron beam is focused to a small spot on the face of a phosphor-coated glass plate (Figure 2-3), light is emitted from that spot. The phosphor that coats the face of the cathode-ray tube (CRT) is a crystalline compound doped with certain impurities. The phosphor is deposited on the face of the tube over a transparent aluminum film. This film is positively charged and forms an anode that attracts the electron beam. The impact of the energetic electrons excites the atoms of the host phosphor, raising some of their electrons to high-energy states. As these electrons decay back to their normal state, light is emitted. The spectrum (color) and persistence (decay rate) of the light generated can be controlled in the manufacture of the phosphor. A wide variety of emission spectra and persistence times from less than 1 microsecond to several seconds is available.

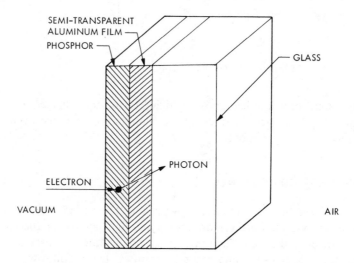

Figure 2-3 CRT target construction

The brightness of the light spot produced by the electron beam is roughly proportional to the average beam current density. The phosphor is made up of granules and is therefore subject to graininess and scattering of light within the phosphor layer. Currently available cathode-ray tubes have a resolution limit of about 30 to 70 line pairs (cycles) per millimeter.

The relatively recent solid-state light-emitting diodes (LEDs) also form compact and convenient light sources. LEDs are typically made of gallium arsenide semiconductor. They emit light at controlled intensity from a spatially small source. This also makes them promising candidates for use in scan-in systems.

Light Sensors

Light sensors produce an electrical signal proportional to the intensity of light falling upon them. Three physical phenomena give rise to three types of light sensors:

photoemissive devices, photovoltaic cells, and photoconductors. Photoemissive substances emit electrons when irradiated with light; photovoltaic substances, such as silicon and selenium solar cells, generate an electrical potential when exposed to light; and photoconductors, such as cadmium sulfide and cadmium selenide, show a drop in their electrical resistance when exposed to light. Photodiodes and phototransistors change their junction characteristics under the influence of light.

Photoemissive Devices: Phototubes and Photomultiplier Tubes. The phototube (Figure 2-4) has a positively charged anode and a negatively charged cathode coated with layers of oxides of the alkaline metals (silver, cesium, antimony, sodium, bismuth,

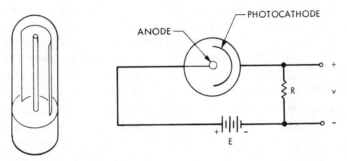

Figure 2-4   The phototube

and rubidium). When photons of sufficient energy ($\lambda < 1$ micron or so) strike the photocathode, electrons are freed from the surface. Under the influence of the electric field, they migrate to the anode, producing a current flow through the device. This current is proportional to the photon flux incident on the photocathode. The current is sensed by the external circuit and may be sampled and quantitized.

The photomultiplier tube (Figure 2-5) has a photoemissive face that forms a semitransparent photocathode. Behind the cathode are a series of dynodes charged with progressively higher positive voltages. Primary electrons freed from the pho-

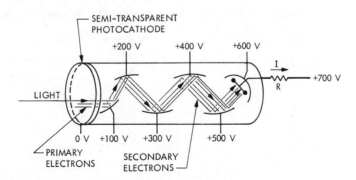

Figure 2-5   The photomultiplier tube

tocathode by photons are accelerated toward the first dynode. The impact on the first dynode frees secondary electrons, producing a multiplying effect. The resulting electrons are then attracted toward the second dynode, and the process continues until the electrons from the last dynode are collected on the anode to produce a current in the external circuit. The photomultiplier tube behaves similarly to the phototube except that it is more sensitive because of the multiplying effect of the dynodes. One primary electron may give rise to as many as a million electrons in the external circuit. This high sensitivity makes the photomultiplier tube useful for digitizing at low light levels.

Photovoltaic Cells. The so-called solar cells are semiconductor junction devices made of silicon or selenium. A junction potential exists at the interface between the *P*-type and *N*-type semiconductor. Impinging photons mobilize carriers near the junction, which move under the influence of the junction potential to produce a current flow in the external circuit. The open-circuit voltage of the cell is primarily due to the junction potential of the semiconductor. The short-circuit current, however, is roughly proportional to light intensity. Photovoltaic cells are useful for solar power applications, but their slow response makes them undesirable for image digitizing applications.

Photoconductive Devices. The common photoconductive cells are made of cadmium sulfide or cadmium selenide semiconductor doped with *N*-type impurities. They are not junction devices but, when connected to an externally supplied field, they conduct under the influence of light. Photon impact frees from the atoms electrons that move under the influence of the externally applied electric field. These devices exhibit resistance dependent on light intensity. Over a wide range, the logarithm of resistance is inversely proportional to the logarithm of light intensity measured in footcandles. These devices are stable and relatively accurate and have found wide use in photographic lightmeters. Their slow response, however, limits their use for image digitizing purposes.

Photodiodes and Phototransistors. The photodiode (Figure 2-6) is a solid-state *P-N* junction that can be exposed to light. In operation, the junction is reverse-biased and exhibits a high impedance. One layer of the device (for instance the *P*-layer) is made very thin so that light can penetrate to the junction. Impinging photons release electron-hole pairs. In the depletion layer, where the electric field is strong, these mobilized carriers do not recombine but migrate under the influence of the field to create a

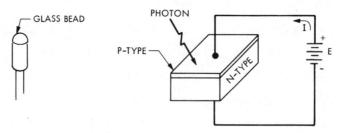

Figure 2-6 The photodiode

current in the external circuit. This current is proportional to the photon flux. Since the reverse-biased junction presents a high impedance, the current is controlled by the light intensity and is relatively independent of the externally applied voltage. The depletion layer is made comparatively thick to capture long wavelength photons.

The avalanche photodiode achieves higher sensitivity than the ordinary photodiode through an electron multiplication effect. The avalanche photodiode is back-biased almost to the point of breakdown. Electrons freed by impinging photons are accelerated by the intense field to velocities at which they have ionizing collisions, freeing more electrons. This effect can produce gain factors as high as 1000, considerably increasing the sensitivity of the device.

The phototransistor (Figure 2-7) is mounted in clear plastic or in a can with a lens on top to permit light to access the transistor junction. Impinging photons release

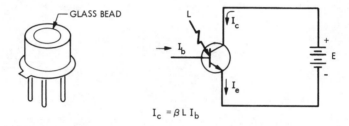

Figure 2-7 The phototransistor

electron-hole pairs in the collector-base junction. The movement of these carriers constitutes base current in the transistor. The collector current is proportional to the base current but is amplified by the current gain factor of the transistor. Externally, the phototransistor behaves like the photodiode except with higher sensitivity. Design requirements for speed and linearity, however, dictate compromises in transistor design and limit available current gains.

In the previous discussion, photodiodes were considered as producing a steady-state current proportional to the incident photon flux. Alternatively, they can operate in the "integrating mode." Since the photodiode junction exhibits capacitance, it will hold a charge of the reverse-biased polarity. Subsequently, photoconduction bleeds off the charge at a rate proportional to incident photon flux. If the photodiode is periodically recharged to some reference voltage, the charge (number of electrons) required is proportional to the integral of incident photon flux over the period between charges. Thus, in the integrating mode, the photodiode senses not instantaneous photon flux but photon flux integrated over a certain period of time.

Two factors limit the dynamic range of photodiodes operating in the integrating mode. First, the small junction capacitance limits the initial charge that can be stored. Secondly, the "dark current," which flows even without incident light, gradually discharges the photodiode. These factors limit the integration period to a few milliseconds and the dynamic range to about 100 to 1 in normal operation. Since the dark current is temperature-sensitive, cooling the photodiode significantly increases practical integration times. Both the photodiode and the phototransistor have fast and stable response to light intensity variations, and they make excellent light sensors for digitizing images.

Scanning Mechanisms

In this section, we discuss techniques that may be used to move the scanning or illuminating spot about the image. In the following section, we consider light sources, sensors, and scanning mechanisms together in complete image digitizing systems.

The Moving-Mirror Scanner. In scan-in systems, the illuminating beam may be reflected off a mirror mounted on a galvanometer motor (Figure 2-8). Current through the motor rotates the mirror, thereby deflecting the illuminating beam. A multifaceted

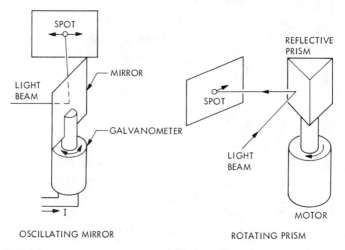

Figure 2-8 Moving-mirror scanning mechanisms

prism driven by a constant-speed motor can produce the same effect. In both cases, the motion of the mirror causes the scanning spot to move across the image. The major disadvantage of the oscillating mirror is low speed of operation. The mass of the mirror and the acceleration required to repeatedly reverse directions severely constrain the speed at which the device can operate. The rotating prism is not so limited by speed. To achieve small scanning angles, however, the prism must have many precisely aligned facets and therefore must be large. This makes the rotating mass bulky.

Moving-mirror scanners are subject to geometric distortion from the scanned image. The linear displacement of the scanning spot is proportional, not to the deflection angle of the mirror but to the tangent of that angle (Figure 2-9). With the rotating prism, this angle increases linearly with time. For the galvanometer, the deflection angle is linear with drive current. Drive current is most easily made linear with time,

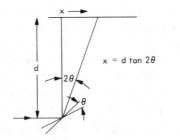

BARREL DISTORTION

Figure 2-9 Scanning angle distortion

and the brightness signal is most conveniently sampled at uniform intervals in time. This produces nonuniform pixel spacing, resulting in pincushion distortion when the image is displayed.

For some applications of narrow-angle scanners, slight geometric distortion is tolerable. It may be partially or totally eliminated by complex galvanometer current waveforms or nonuniform sampling techniques. Both of these refinements entail additional complexity. Furthermore, acceleration and deceleration during the line scan limit the speed at which a moving mirror scanner can operate satisfactorily. Faster operation is obtained when the mirror accelerates before the line scan begins, maintains constant velocity during the line scan, and decelerates and reverses after it is over. Even faster operation is possible using a Perkin Elmer design. In this system, the mirror is driven sinusoidally at its frequency of mechanical resonance. This is several times faster than the highest scan rates that can produce linear sweeps. Light reflected off a grating on the backside of the mirror and detected by a photodiode triggers sampling at uniform angular spacing.

Mechanical Scanning Devices. Figure 2-10 shows two mechanical methods for image scanning, the rotating drum and the lead screw. A photographic image is wrapped, partially or completely, around a cylindrical drum, and the drum is rotated to pull the

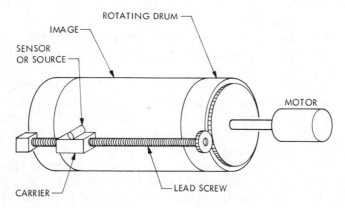

Figure 2-10 Mechanical scanning mechanisms

image past a stationary aperture. This effects scanning in one direction. The scanning aperture may be mounted on a lead screw that rotates to move the aperture across the image. In Figure 2-10, a rotating drum and lead screw have been combined to produce a two-dimensional image scanner. If the lead screw turns continuously rather than in steps, the scan is helical, but this is usually an adequate approximation to a rectilinear scan. Mechanical scanning devices such as these are somewhat limited in speed of operation but provide good geometric stability at relatively low cost.

Electron Beam Scanning. Several electronic devices useful in image digitizing and display scan images with an electron beam. Figure 2-11 illustrates two means of deflecting an electron beam to scan a target. An electron beam, generated by an electron

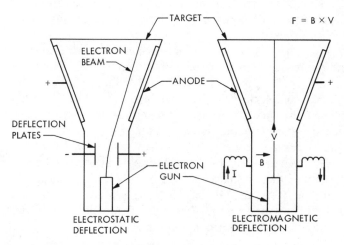

Figure 2-11 Electron beam deflection

gun in the base of the tube, is attracted toward the target by the positively charged anode. As the electron beam passes between the electrostatic deflection plates, the electric field exerts a force on the electrons, changing their direction of travel. The deflection angle is dependent on the beam velocity and the electrical potential between the plates. By controlling the deflection potential, one can cause the electron beam to impact any point on the target.

A transverse magnetic field can also be used to deflect an electron beam. The force on a charged particle in a magnetic field is the vector cross product of the particle velocity and the magnetic field. The deflecting force acts at right angles to both the beam and the magnetic field. Thus, in Figure 2-11, the negatively charged electrons will be deflected downward.

Electron beams must be focused as well as deflected. Figure 2-12 shows one method for focusing an electron beam. The electrostatic lens, a cylindrical structure shown in cross section in Figure 2-12, is arranged to produce a convex pattern of equipotential lines. As an electron moves through the electrostatic lens, it is acted

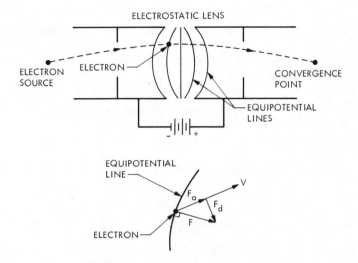

Figure 2-12   Electrostatic focus

upon by a force normal to the equipotential lines. This force may be resolved into an accelerating component and a deflecting component. The deflecting component changes the electron's path, deflecting it toward the convergence point. In a well-designed electrostatic lens, all electrons from the source reaching the lens will be deflected to arrive near the convergence point.

Figure 2-13 illustrates the use of an axial magnetic field to converge an electron beam. Electrons exit the anode aperture with approximately equal axial components

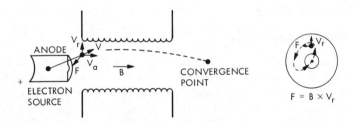

Figure 2-13   Electromagnetic focus

of velocity but differing in radial velocity. Since the force on an electron is the cross product of its velocity and the magnetic field vector, the axial component of velocity produces no deflecting force. The radial component, however, interacts with the magnetic field to produce a tangential accelerating force. This causes the electrons to move in a spiral or helical path about the axis of the tube. The magnetic field strength can be adjusted to cause all electrons leaving the aperture to make exactly one complete spiral before arriving at the convergence point.

Figure 2-14 illustrates the construction of an electron gun. Electrons boiling off the heated cathode exit through an aperture in the control grid. The negative potential on the control grid determines the rate at which electrons exit the aperture. The electrons are attracted by the positively charged components farther down the tube. The first electrostatic lens images the control grid aperture on the second lens. The second lens, in turn, images the first lens at infinity, producing a parallel exit beam of electrons. The positive voltages on the first and second anodes and the target establish the velocity of the electron beam.

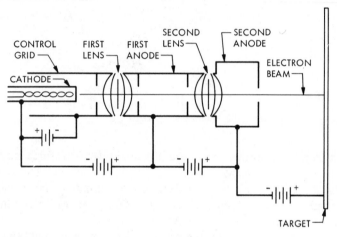

Figure 2-14   Electron gun construction

ELECTRONIC IMAGE TUBES

The Vidicon Camera Tube

Figure 2-15 illustrates the construction of the vidicon, an important type of television image sensing tube. The vidicon is a cylindrical glass envelope containing an electron gun at one end and a target and faceplate at the other. The tube is surrounded by a yoke containing electromagnetic focus and deflection coils. The faceplate is coated on the inside with a thin layer of photoconductor over a thin transparent metal film that forms the target. Behind the target is a positively charged fine wire mesh. Electrons are decelerated after passing through the mesh and reach the target with approximately zero velocity. A small positive charge is applied to the metal coating of the target. In darkness, the photoconductor behaves as a dielectric, and the electron beam deposits a layer of electrons on the inner surface of the photoconductor to balance the positive charge on the metal coating. Thus, after a complete scan by the electron beam, the photoconductor appears as a capacitor with a positively charged plate on one side and a surface charge of electrons on the other side.

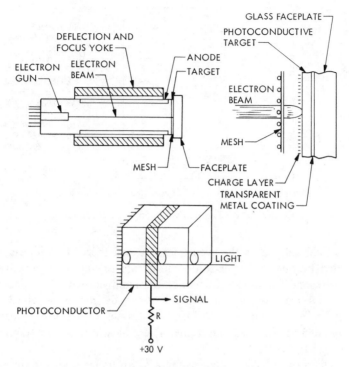

Figure 2-15 The Vidicon camera tube

When light strikes a small area of the photoconductor, it allows the electrons to flow through, locally depleting the electron charge layer. Thus, if an optical image is formed on the target, the photoconductor will form an identical electron image on the back of the target, that is, electrons will be present in dark areas and absent in light areas. As the electron beam scans the target, it replaces the lost electrons, restoring the uniform surface charge. As the electrons are replaced, a current flows in the external target circuit. This current is proportional to the number of electrons required to restore the charge and, therefore, to the light intensity at that point. It is also proportional to the beam velocity, which, in turn, determines the time available for the charge to flow. Current variations in the target circuit produce the video signal of the vidicon. The electron beam repeatedly scans the surface of the target, replacing the charge that bleeds away during the interval between scans. The vidicon target is thus an integrating sensor.

Figure 2-16 illustrates the RETMA (Radio–Electronics–Television Manufacturers Association) scanning convention, which has become a standard for broadcast television in the United States. The beam scans the entire target surface in 525 lines, 30 times each second. Each frame is made up to two interlaced fields, however, each consisting of 262.5 lines. The first field of each frame scans all the odd lines, while

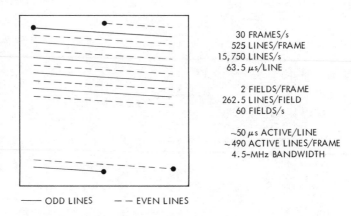

30 FRAMES/s
525 LINES/FRAME
15,750 LINES/s
63.5 μs/LINE

2 FIELDS/FRAME
262.5 LINES/FIELD
60 FIELDS/s

~50 μs ACTIVE/LINE
~490 ACTIVE LINES/FRAME
4.5–MHz BANDWIDTH

——— ODD LINES   — — EVEN LINES

Figure 2-16   The RETMA scanning convention

the second field scans the even lines between them. This interlaced convention was adopted to yield a 60-per-second field rate to minimize perceived "flicker" and a 30-per-second frame rate to reduce the bandwidth of the transmitted signal. Each horizontal line scan requries 63.5 microseconds (μs), of which 83% or approximately 50 μs are active. Of the 525 lines per frame, 16 are lost in each vertical retrace, leaving 483 active lines per frame. The bandwidth of the standard video signal is 4.5 megahertz (MHz).

One can use the vidicon camera as an image digitizer simply by sampling the video signal. To obtain approximately 500 points per line, however, one must sample the video every 100 nanoseconds. While currently available electronic circuitry can operate at this speed, most bulk data storage devices cannot accept data at this rate. For image transfer to tape or disk storage devices, the data rate must be reduced. This can be done by reducing the scanning rate of the vidicon and sampling the video signal or by pulsing the beam. Special-purpose vidicon digitizing cameras have been designed for noninterlaced scanning rates of 10 frames per second. This reduces the video bandwidth but produces flicker in a real-time display.

Another reduced-data-rate scanning method uses a standard vidicon camera with a one-line data buffer. In each field, one line is digitized at a 10-MHz sampling rate and the data stored in a high-speed solid-state memory. During the remainder of the field, the data is buffered to storage on tape or disk. During the next field, the line directly below is digitized as before. This technique can digitize an entire 500-line image in $8\frac{1}{3}$ seconds. It produces data at a rate convenient for transfer but requires a relatively large solid-state memory, typically 500 8-bit bytes.

Another way to reduce the data rate is to sample one pixel per line. This technique is called *column digitizing*. On each frame, one vertical column of pixels is sampled. The digitizing column moves from left to right, from frame to frame, completing the entire image in $16\frac{2}{3}$ seconds. Since the fields are interlaced, one field will give all the odd-numbered pixels on a vertical column and the following field will give the even-numbered pixels.

The Plumbicon and the Silicon Diode Vidicon

The photoconductive target of the standard vidicon is made of selenium photoconductor. A similar tube, the plumbicon, has a lead oxide target. A more recently developed tube, the silicon diode vidicon, has a silicon photoconductor target doped with *N*-type impurities. An array of small *P*-type islands is diffused into the silicon target. This produces a rectangular array of approximately 400,000 diodes in a $\frac{1}{2}$-in. square. The operation of the plumbicon and the silicon diode vidicon is similar to that of the standard vidicon.

The plumbicon is somewhat more sensitive than the vidicon and thus applicable to low-light-level situations. The plumbicon also has faster transient response. It exhibits less lag than the vidicon when responding to a rapidly changing image. The silicon diode vidicon shows superior sensitivity in the long wavelengths, particularly the infrared. Nonuniformities in the diode array, however, can introduce blemishes in the scanned image.

The Image Dissector

Another electronic camera tube that has found considerable application in image digitizing is the image dissector tube shown in Figure 2-17. This tube has a photoemissive surface coated on the back of a glass faceplate. Coils in the electromagnetic yoke

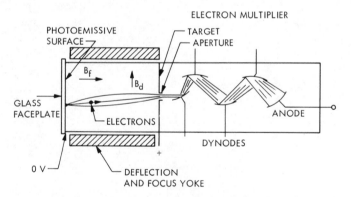

Figure 2-17 The image dissector camera tube

image the photoemissive coating on the positively charged target further back inside the tube. An optical image on the faceplate produces an electron image on the photoemissive surface. This image drifts rearward under the influence of the positively charged target. The electromagnetic yoke contains deflection coils so that electrons from any point in the image may be made to pass through the target aperture. These electrons then pass through an electron multiplier section and impact the anode, producing a current in the external circuit.

Unlike the vidicon, the image dissector can "dwell" on any point of the image for an arbitrary length of time. It does not depend on maintaining a constant scanning beam velocity (or pulsing the beam) to produce an output signal. The image dissector is fundamentally less sensitive than the vidicon, however, because it is a point integrating sensor. Light energy is wasted if it strikes points other than the one being sampled, whereas the vidicon integrates light intensity at every point and stores it until the electron beam arrives. The electron multiplier of the image dissector partially compensates for its lack of sensitivity. As a digitizing sensor, the image dissector is essentially "shot noise limited," since, in effect it counts photons from one image point at a time. It can also suffer from blemishes in the photoemissive surface.

OTHER DIGITIZING SYSTEMS

The Dual Moving-Mirror Image Plane Scanner

Figure 2-18 diagrams the dual moving-mirror type of image plane scanner. This device images the object on an image plane containing a small aperture. A photomultiplier tube measures the light transmitted through the aperture. Two galva-

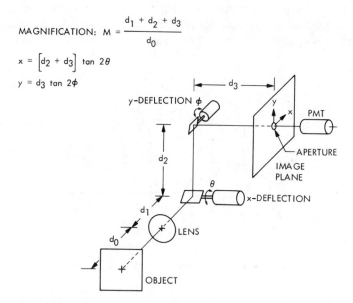

Figure 2-18 The dual moving-mirror image plane scanner

nometer mounted mirrors provide *x*- and *y*-deflection. This system can produce high-quality digitized images but is restricted in speed because of mirror deflection dynamics and the fact that it is a point integrating system. The object must be illuminated at high intensity, since the fraction of light coming through the aperture is low. This design is also subject to scanning angle distortion, particularly in the direction of widest scanning angle.

Self-Scanning Arrays

A promising new type of image sensor is the electronic self-scanning solid-state array. These devices have a linear or rectangular array of photodiodes on a single integrated circuit chip, complete with clocking and scanning logic. The photodiodes operate in the integrating mode. The self-scanning photodiode array (Figure 2-19) contains, on one chip, an array of sensors, a series of switches, and associated control circuitry. Responding to externally supplied clock pulses, the circuit closes the switches, one at a time, to allow the junction capacitance to be recharged by the external circuit. The charging current I_c reflects light intensity.

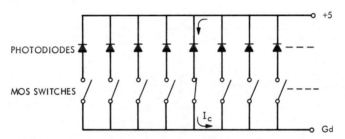

Figure 2-19 The self-scanning photodiode array

Charge-coupled devices (CCDs) also have an array of photodetectors. The charge developed in each sensor is shifted as a "packet" down a series of internal capacitors until it reaches an external terminal. These devices are currently somewhat limited in spectral sensitivity, integration time, noise level, and pixel-to-pixel uniformity of sensitivity. With continued development, however, they promise a line of compact and rugged solid-state cameras for television and for image digitizing.

FILM SCANNING

Photography plays an important role in digital image processing, both before digitizing and after display. In this section, we discuss the photographic process and some considerations that apply to film scanners.

Transmittance and Density

When an object passes some but not all of the light incident upon it, it is neither transparent nor opaque. We use two measures of this partial light-transmitting property, transmittance and optical density (OD). The transmittance of the left-hand object in Figure 2-20 is given by

$$T_1 = \frac{I_2}{I_1} \qquad 0 \le T_1 \le 1 \tag{1}$$

where I_1 is the incident and I_2 the transmitted photon flux density. Transmittance is merely the factor by which an object attenuates light intensity. The optical density of the left-hand object is given by

$$D_1 = \log \frac{I_1}{I_2} = \log \frac{1}{T_1} = -\log T_1 \qquad 0 \le D_1 \le \infty \tag{2}$$

OD is not confined to a convenient range and approaches infinity for opaque objects.

The two objects in Figure 2-20 are serially arranged in the light path and thus superimposed. The transmittance of the combination is

$$T_3 = \frac{I_3}{I_1} = \frac{I_3}{I_2}\frac{I_2}{I_1} = T_1 T_2 \tag{3}$$

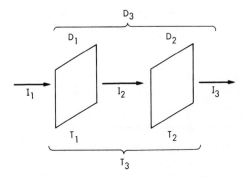

Figure 2-20 Density, transmittance, and superimposition

while the OD of the combination is given by

$$D_3 = \log \frac{I_1}{I_3} = \log \frac{1}{T_1 T_2} = \log \frac{1}{T_1} + \log \frac{1}{T_2} = D_1 + D_2 \tag{4}$$

Thus, when light-absorbing objects are superimposed, their optical densities add and their transmittances multiply. We shall use these properties in Chapter 7 to remove undesirable superimposed information by computer processing.

The Photographic Process

The construction of photographic film is illustrated in Figure 2-21. The film base is either glass or a flexible transparent acetate sheet that provides the film its mechanical stability. The base is coated with a 5- to 25-micron-thick emulsion made up of silver salt grains embedded in gelatin. The grains are halides of silver—silver chloride, silver bromide, or silver iodide crystals. During the manufacture of the film, the silver halide grains are activated to make them photosensitive. During exposure, different parts of the emulsion receive light at varying intensities. When a silver halide grain absorbs a photon, one or more molecules are reduced to silver, and the grain becomes "exposed."

The development process reduces the silver halide grains to silver; however, the reduction reaction proceeds much more rapidly on the exposed grains. After a suitable period of time, most of the exposed grains and only a few of the unexposed grains have

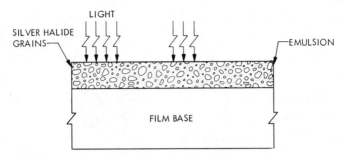

FILM DURING EXPOSURE

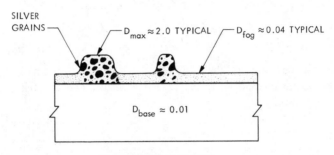

FILM AFTER DEVELOPMENT

Figure 2-21   The photographic process

been reduced. Finally, the unreduced grains are washed off the base. Thus, the developed film shows a silver coating of varying thickness. In areas that have been heavily exposed, the entire emulsion thickness is maintained, giving a maximum density. In unexposed areas, the silver halide grains are almost completely removed, leaving only a "fog level" of approximately 0.04 optical density.

Figure 2-22 shows a convenient means for characterizing the light response of an emulsion. It is called the "D/log E curve" or the "H and D curve" after its original proponents, Hurter and Driffield. It shows the density of the developed film as a function of the logarithm of exposure. For reasonable exposure times from milliseconds to seconds, exposure may be taken as the product of incident radiant energy flux density times duration. This equivalence between intensity and duration is called the *reciprocity law*. The breakdown of this relationship at extremely long or short exposure durations is called *reciprocity failure*.

The gross fog level (emulsion fog level plus base density) sets the minimum density for unexposed film. The maximum density is also limited. Over a relatively wide range of exposure, the relationship between density and log exposure is approximately linear. This is the normal working range of a photographic emulsion. The abscissa length of the linear portion is the *latitude* of the emulsion. The slope of the curve in the straight portion is called the *gamma* and represents the contrast of

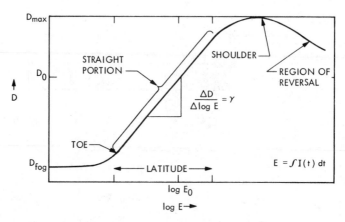

Figure 2-22 The emulsion characteristic curve

the emulsion. Beyond the shoulder of the curve is a region of reversal, where continued exposure brings about a decrease in density. The chemistry of this phenomenon is not well understood.

The emulsion thickness and grain size determine several important characteristics of the film. For example, a high maximum density is possible only with a thick emulsion. High resolution, however, requires small grains in a thin emulsion to avoid light scatter within the emulsion. Highly sensitive films that must work at low light levels require a thin emulsion containing relatively few grains. Therefore, any film is a compromise among the opposing constrains of resolution, maximum density, and sensitivity. So many different emulsions are available, however, that a suitable compromise usually can be found. In general, the lower the speed rating (sensitivity) of an emulsion, the higher its resolution and gamma will be, while its granularity and latitude will be lower. Granularity is due to the random but nonuniform grain distribution within the emulsion. It results in the subjective phenomenon of graininess and becomes more pronounced as density increases.

Maximum resolution of a low-contrast image is obtained when it is exposed and developed to lie between about 0.8 and 1.2 OD on the film. Below that range, the toe of the H and D curve reduces the contrast, while granularity becomes more of a problem at higher densities.

The H and D curve for a particular emulsion varies with the parameters of the development process. For example, overdevelopment tends to shift the curve to the left and increase the gamma. Obtaining predictable, reproducible results requires careful control of exposure and development parameters.

While the H and D curve illustrates an emulsion's response to light, it says nothing about the resolution of the film. The modulation transfer function (Figure 2-23) is a common way to specify the resolution characteristics of an emulsion. Suppose we expose an emulsion with a spatially periodic pattern of light intensity given by

$$\log E = \log E_0 + \sin (2\pi f x) \tag{5}$$

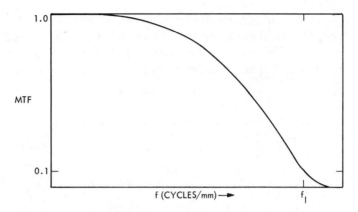

Figure 2-23   The modulation transfer function

where $\log E_0$ falls in the central part of the straight portion of the H and D curve. From the H and D curve, one would expect the density to be

$$D(x) = D_0 + \gamma \sin (2\pi f x) \tag{6}$$

When the spatial frequency f is high, however, grain size and light scatter within the emulsion reduce the contrast of the sinusoidal density variations. Thus, the observed density is

$$D(x) = D_0 + \gamma M(f) \sin (2\pi f x) \qquad 0 \le M(f) \le 1 \tag{7}$$

where $M(f)$ represents the loss of image contrast as a function of spatial frequency. To further simplify the specification of film resolution, manufacturers often refer to the "frequency of limiting resolution," f_L. This is the spatial frequency at which the modulation transfer function falls to 0.1, and it corresponds roughly to the limit of visibility.

Photocopying

Frequently, one must work with film images that are not originals but photographic copies of other film images. Figure 2-24 illustrates the setup for photocopying and

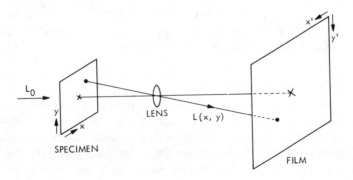

Figure 2-24   Photocopy configuration

photomicrography. Let the density of the original be given by $D_s(x, y)$ and assume that the copy film has the characteristic shown in Figure 2-25. The specimen is illuminated from behind with intensity I_0 for duration T. This means that the amount of exposure coming from (x, y) and reaching point (x', y') on the film is given by

$$E(x', y') = I_0 T\, 10^{-D_s(x,y)} \tag{8}$$

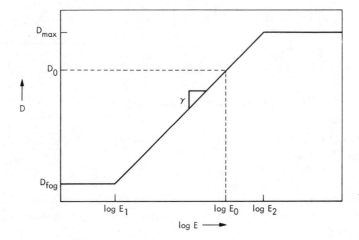

Figure 2-25 Idealized emulsion characteristic curve

Letting $E_0 = I_0 T$ and taking the log of both sides yields

$$\log [E(x', y')] = \log E_0 - D_s(x, y) \tag{9}$$

In the linear region of Figure 2-25, density is given by

$$D(\log E) = D_{\text{fog}} + \gamma(\log E - \log E_1) \tag{10}$$

Combining with Eq. (9) produces

$$D(x', y') = D_{\text{fog}} + \gamma[\log E_0 - D_s(x, y) - \log E_1] = D_0 - \gamma D_s(x, y) \tag{11}$$

where

$$D_0 = D_{\text{fog}} + \gamma(\log E_0 - \log E_1) = D(\log E_0) \tag{12}$$

Equation (11) illustrates that the copy is a negative image with density falling below D_0 and with contrast modified by the factor γ.

Electronic Log Conversion

In order to digitize with gray levels proportional to optical density in a film scanner, it is necessary to quantize a signal that is proportional to the negative of the logarithm of transmittance. Figure 2-26 shows an electronic circuit for logarithmic conversion of the output of an intensity sensor. Each pixel location on the film is illuminated with a beam of intensity L_i. The film attenuates the beam intensity to L_t before it strikes

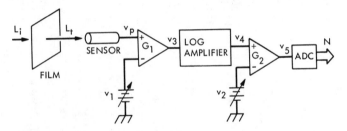

Figure 2-26 Electronic log conversion circuit

the photomultiplier tube. The sensor generates an output voltage given by

$$v_p = v_d + kL_t \tag{13}$$

where v_d is the dark current component and k is the sensitivity coefficient. The signal passes into a differential amplifier whose output is

$$v_3 = G_1(v_p - v_1) \tag{14}$$

where the gain G_1 and the offset voltage v_1 are adjustable. The logarithmic amplifier produces an output

$$v_4 = \log v_3 \tag{15}$$

Finally, a second differential amplifier produces

$$v_5 = G_2(v_4 - v_2) \tag{16}$$

where G_2 and v_2 are again adjustable. The analog-to-digital converter can be adjusted to produce an integer N in response to v_5 such that

$$N = \text{integer}(25.5v_5) \tag{17}$$

That is, N goes from 0 to 255 linearly as v_5 goes from 0 to 10 volts (v).

Suppose we want the digital output N to be proportional to film density over the range from D_1 to D_2, where $D_1 < D_2$. We now demonstrate that the circuit in Figure 2-26 is sufficient for the task.

We can combine Eq. (13) through (16) to express

$$v_5 = G_2\{\log [G_1(v_d + kL_t - v_1)] - v_2\} \tag{18}$$

In order for N to go from 0 to 255 as film density goes from D_1 to D_2, we must have

$$v_5 = \frac{10}{D_2 - D_1}(D - D_1) = \frac{10}{D_2 - D_1}(\log L_i - \log L_t - D_1) \tag{19}$$

We can begin to work Eq. (18) into the proper form if we let

$$G_2 = \frac{-10}{D_2 - D_1} \tag{20}$$

Then we can write Eq. (18) as

$$v_5 = \frac{10}{D_2 - D_1}\{v_2 - \log [G_1(v_d - v_1 + kL_t)]\} \tag{21}$$

Further progress results from letting

$$v_1 = v_d \qquad (22)$$

and

$$G_1 = \frac{1}{k} \qquad (23)$$

Now

$$v_s = \frac{10}{D_2 - D_1}(v_2 - \log L_t) \qquad (24)$$

Clearly, setting

$$v_2 = \log L_t - D_1 \qquad (25)$$

produces Eq. (19), as desired. Notice that we are letting v_1 compensate for v_d [Eq. (22)] and G_1 compensate for k [Eq. (23)]. Furthermore, v_2 offsets L_t [Eq. (25)] and, with G_2 [Eq. (20)], sets the density range of the digitizer.

The preceding exercise shows that the circuit in Figure 2-26 is adequate to produce integers that are proportional to optical density if the two gains and the two offset voltages are properly adjusted.

BIBLIOGRAPHY

D. G. FINK, ed., *Television Engineering Handbook*, McGraw-Hill Book Company, New York, 1957.

D. H. SHEINGOLD, ed., *Nonlinear Circuits Handbook*, Analog Devices Inc., Norwood, Massachusetts, January, 1976.

HEWLITT PACKARD OPTOELECTRONICS STAFF, *Optoelectronics Application Seminar Handbook*, Hewlett Packard, Palo Alto, California, May 1974.

TEXAS INSTRUMENTS ENGINEERING STAFF, *The Optoelectronics Data Book*, Texas Instruments Incorporated, Dallas, Texas, 1976.

F. POULIOT, "Test Your Logarithmic Amplifier IQ," *EDN*, **21**, No. 4, February 20, 1976.

C. E. K. MEES and T. N. JAMES, *The Theory of the Photographic Process*, The MacMillan Company, New York, 1966.

DIGITAL IMAGE DISPLAY

llllllllllllllllllllllllllllllllllllllllllllllllllllllllllllllllllllllllllllllllllllllllllllllllllllllllllllllllllll

INTRODUCTION

Image display is the final link in the digital image processing chain. After all processing is completed, the display transforms the digital image into a form suitable for human consumption. Strictly speaking, digital image display is not required for digital image analysis that produces its output in the form of numerical data or decisions. The display is useful in image analysis, however, and is required for digital image processing, which produces gray level images as output.

In this chapter, we consider the construction and characteristics of digital image display systems and, in particular, the factors that determine the quality of display systems. There are several display pitfalls that should be avoided if one is to produce images that do not call attention to themselves as being computer processed. In view of our philosophy that computer processing *per se* should not degrade image quality, we must not allow a carefully digitized and accurately processed digital image to be degraded by a noisy or inaccurate display system.

The human eye/brain is the ultimate consumer of the fruits of digital image processing. The eye is capable of resolving only about 40 gray levels. This means that if the range between black and white were divided into more than 40 equal intervals, adjacent gray levels would appear identical to the human eye. However, there is a built-in edge enhancement process on the retina. This makes it possible for the eye to detect gray level transitions much smaller than $\frac{1}{40}$ of the total range. For example, consider a gray scale target consisting of an array of 256 squares ranging in gray level from black to white. The normal human observer can easily see the boundary between adjacent squares even though the gray level difference is only one step in a range of 256 steps. However, if the boundary between adjacent squares is obscured by a narrow white strip, then adjacent squares appear to have equal brightness.

There are two basic types of displays, permanent and volatile. Permanent displays produce a "hard-copy" image on paper, film, or other permanent recording medium. Volatile displays produce a temporary image on a display screen. These are commonly used in interactive processing, since the image is visible only until it is replaced by another image or the system is deactivated.

The basic components of a display system are similar to those of a scan-in digitizer. The display differs in that it requires no light sensor, and the scanning spot intensity is controlled by the gray level values of the digital image being displayed. The similarity between scan-in digitizers and display systems is so great that many CRT devices can be used as both film scanners and film recorders.

Ordinarily, a display system produces an image in which either the brightness or the density of each display pixel is controlled directly by the gray level of the corresponding pixel in the digital image. However, the primary function of the display is to allow the human observer to understand and interpret the content of the image. Thus, in some cases, it is helpful to match the display to the characteristics of the human eye. For example, the human eye has considerable acuity in discriminating fine detail (high-spatial-frequency information), although it is not particularly sensitive to low-frequency (slowly varying) image information. Some images may be more easily understood if they are displayed indirectly by using contour lines, derivatives, or some graphical representation. Examples of such displays appear later in this book.

DISPLAY CHARACTERISTICS

In this section, we discuss those characteristics which, taken together, determine the quality of a digital image display system and its suitability for particular applications. The primary characteristics of interest are the size, photometric and spatial resolution, low-frequency response, and noise characteristics of the display.

Displayed Image Size

The image size capability of a display system has two components. First is the physical size of the display itself, which should be large enough to permit convenient examination and interpretation of the image. The second characteristic is the size of the largest digital image that the display system can handle. Clearly, the display should be adequate for the number of lines and the number of samples per line in the largest image to be displayed. Such operations as displaying images in pieces and pasting them together prove cumbersome and usually impractical. The trend in contemporary image processing is toward images with more and more pixels. Inadequate display size can significantly reduce the effectiveness of an image processing installation.

Photometric Resolution

For display systems, photometric resolution refers to the accuracy with which the system produces the correct brightness or density value at each pixel position. Of particular interest is the number of discrete gray levels that the system can generate. This is partially dependent on the number of bits used to control the brightness of each pixel. Some displays are capable of handling only 4-bit data, therefore producing only 16 distinct gray levels, while others handle 8-bit data for 256 gray levels. However, it is one thing to design a display with 8-bit data capacity and quite another to produce a system that can reliably display 256 distinct levels of gray. The effective number of gray levels is never more than the number of gray levels in the digital data. If electronic noise within the display system occupies more than 1 gray level, then the effective number of gray levels is reduced. As a rule of thumb, the root-mean-square (RMS) noise level represents a practical lower limit for gray scale resolution. For example, if the RMS noise level is 1% of the total display range from black to white, then the display can be assumed to have a photometric resolution of 100 shades of gray. If the display system accepts 8-bit data, it still has only 100 effective gray levels. If it is a 6-bit display system, it has 64 gray levels. The RMS noise level is used because, if the noise is assumed to have a normal distribution, then it will stay between ± 1 standard deviation about 68% of the time.

Gray Scale Linearity

Another important display characteristic is the linearity of the gray scale. By this we mean the degree to which brightness or density is proportional to input gray level. Any display device has an input gray level to output density or brightness transfer curve. For proper operation, this curve should be reasonably linear and constant from one usage to the next. With permanent displays involving a film recorder and subsequent development and enlargement, careful quality control is required for reproducible results. Fortunately, the human eye is not a very accurate photometer. Slight nonlinearities in the transfer curve, as well as 10 to 20% intensity shading across the image, are hardly noticed. If the transfer curve has a noticeable shoulder or toe at one end or the other, however, then information may be lost or degraded in the light or dark gray areas. On volatile displays involving television monitors, the transfer curve depends in part on the brightness and contrast control settings on the monitor. Thus it is possible for the user to alter the transfer curve to suit his displayed image and personal taste. In most cases, however, the processing should be done by the computer and not the display system, which should merely present the data to the operator without "enhancement."

Low-Frequency Response

In this section, we consider the ability of a display system to reproduce large areas of constant gray level, or "flat fields." This ability depends primarily on the shape of the display spot, the spacing between the spots, and the amplitude noise and position noise characteristics of the display system.

Flat Field Response. An important component of display quality is the accuracy with which the system can display a "flat field" or region of equal amplitude pixels. This depends primarily on the amount of shading present and on display spot shape and pixel spacing in relation to spot width. Flat field response is somewhat a matter of taste since some image interpreters prefer to have the pixels visible in the displayed image while others do not. If our goal is to minimize the visible effects of digital processing, we prefer flat fields to be displayed with uniform intensity.

A flat field can, of course, be displayed at any shade of gray between black and white. On a cathode-ray tube display, for example, a high-intensity pixel is displayed as a bright spot on an otherwise dark tube face. Zero-intensity pixels leave the tube face in its intrinsic dark state. In a CRT film recorder, a high-intensity pixel leaves a black spot on otherwise transparent film. Zero-intensity pixels leave the film transparent. Thus any display system has a characteristic pixel polarity. No matter what the display polarity, zero-intensity flat fields are displayed uniformly flat. Therefore, flat field performance becomes a problem only at intermediate- and high-intensity gray levels. These may be either black or white, depending on the display system polarity.

Flat field performance depends primarily upon how well the pixels "fit" together. Certain rotating drum film recorders image a rectangular aperture on the sensitized film, producing sharp-edged rectangular pixels that fit together accurately, creating excellent field flatness. However, CRT devices, which are common for digital image display, use a rectangular array of circular spots. In this section, we examine the factors that affect field flatness with circular spots. We model these circular display spots with the Gaussian function. Physically realizable CRT spots may not be exactly Gaussian in shape, and the exact shape of a CRT spot is rather difficult to measure. Nevertheless, the Gaussian provides a mathematically tractable model and, for accuracy to a few percent, a reasonable model for the study of display spot interaction.

When using a CRT display, it is customary to focus the electron beam to produce the smallest possible display spot on the tube face. This is usually not an operator function but is performed periodically by maintenance personnel. The selection of display spot spacing is much simpler and usually can be done by the operator to suit his individual needs or taste. Therefore, we address the problem of selecting display spot spacing in terms of the spot radius. Display using a defocused (enlarged) spot is possible but uncommon in practice.

The Gaussian Display Spot. Assume that the display spot has a two-dimensional Gaussian intensity distribution of the form

$$p(x, y) = e^{-(x^2+y^2)} = e^{-r^2} \tag{1}$$

where r is radial distance measured from the center of the spot. If we define R as the radius at which the intensity drops to one-half its maximum value, we can write the spot profile function as

$$p(r) = e^{-(r/R)^2 \ln(2)} \tag{2}$$

since this has the form of Eq. (1) and takes the value 0.5 at $r = R$. We may rearrange the exponent to yield

$$p(r) = e^{\ln(2^{-r^2/R^2})} \tag{3}$$

which is simply

$$p(r) = 2^{-(r/R)^2} \tag{4}$$

Thus the Gaussian profile can be expressed as a negative squared exponent of 2, as illustrated in Figure 3-1. The intensity distribution of a single spot becomes

$$p(x, y) = 2^{-[(x^2+y^2)/R^2]} \tag{5}$$

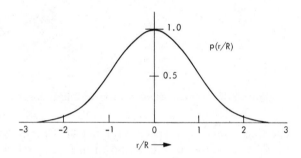

Figure 3-1   The Gaussian spot profile

which is depicted in Figure 3-2. We denote the display intensity by $D(x, y)$, which reflects the contributions of all the spots.

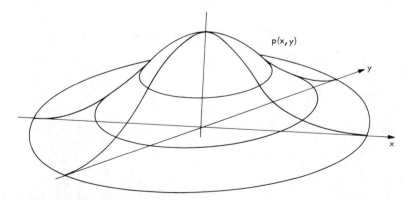

Figure 3-2   The intensity distribution of the Gaussian spot

Display Spot Interaction. Since the Gaussian spot does not fall below 1% of its peak amplitude until a distance of about $2\frac{1}{2}$ radii from center, display spots overlap unless they are rather widely spaced. Figure 3-3 illustrates the density distribution along a line connecting two adjacent equal-amplitude Gaussian spots separated by a distance $d = 2R$. Notice that there is a 12.5% variation in intensity between the spot centers and the midway position. Thus, $d = 2R$ cannot yield microscopically flat fields.

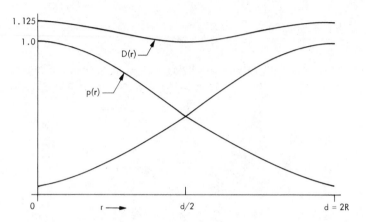

Figure 3-3 Overlap between adjacent Gaussian spots

In selecting display spot spacing for flat field performance, we are concerned with the three "worst-case" positions shown in Figure 3-4. These positions are pixel center, mid-pixel (midway between two pixels), and mid-diagonal (midway between four pixels). Ideally, pixel spacing would be chosen to make $D(x, y)$ equal at all three positions. We may write the display density at pixel center in a flat field of unit amplitude spots as

$$D(0, 0) \approx 1 + 4p(d) + 4p(\sqrt{2}\,d) \tag{6}$$

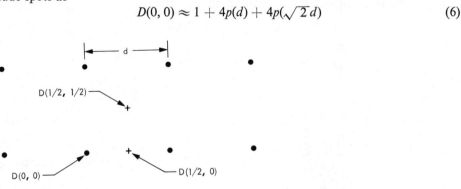

Figure 3-4 Critical positions for flat field display

since only the eight nearest neighbor spots contribute 1% or more to the density for $d \geq \sqrt{2} R$. Similarly, we can write

$$D\left(\frac{1}{2}, 0\right) \approx 2p\left(\frac{d}{2}\right) + 4p\left(\sqrt{5}\,\frac{d}{2}\right) \tag{7}$$

for the mid-pixel position, accounting for six neighbors. Finally,

$$D\left(\frac{1}{2}, \frac{1}{2}\right) \approx 4p\left(\sqrt{2}\,\frac{d}{2}\right) + 8p\left(\sqrt{10}\,\frac{d}{2}\right) \tag{8}$$

accounts for 12 spots surrounding the mid-diagonal position.

Figure 3-5 shows a plot of Eq. 6, 7, and 8 in the range $2R \leq d \leq 3R$ for the Gaussian spot. Notice that no choice of d makes the density equal at all three points. The best field flatness falls in the range $1.55R \leq d \leq 1.65R$. At the intuitive choice of $d = 2R$, there is a 26% intensity variation. At $d = 3R$, the pixels are clearly visible in high-intensity areas of the displayed image.

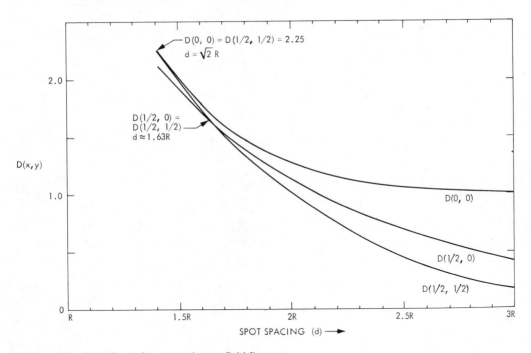

Figure 3-5 The effect of spot overlap on field flatness

High-Frequency Response

At the opposite end of the image content spectrum from flat fields lies fine detail. How accurately a display system can reproduce fine detail again depends on spot interaction and, hence, spot shape and spacing. We now consider the effect of spot spacing on the fidelity of reproduction of two "worst-case" images containing fine detail.

The High-Frequency Line Pattern. A common high-frequency test pattern consists of alternating light and dark vertical (or horizontal) lines spaced one pixel apart. These are sometimes referred to as "line pairs," where a pair consists of one dark and one adjacent light line. Every second column contains high-intensity pixels, while the columns in between contain zero-intensity pixels. How well a display system can reproduce a line pattern gives an indication about its performance on fine image detail.

Figure 3-6 shows the positions of interest when a high-frequency vertical line pattern is displayed. Bold dots represent pixels of unit amplitude, and small dots imply zero amplitude. We can write the pixel center density on the lines as

$$D(0, 0) \approx 1 + 2p(d) + 2p(2d) \tag{9}$$

and between the lines as

$$D(1, 0) \approx 2p(d) + 4p(\sqrt{2}\,d) \tag{10}$$

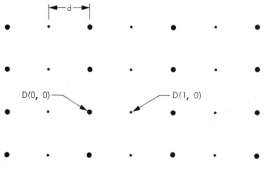

Figure 3-6 Critical positions for the vertical line pattern

Subtracting Eq. (10) from Eq. (9) yields

$$D(0, 0) - D(1, 0) \approx 1 + 2p(2d) - 4p(\sqrt{2}\,d) \tag{11}$$

the contrast of the displayed line pattern. The modulation factor

$$M = \frac{D(0, 0) - D(1, 0)}{D(0, 0)} \tag{12}$$

is graphed in Figure 3-7 as a function of spot spacing. Notice that the modulation depth falls off rapidly as spot spacing decreases below $2R$.

The Checkerboard Pattern. Another worst-case high-frequency display pattern is the single-pixel checkerboard. Here pixel intensity alternates both horizontally and vertically. The critical positions for this pattern are shown in Figure 3-8. The maximum density is given by

$$D(0, 0) \approx 1 + 4p(\sqrt{2}\,d) \tag{13}$$

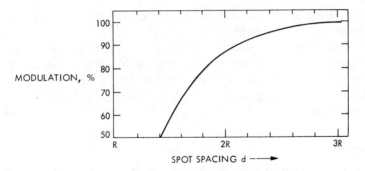

Figure 3-7 Spot spacing effect on the vertical line pattern

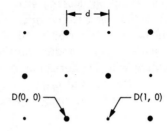

Figure 3-8 Critical positions for the
checkerboard pattern

and the minimum density by

$$D(1, 0) \approx 4p(d) + 8p(\sqrt{5}\,d) \tag{14}$$

The modulation factor, again given by Eq. (12), is plotted in Figure 3-9. The loss of

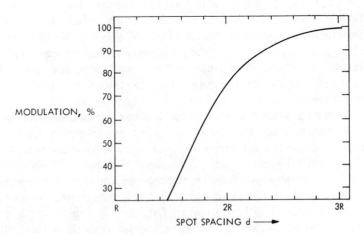

Figure 3-9 Spot spacing effect on the checkerboard pattern

modulation depth with decreasing spot spacing in the checkerboard pattern is even worse than in the line pattern.

Obviously, the goals of field flatness and high-frequency response place opposing constraints upon the selection of spot spacing. The actual compromise depends on the relative importance of high- and low-frequency information in each individual image. Spot spacing can be considered a display variable that must be matched to each image.

Noise

Electronic noise in a display system produces variations in both the intensity and position of the display spot. Random noise in the intensity channel can produce a salt-and-pepper effect particularly visible in flat fields. The previously stated rule of thumb indicates that the effective quantizing level is roughly equal to the RMS noise amplitude. If the noise is periodic and of reasonably high intensity, it can produce a herringbone pattern superimposed on the displayed image. If the noise is periodic and synchronized with the horizontal or vertical deflection signals, it can produce a pattern of bars. The general display quality is adequate if all noise, random and periodic, is kept at or below one gray level in amplitude. In many systems, it is much worse.

Spot Position Noise. A more serious effect is produced by noise in the display spot deflection circuits. This produces nonuniformity in display spot spacing. Display position noise, unless extremely severe, will not have a noticeable geometric effect upon the image. However, the effects of spot interaction combine with position noise to produce considerable amplitude variation. Recall from Figure 3-5 that variations in spot spacing cause considerable change in the mid-pixel, pixel center, and mid-diagonal intensities of flat fields. As an example, suppose that a 1000 by 1000 pixel display uses a spot spacing equal to twice the spot radius. Notice from Figure 3-5 that, as spot spacing goes from $1.9R$ to $2.1R$, mid-diagonal intensity increases from about 0.87 to 1.16. This reflects a 19% change using 1.0 as an intensity reference. However, a spot spacing variation of $0.2R$ is only 0.01% of full-scale deflection. Thus 0.01% peak-to-peak noise in the deflection circuit produces a 19% variation in mid-diagonal amplitude. Pixel center and mid-pixel amplitudes are also affected but to a lesser degree. At spot spacings less than $2R$, the effect of position noise is even more pronounced.

When position noise is random, it produces a salt-and-pepper effect throughout the displayed image. Position noise is most visible in high-intensity flat fields, where spot interaction is most obvious. Frequently, nonrandom position noise is introduced by inaccurate digital-to-analog converters. The analog deflection signal is often produced by digitally switching resistances into and out of a resistive voltage divider network. If these resistance values are not precise, the conversion, and thus the deflection signal, will be inaccurate. Digital-to-analog converter noise produces fine vertical and/or horizontal lines at regular intervals throughout the image. Because spot interaction effects amplify position noise, careful display design requires precise pixel position control.

Numerous other techniques exist to improve the flat field and high-frequency response of display systems. For example, a hexagonal sampling grid may be used for some improvement of flat field response. In some cases, it is possible to control pixel shape for better pixel overlap characteristics. In many systems, however, the pixel shape is beyond the operator's control. In Part II, we develop analytical techniques to describe the effects of pixel shape and spot spacing.

DISPLAY TECHNOLOGIES

Display systems come in two types, volatile displays and permanent displays or hard-copy devices. Hard-copy devices produce an image by permanently altering the light-absorbing characteristics of a sensitized medium such as paper or film. Volatile displays produce an image on a display screen, but that image remains visible only while the display is active.

Volatile Displays

The most common volatile display uses a cathode-ray tube, scanned in raster fashion, while the pixel spot intensity varies with position to produce the image. An ordinary television monitor can act as a digital image display if it is provided with a suitable video signal. Since the display spot continuously scans the image, the display must be continually "refreshed." Refreshing can take place from a stored digital or analog image. Volatile displays can be refreshed digitally from a digital image stored in a dedicated core memory, a solid-state shift register or random access memory, or a digital disk system. Display systems can be analog refreshed either from an analog video signal stored on disk or an electron image stored on a scan converter tube. The scan converter tube is similar to the vidicon except that an electron image is first written on the target under digital control and then scanned repeatedly to generate the analog display signal. Well-designed scan converter tubes can refresh a CRT display for up to $\frac{1}{2}$ hour before the electron image on the target is significantly degraded.

Photochromic displays use a cathode-ray tube whose target is coated with a photochromic substance rather than an electroluminescent phosphor. The electron beam changes the light-absorbing characteristics of the photochromic coating so that the displayed image may be viewed under external illumination. Once the image is written on the photochromic target, it remains until it is erased by the electron beam.

Laser displays can be built using moving mirrors or other means for beam deflection and a Kerr cell for intensity modulation of the beam. Gas discharge displays are made by sandwiching a fine mesh between two sheets of glass, leaving a rectangular array of cells containing an ionizable gas. By using coincident horizontal and vertical addressing techniques, the cells can be made to glow under the influence of a permanent sustaining electrical potential. Currently, these displays are limited in their available number of gray levels.

Several promising new types of solid-state displays are on the horizon. These promise to be compact and relatively inexpensive, using liquid crystal and light-emitting diode technology.

Permanent Displays

The most common permanent display is the CRT film recorder. This is basically a film camera, mounted in front of a CRT display. With the shutter open, the entire image is displayed, pixel by pixel, to expose the film. Since only one pass is needed, no refreshing is required. Pixel intensity can be modulated by controlling either the brightness of the spot or the time duration for which each pixel is displayed. Pixel exposure on the film is nominally proportional to the product of exposure intensity and exposure time. If spot intensity modulation is used, the phosphor brightness must be a linear function of beam current or any nonlinearities must be compensated. Since the CRT is subject to geometric distortion and to loss of spot focus at the periphery, high-quality film recorders incorporate dynamic geometric distortion correction and dynamic focus correction circuits to maintain the accuracy of spot position and focus.

Drum feed display devices use a slowly rotating drum to pull roll paper or film past a linear scanning mechanism operating perpendicular to the direction of paper motion. The linear scanning element can be a cathode-ray tube performing a single line scan. The scan line is imaged on the paper to effect exposure. The scanning mechanism might also be a laser beam exposing photosensitive paper or an electric current exposing electrolytic paper. Electrolytic paper is sensitized so that a localized current through the paper causes that portion to darken. In general, the degree of darkening is proportional to the current, and this provides a means to modulate pixel intensity. Better results are obtained by varying the area of a solid black pixel, since the electrolytic process is considerably more repeatable if carried to saturation. An electrostatic charge image can be written on the paper and used to attract a powdered toner.

The rotating drum display uses a rotating drum and lead screw arrangement to expose a single sheet of film (recall Figure 10 in Chapter 2). Rotating drum film recorders use an objective lens to image an aperture upon the unexposed film. The aperture is illuminated by light source, typically a light-emitting diode. Light source intensity is modulated by the gray level of the digital image. Frequently, the size and shape of the aperture as well as the horizontal and vertical pixel spacing are adjustable.

Other display technologies can be combined to produce systems not enumerated here. We have covered the fundamentals of digital image display at the current state-of-the-art, however, and this should provide a basis upon which to evaluate the adequacy of individual display systems for particular image display tasks.

Future developments in the rapidly expanding field of solid-state electronics promise to improve display technology in the near future. We can look forward to compact, efficient, high-quality digital image display devices at reasonable cost.

BIBLIOGRAPHY

H. R. LUXENBERG and R. L. KUEHN, eds., *Display Systems Engineering*, McGraw-Hill Book Company, New York, 1968.

D. FINK, ed., *Television Engineering Handbook*, McGraw-Hill Book Company, New York, 1957.

H. POOLE, *Fundamentals of Display Systems*, MacMillan and Company, London, 1966.

J. H. HOWARD, ed., *Electronic Information Display Systems*, Spartan Books, Inc., Washington D.C., 1963.

D. M. AVEDON, *Computer Output Microfilm*, National Microfilm Association, Annapolis, Maryland, 21401, 1969.

F. C. BILLINGSLEY, "Noise Considerations in Digital Image Processing Hardware," in T. S. Huang, ed., *Picture Processing and Digital Filtering*, Springer-Verlag, New York, 1975.

IMAGE PROCESSING SOFTWARE

IIIIIIIIIIIIIIIIIIIIIIIIIIIIIIIIIIIIIIIIIIIIIIIIIIIIIIIIIIIIIIIIIIIIIIIIIIIIIIIIIIIIIIIIIIIIIIIIIIIIIIIIIIIIIIIIIIIIIIIIIIIIIIIIIIIIIIIIIIIIIIIIIIIIIIIIIIIIIIIIIIIIII

INTRODUCTION

Software to implement digital image processing can be organized in many different ways. In the early stages of a developing system, the emphasis is usually placed on making the first few programs work, with little thought given to overall software organization. However, the software requirements of image processing are somewhat different from those of general computing. Available compilers and operating systems gradually become awkward as the user is bogged down in data handling details. At that point, it becomes expedient to devote some effort to the software that handles the myriad of details accompanying the execution of image processing programs.

In this chapter, we discuss image processing software organization principally by describing one system—the VICAR (Video Image Communication and Retrieval) system, developed at JPL and used extensively at several other institutions (Ref. 1). This illustrates how a well-designed software system can ease the burden of those who write and use image processing programs. VICAR was designed to support general-purpose image processing, and it illustrates a straightforward and effective approach to image processing software organization.

A well-designed image processing software system simplifies both program development and production jobs. It should be simple for the uninitiated user to learn, and should rescue the user from the repetitious tasks of parameter and label processing, tape positioning, file management, and data buffering.

It should be pointed out that a special software system is not strictly necessary, since the operating system supplied with most computers includes utility routines for call-up and execution of programs stored in a user library. However, a system that significantly simplifies this task can do much to promote image processing by users from diverse backgrounds outside computer science.

Three groups of software are involved in image processing with the VICAR system. They are the system supervisor or operating system, the VICAR control programs, and the VICAR application programs. The system supervisor is a manufacturer-supplied program that controls the input and execution of jobs. The VICAR control programs work with the system supervisor to simplify data management and program execution for the user. The application programs perform the actual image processing operations.

For example, the five statements in Figure 4-1 are all that are required (excluding system supervisor commands to get into and out of VICAR) to define an input tape and an output tape, reserve a disk data set for temporary image storage, and execute two library programs (STRETCH and ROTATE) on an image from the first tape, leaving the output image on the second tape. The statements in Figure 4-1 are considerably simpler than those required to do the same job with utility routines.

```
TAPE,080,, IN
TAPE,081,,OUT
RESERVE, 2, 1000, 1000, 190, TEMP
EXECUTE, STRETCH, IN, TEMP
EXECUTE, ROTATE,TEMP,OUT
```

Figure 4-1   A VICAR example

Application Programs

The programs that actually perform the image processing operations are the application programs stored in the VICAR library (Ref. 2). These are written in either FORTRAN or assembly language. Application programs are written, debugged, documented, and then loaded permanently into the VICAR library (Ref. 3). Using the VICAR documentation (Ref. 2), the user selects the programs required to perform a desired function (see Appendix II). The program documentation explains what the program does and how it is used. Many programs have a variety of options that may be selected by appropriate keywords in the execution parameter set.

Figure 4-2 illustrates the execution of a VICAR application program. One or more input images reside in tape or disk data sets. The program reads the input images, line by line, into memory as needed. Using this data, it computes the output image, line by line, and writes it on the output data set. Certain complex programs produce more intermediate data storage than can be accommodated in memory. These programs use an intermediate data set for temporary storage during execution.

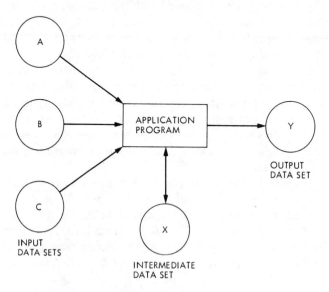

Figure 4-2 Execution of a VICAR task

Notice that the processing is nondestructive, since the images in the input data sets are preserved. Rather than transforming an input image into an output image, the application program uses the input image to guide the generation of an output image. An image is destroyed only when its data set is used for output or is deleted at the end of a job.

Jobs and Tasks

Figure 4-3 shows the configuration of a computer system on which VICAR might run. The user submits a VICAR job consisting of several tasks. Each task involves the loading and execution of a previously written application program. In the job, the user specifies where each input image is located, which application programs are to be executed, and where the computer should place the output images. The VICAR system reads the job deck and executes programs from the library as requested. Typically, the input images are supplied on digital magnetic tape. The VICAR job copies the images into temporary disk data sets. The processing tasks involve accessing one or more input images on disk and generating an output image that is stored in a

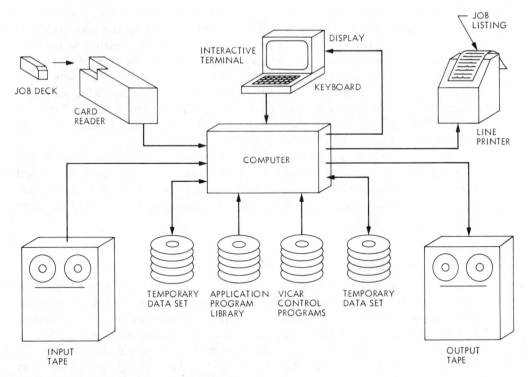

Figure 4-3 An image processing system

separate disk data set. At the end of the job, the final processed images are copied to digital tape for long-term storage. The line printer lists the VICAR job deck and the execution details of each task. In cases of abnormal task or job termination, the listing becomes a valuable diagnostic aid.

For interactive processing, the user sits at a console consisting of a keyboard and an image display device. The VICAR control statements are entered through the keyboard rather than the card reader. Processed images may be displayed at the terminal. Output images to be saved are written on tape.

Image Format and Labels

In the VICAR system, each image is stored in a tape or disk data set. Each data set is a file containing one image. End-of-file marks separate the images. The last image on a tape is followed by two end-of-file marks. Each line is stored as one logical record. Each image is preceded by one or more 360-byte VICAR label records. Data sets may be "blocked"—written with several logical records per physical record—to reduce input/output time.

The VICAR label has three components: the system label, the user labels, and the history labels. The system label contains such information as the number of lines in the image and the number of samples per line. An application program can read the system label, obtain the size information, and then set up the parameters for line-by-line input. The user labels are strings of alphameric text supplied by the user. They generally contain descriptive text designed to aid identification of the image. Each time the image is displayed, the user labels appear with it. This aids in the identification of similar images. The history labels are also alphameric but are added automatically. They contain the names of application programs that have been executed on the image. By reading the history labels on a displayed image, the user sees a convenient summary of its processing history.

Each time before an application program is executed, the labels from the primary input data set are copied to the output data set. The system label is modified if the size of the image is to be changed by the process. New user labels, if specified, are added. Finally, a new history label is added to indicate the name of the application program about to be executed.

Execution Parameters

Generally, an application program requires a set of execution parameters. This user-supplied information specifies the details of the processing step. For example, the application program STRETCH performs contrast enhancement, but a set of parameters must be supplied to specify the exact contrast enhancement desired. The user may elect not to specify some or all of the parameters. In this case, the program uses preestablished "default" values. The documentation for each application program describes its parameters and lists their default values.

Many of the VICAR application programs have numerous parameters, most of which are defaulted at any one time. These programs allow the user to specify parameters in the format of a keyword, followed by a numerical value. The keyword designates which of the parameters is to be specified, and the subsequent number establishes its value. This format allows the user to specify only those parameters that he chooses not to default. Keywords are also useful for selecting among several major execution options offered by some application programs. The keywords for each program and their functions are described in the application program documentation (Ref. 1).

Control Programs

In addition to the application program library, there are three programs which comprise the VICAR system: VTRAN, VMAST, and VMJC (Refs. 1, 3). Their functions are summarized in Table 4-1.

VTRAN is the VICAR language translator. It reads the VICAR control commands and uses them to generate job control language (JCL) that the system supervisor uses to control the processing. Its operation is diagrammed in Figure 4-4. VTRAN generates a disk data set called JCFILE. This data set contains the JCL for allocation

Table 4-1. **VICAR control programs**

Program	Functions
VTRAN	VICAR language translator. Converts VICAR control statements into JCL and generates task queue.
VMAST	VICAR resident supervisor. Loads VMJC between tasks, handles I/O during application program execution.
VMJC	VICAR transient supervisor. Loaded by VMAST and executed between tasks. Prepares for next task by translating parameters, processing labels, and positioning data sets. Loads application program for next task.

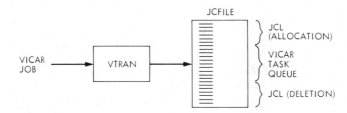

Figure 4-4 The operation of VTRAN

of the required data sets, the task queue, and the JCL required for deletion of data sets at the end of the job. The task queue specifies the application programs to be executed, their input and output data sets, and all execution parameters.

VMAST is the VICAR resident supervisor. It is loaded at the beginning of a job and remains resident until the job terminates. A portion of VMAST, called VMIO, handles input and output during the execution of each application program. VMAST loads another program, VMJC, before and after each application program execution.

VMJC is the transient portion of the VICAR supervisor (Figure 4-5). It prepares the system for each upcoming application program. This includes translating the parameters in JCFILE into a form acceptable by the application program, generating the label for the output data set, and positioning input and output data sets as required for the task. After this preparation, it is overlayed with the application program that is rolled in.

THE PROCESSING SEQUENCE

The three programs, VTRAN, VMAST, and VMJC, work together to effect the execution of a VICAR job. Table 4-2 shows the steps involved in the execution of a VICAR job.

instrx9sacapteosobpnaotnoregte

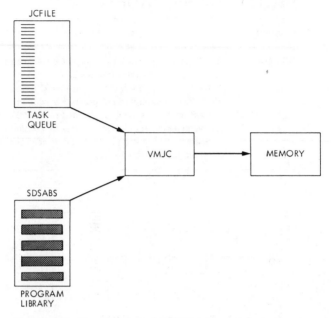

Figure 4-5 The operation of VMJC

Table 4-2. VICAR processing sequence

Steps	Actions
1	VICAR input deck requests VTRAN.
2	VTRAN translates VICAR statements, generating a task queue in JCFILE.
3	The system supervisor reads JCFILE and allocates data sets as specified.
4	The supervisor loads VMAST.
5	VMAST loads VMJC.
6	VMJC updates labels, translates parameters, and positions data sets for the upcoming task.
7	VMJC loads application program, overlaying VMJC, and initiates execution.
8	The application program executes, using VMIO, the I/O package in VMAST.
9	The application program terminates with a return to VMAST.
10	VMAST loads VMJC.
11	If there are more tasks in the queue, VMJC proceeds with step 6.
12	The supervisor reads JCFILE, deletes data sets, and terminates the job.

Control Statements

A VICAR job consists of a series of control statements. A partial list of the VICAR control statement functions is shown in Table 4-3. A complete list appears in Ref. 1. The statements are divided into five functional groups: those which allocate disk

Table 4-3. VICAR statement functions

Function Code	Purpose
RESERVE	To establish a disk data set for temporary image storage; it specifies the symbolic name and size of the data set and the physical device (disk drive) to be used.
TAPE	To establish a tape drive for input, output, or both; it specifies the symbolic name and physical device (tape drive) to be used.
EXECUTE	To define and effect execution of a VICAR task; it specifies the program to be executed, the input and output data sets, and the execution parameters to be used.
PARAMS	To define a set of execution parameters; it specifies the symbolic name of the parameter set.
LABEL, RELABEL	To append user-specified alphameric text to the label record of output data sets.

data sets, those which define tape data sets, and statements for program execution and parameter and label processing.

Disk data sets are commonly used for temporary image storage during the execution of a job. Each disk data set is referred to by a user-assigned symbolic name. The RESERVE statement allows the user to name a data set, define its size, and specify the disk drive to which it will be allocated. The TAPE statement similarly defines a tape data set.

The EXECUTE statement causes an application program to be called up from the library and executed. This statement specifies the program and the input and output data sets. If execution parameters are required, they may be listed in the EXECUTE statement or in a PARAMS statement to which the EXECUTE statement points.

A major convenience of the VICAR system is the ease with which the user can maintain alphameric labels attached to his images. These labels always appear when an image is displayed, and they can prevent considerable confusion. The LABEL and RELABEL statements allow the user to attach labels to images.

VICAR statements consist of up to 10 "fields" delineated by commas. Blanks are ignored. Each field may have up to 10 subfields also separated by commas. If a field contains more than one subfield, it must be enclosed in parentheses. The first field of a VICAR statement always contains the statement function code.

RESERVE Statement. The format of the RESERVE statement is illustrated in Table 4-4. Field 2 specifies the number of data sets to be reserved, and Fields 3 and 4 specify

Table 4-4. RESERVE statement format

Field	1		2		3		4		5		6
Content	Function		No. of Data Sets		Record Length		No. of Records		Device Address		Data Set Names
Example 1	RESERVE	,	1	,	900	,	605	,	190	,	A
Example 2	RESERVE	,	3	,	500	,	505	,	*	,	(A, B, C)
Example 3	RESERVE	,	3	,	600	,	705	,	PAK027	,	C

their size. Field 3 contains the record length in bytes and generally equals the number of pixels per line. Field 4 specifies the number of lines reserved. The size of a disk data set must always be at least as large as the largest image to be stored there during the job. Furthermore, since the VICAR system label occupies one or more lines, it is necessary to reserve a few extra lines.

Field 5 is used to specify the device (disk drive) to be used. In Example 1, the unit with system address 190 is specified. One may code Field 5 with an asterisk, as in Example 2, and the system supervisor will make the assignment. The user may specify the volume serial number of a particular disk pack, as in Example 3. In this case, the system selects a device and notifies the operator to mount the correct pack there.

In Field 6, the user may specify the symbolic names of the data sets reserved. Example 1 reserves one data set on device 190 and assigns it the symbolic name A. Example 2 reserves three data sets with the names A, B, and C. Example 3 also reserves three data sets, but only the first is assigned a symbolic name. The other two data sets are reserved unnamed. These "dummy" data sets may be used by the system as described later.

TAPE Statement. The format of the TAPE statement is shown in Table 4-5. Field 2 contains the device address which may be specified (Example 1) or left to the system

Table 4-5. TAPE statement format

Field	1		2		3		4		5
Content	Function		Device Address		Reel ID and Output DS Names		Symbolic Tape Name		Format Code
Example 1	TAPE	,	080	,		,	TAP	,	8
Example 2	TAPE	,	*	,	(MARS21,C)	,	OUTAPE	,	8F

supervisor (Example 2). In Field 3, the user may specify the reel identification label of his tape and the names of output data sets that he wishes to have saved on this tape. The user may leave this field blank (Example 1), specify either, or specify both (Example 2). In Example 2, the tape in use would have a label marked "MARS21" attached to the reel. Also, the disk data set C is specified as an output data set in Example 2. This means that after every task in which data set C is used for output, its contents are automatically copied to this tape.

Field 4 is used to specify the symbolic name of the tape. Throughout the job, the tape is referred to by this name. Field 5 specifies the data format of the tape. A partial listing of the acceptable VICAR tape formats is shown in Table 4-6. VICAR

Table 4-6. VICAR tape formats

Format Code	Format	Density	Bits per Pixel	VICAR Labels
6	9-track	1600 BPI	8	Yes
9	9-track	800 BPI	8	Yes
8	7-track	800 BPI	8	Yes
8A	7-track	800 BPI	8	No
8F	7-track	800 BPI	6	No

tapes are all written with one image per file and one line per logical record. The end of the last image on a tape is signified by two end-of-file marks. This format allows efficient packing of image data on a tape and facilitates input and output. The tape formats differ, however, in the number of tracks employed, the packing density, the number of bits per picture element, and the presence or absence of VICAR system labels. Most VICAR application programs require labeled input data sets.

In the TAPE statement, the user must specify the format code corresponding to the data format on his tape. In Table 4-5, Example 1, the seven-track input tape has 8 bits devoted to each pixel, and VICAR system labels are present. The output tape in Example 2 is a 6-bit tape with no VICAR labels. This is typical of tapes prepared for off-line display devices.

EXECUTE Statement. Table 4-7 illustrates the format of the EXECUTE statement. Field 2 contains the name of an application program residing in the library. Fields 3 and 4 specify the symbolic names of the input and output data sets, respectively. In Example 1, the program GUDSTF accesses input data sets A and B and writes its output in data set C. In Example 2, the input data set is the tape defined by Example 1 of Table 4-5. The number 3 following the tape name indicates that the third file on the tape will be used as input. Before execution of program HOTWUN, the tape named TAP will be positioned at the beginning of file 3. In Example 2, the asterisk in Field 4 indicates that program HOTWUN does not require an output data set. This is typical of programs that produce their output on a line printer or real-time display device.

Table 4-7. EXECUTE statement format

Field	1	2	3	4	5	6
Content	Function	Program Name	Input Data Sets	Output Data Sets	Output Size	Parameters
Example 1 ,	EXECUTE ,	GUDSTF ,	(A,B) ,	C ,	(1,1,500,600) ,	(MIN,0,MAX,255)
Example 2 ,	EXECUTE ,	HOTWUN ,	TAP3 ,	* ,	,	P1
Example 3 ,	EXECUTE ,	HEAVY ,	*TAP3 ,	C ,	(NL = 200) ,	
Example 4 ,	EXECUTE ,	BIGDIG ,	(TAP/1-4,6) ,	OUTAPE		

The VICAR system incorporates a feature known as *indirect copy*. The EXECUTE statement in Example 2 actually specifies that the image on file 3 of TAP be copied into an unnamed disk data set before HOTWUN is executed. This feature is useful when the same file is to be used as input for several tasks. On the first such EXECUTE statement, the image will be copied from tape to an unnamed disk data set. On subsequent executions involving file 3, the disk data set is used as input. This prevents repeated backspacing and rereading of the tape. Also, if TAP had been in an unlabeled format (i.e., 8F), this indirect copy step would have allowed VICAR to supply a system label, thus making the image acceptable as input to an application program. If the indirect copy is unnecessary or undesirable, it may be suppressed by coding an asterisk in Field 3, as illustrated in Example 3.

Frequently, it is desirable to execute a program on several files on a tape. While one could accomplish this with several EXECUTE statements, VICAR allows it to be done with one. Field 3 in Example 4 specifies that program BIGDIG is to be executed on files 1, 2, 3, 4, and 6 of TAP. This multi-file EXECUTE statement is the equivalent of five single-file EXECUTE statements.

Field 5 of the EXECUTE statement allows the user to specify the size of the output image in terms of the input image size. The convention for size specification is illustrated in Figure 4-6. The upper left-hand corner of the output image is specified by the address of the corresponding input pixel. These are referred to as the starting line (SL) and starting sample (SS) of the output image. The size of the output image is specified by its number of lines (NL) and number of samples (NS). In Example 1, the output image will start at line 1, sample 1 of the input image and will consist of 500 lines of 600 pixels each. In Example 2, Field 5 is defaulted (left blank). In this case, the starting line and starting sample are both assumed to be 1, and the numbers of lines and samples are taken to be the same as the input image. Thus the default case makes the output image the same size as the input image. In Example 3, the number of lines in the output image is set to 200, while the other three parameters are defaulted as before. This will produce an output image that has the same number of samples per line as the input but only the first 200 lines are processed. In Example 4, Field 5 and all subsequent fields are defaulted, and delineating commas are not required.

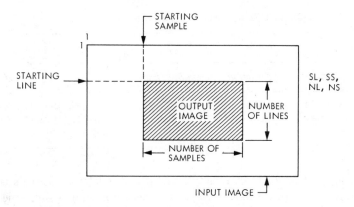

Figure 4-6 Output image size specification

    Execution parameters may be listed or referred to in Field 6. In Example 1, two parameters are given, and their format is a keyword followed by a value. In this case, the keywords are MIN and MAX, and the values are 0 and 255, respectively. Parameters may be numeric or alphameric but must be in a format acceptable to the particular application program. The documentation for each application program specifies its parameter format (Ref. 2).

    Since a VICAR statement field may contain no more than 10 subfields, this may be insufficient for parameter specification. In Example 2, Field 6 is coded with the name of a parameter set. Elsewhere in the program will be defined a set of parameters with the name P1.

PARAMS Statement. The statement that defines a parameter set is illustrated in Table 4-8. The PARAMS statement has only two fields. Field 2 contains the symbolic name of the parameter set. The parameters themselves are entered on lines following the PARAMS statement. All data between the PARAMS statement and the next valid VICAR control statement constitutes the parameter set P1.

Table 4-8. PARAMS statement format

Field	1	2
Content	Function	Parameter Set Name
Example	PARAMS	P1

LABEL Statement. The LABEL statement (Table 4-9) allows the user to attach alphameric label information to his output images as they are produced. Only three fields are required. Field 2 specifies the output data set name, and Field 3 contains

Table 4-9. LABEL and RELABEL statement format

Field	1	2	3
Content	Function	Data Set Name	Text
Example 1	LABEL ,	,	ENHANCED CHEST X-RAY
Example 2	LABEL ,	C ,	LANDSAT IMAGE OF CENTRAL TEXAS
Example 3	LABEL ,	* ,	CLOSEUP OF VICTIM

the alphameric text. In Example 1, Field 2 is left blank. This is adequate for the common case of a single LABEL statement following a single-file EXECUTE statement. Field 2 may again be left blank when a multi-file EXECUTE statement is followed by a series of LABEL statements, one for each file. If more than one label is to be added to the data set produced by a single-file EXECUTE statement, the data set name is coded in Field 2, as shown in Example 2. If a single label is to be added to all the outputs of a multi-file EXECUTE statement, an asterisk must be coded in Field 2 (Example 3). The RELABEL statement has the same effect as LABEL, except that all previous user labels are deleted before the new label is added.

Miscellaneous Statements. The VICAR language contains other statements besides those just discussed. For instance, it may be important for one VICAR job to communicate with a subsequent job. Ordinarily, data sets created in one job are deleted at the end of that job. The SAVE statement can prevent the deletion and allow a data set containing parameters or an image to be passed to a subsequent job. This job-to-job communication may be accomplished by using a SAVE statement in the first job to prevent the deletion of one of its disk data sets. The second job can use a FIND statement to access the previously saved data set. Finally, the second job may use the RELEASE statement to release the data set when it is no longer needed. The statements ALLOCATE and BLOCK are similar to RESERVE, and READ and WRITE are synonyms for TAPE. The END statement (optional) marks the end of a VICAR job.

DO-Groups. The multi-task feature of the EXECUTE statement allows one operation to be performed upon several files. Frequently it is desirable to have a series of two or more operations performed upon several files. VICAR has a loop option called a "DO-Group" that allows a series of programs to be executed on each of a series of images. An example is shown in Figure 4-7.

A DO-Group must begin with a multi-file EXECUTE statement. If such a statement begins a DO-Group, it is coded with a leading asterisk. All subsequent EXECUTE statements in the DO-Group, except the last, are also coded with leading asterisks. The last EXECUTE statement in the DO-Group is the first one without a leading asterisk.

In Figure 4-7, the DO-Group consists of three EXECUTE statements. The DO-Group operates on files 1 through 5 of tape TA. For each file, program SAR, after

```
READ, 080, , TA, 8A

WRITE, 081, , TB, 8F

RESERVE, 3, 190, 600, 505, (A, B)

*EXECUTE, SAR, (TA/1-5), A

*EXECUTE, STRETCH, A, B          DO-GROUP

EXECUTE, MASK, B, TB

EXECUTE, VUNLOAD, TA, *
```

Figure 4-7 VICAR DO-Group example

an indirect copy operation, is executed into data set A. Then program STRETCH is executed from A into B, and finally, program MASK is executed from data set B to tape TB. This four-step process (including the indirect copy) is repeated for each of the first five files of tape TA. At the end of the process, tape TB contains five files of processed images from tape TA. The final EXECUTE statement is not in the DO-Group and is executed only once.

Procedures. Frequently the VICAR user finds himself using the same series of VICAR statements several times in one job or in many different jobs. In these cases, the VICAR DO-Group cannot replace the repetitious statements, but an extension to VICAR, called "EVIL," can.

EVIL (Extensible Video Interactive Language) is a preprocessor that handles the VICAR job deck before it is passed to VTRAN. EVIL is written in TTM (Ref. 4), an interpretive language for character string manipulation. EVIL allows the user to define "procedures," which are series of VICAR control statements, and then insert each procedure, by name, into the VICAR job with a single statement. EVIL also has a procedure library that makes commonly used procedures available to all VICAR users.

EVIL adds three statements to the VICAR language—DEFINE, GET, and CALL. The DEFINE statement specifies the name of a procedure and lists the parameters that are passed to it. The procedure itself consists of the VICAR statements that follow the DEFINE statement, up to another DEFINE statement or an END statement. Wherever the user wants the procedure executed in his job, he merely inserts a CALL statement specifying the procedure name and listing the parameters to be passed to it. If the procedure already exists in the procedure library, the GET statement is used to fetch it and the DEFINE statement is not needed. At run time, EVIL removes the procedure definitions from the job, replaces the CALL statements with the appropriate procedures, and passes the modified job to VTRAN for a normal VICAR job execution.

When defining a procedure, the user must decide which parameters are to be passed from the CALL statement and which should be fixed in the procedure definition. For example, one might place into the procedure library a procedure to RESERVE four disk data sets whose size is fixed at 1000 by 1000 pixels but whose names are specified in the CALL statement in each VICAR job. This affords the users a convenient way to RESERVE and name four data sets with one statement. Also, one might define a procedure consisting of a series of EXECUTE statements, in which the size of the input image (NL, NS) is passed as a parameter. This allows the user to specify compactly the execution of a complex procedure upon a variety of different-sized input images in one job.

By making TTM available to the VICAR user, EVIL allows looping, nesting of procedure CALLs, and conditional executions based on numeric or logical comparisons. It greatly enhances the flexibility of VICAR, especially in interactive processing. Furthermore, a well-developed procedure library is a resource almost as valuable as the application program library.

Application Programs

Application programs are generally written in FORTRAN for convenience of programming. After a program becomes operational, it may be analyzed to determine which portions consume the greatest amount of execution time. Critical portions may be recoded more efficiently as assembly language subroutines to reduce execution time. A partial listing of VICAR application programs appears in Appendix II.

The input/output package available in the FORTRAN language is somewhat inefficient for image processing in terms of subroutine size, data packing density, and execution speed. For this reason, VMJC contains an input/output routine VMIO, which is specially designed for image processing. The applications programmer may use VMIO for all data communication to avoid loading the large FORTRAN input/output package. The application program accomplishes its I/O with a series of calls to VMIO (Ref. 3).

Table 4-10 shows a partial listing of the available calls and their functions. Typically, an application program begins with a CALL PARAM, which brings in the parameters from the appropriate EXECUTE statement or parameter set. Next it uses CALL OPEN to open the input and output data sets. At this point begins the core

Table 4-10. VMIO CALLs

Format	Function
CALL PARAM	Fetch user parameters that have been translated by VMJC.
CALL OPEN	Prepare a data set for use (input or output) and define the buffering scheme.
CALL READ	Read one record (image line) from an input data set to an array in memory.
CALL WRITE	Write one record (image line) from a memory array to an output data set.
CALL PRINT	Print a message on the line printer.
CALL CLOSE	Finalize a data set after I/O is finished.
CALL END	Terminate current task and return control to VMAST.

of the application program composed of the FORTRAN statements that actually accomplish the processing. The CALL READ and CALL WRITE statements rather than the FORTRAN READ and WRITE statements are used for data input and output. CALL PRINT is used to print execution messages on the line printer. When the processing is complete, the program uses CALL CLOSE to write end-of-file marks in the data sets. A CALL END statement terminates the task.

REFERENCES

1. J. B. SEIDMAN, *VICAR Image Processing System Guide to System Use*, Publication 77-37, Jet Propulsion Laboratory, Pasadena, California, May 1977.

2. *Image Processing Laboratory User's Documentation of Applications Programs*, Jet Propulsion Laboratory Internal Document No. 900-670, Pasadena, California, October 11, 1974.

3. "VICAR—Version 3," *JPL Digital Image Processing Manual*, Section 1, Jet Propulsion Laboratory Internal Document No. 68–369, Pasadena, California, August 30, 1968.

4. S. H. CAINE and E. K. GORDON, *TTM: A Macro Language for Batch Processing*, Booth Computing Center Report No. 8, California Institute of Technology, Pasadena, California, May 1969.

THE GRAY LEVEL HISTOGRAM

IIIIIIIIIIIIIIIIIIIIIIIIIIIIIIIIIIIIIIIIIIIIIIIIIIIIIIIIIIIIIIIIIIIIIIIIIIIIIIIIIIIIIIIIIIIIIIIIIIIIIIIIIIIIIIIIIIIIIIIIIIIIIIIIII

INTRODUCTION

One of the simplest and most useful tools in digital image processing is the gray level histogram. This function summarizes the gray level content of an image. While the histogram of any image contains considerable information, certain types of images are completely specified by their histograms. Computation of the histogram is simple and may be done at little apparent cost during the transfer of an image from one data set to another.

Definition

The gray level histogram is a function showing, for each gray level, the number of pixels in the image that have that gray level. The abscissa is gray level and the ordinate is frequency of occurrence (number of pixels). Figure 5-1 shows an example.

There is another way to define the gray level histogram (Refs. 1, 2), and the following exercise yields insight into the usefulness of this function. Suppose we have a continuous image, defined by the function $D(x, y)$, which varies smoothly from high gray level at the center to low gray level at the borders. We may select some gray level D_1 and define a set of contour lines connecting all points in the image having

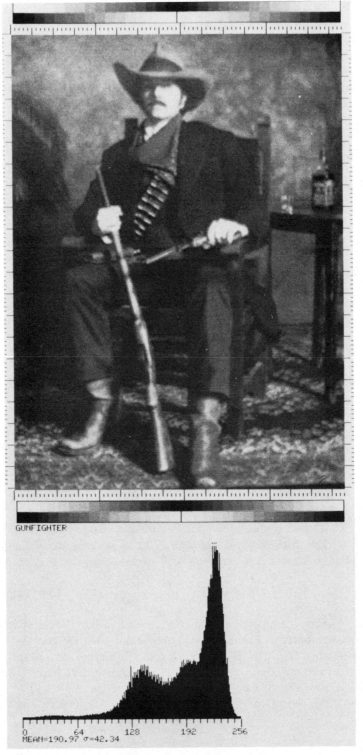

GUNFIGHTER

MEAN=190.97 σ=42.34

Figure 5-1 An image and its gray level histogram

gray level D_1. The resulting contour lines form closed curves that surround regions in which the gray level is greater than or equal to D_1. Figure 5-2 shows an image containing one contour line at the gray level D_1. A second contour line has been drawn at a higher gray level D_2. A_1 is the area of the region inside the first contour line and, similarly, A_2 is the area inside the second line.

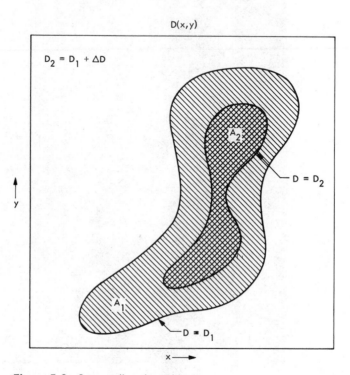

Figure 5-2 Contour lines in an image

The threshold area function A(D) of a continuous image is the area enclosed by all contour lines of gray level D. Now the histogram may be defined as

$$H(D) = \lim_{\Delta D \to 0} \frac{A(D) - A(D + \Delta D)}{\Delta D} = -\frac{d}{dD} A(D) \qquad (1)$$

Thus the histogram of a continuous image is the negative of the derivative of its area function. The minus sign compensates for the fact that $A(D)$ decreases with increasing D. If the image is considered a random variable of two dimensions, the area function is proportional to its cumulative distribution function and the gray level histogram to its probability density function.

For the case of discrete functions, we fix ΔD at unity, and Eq. (1) becomes

$$H(D) = A(D) - A(D + 1) \qquad (2)$$

The area function of a digital image is merely the number of pixels having gray level greater than or equal to D for any gray level D.

The Two-Dimensional Histogram

Frequently one finds it useful to construct histograms of higher dimension than one. Figure 5-3 shows images digitized from a microscope field containing a white blood cell and several red blood cells. The field was digitized in white light and, through colored filters, in red and blue light. At the lower right is the two-dimensional "red versus blue" histogram of the latter two images. The two-dimensional histogram is a function of two variables, gray level in the red image and gray level in the blue image. Its value at the coordinates D_R, D_B is the number of corresponding pixel pairs having gray level D_R in the red image and gray level D_B in the blue image. Recall that a multispectral digital image such as this can be thought of as having a single pixel at each sample point, but each pixel has multiple values, in this case two. The two-dimensional histogram shows how the pixels are distributed among combinations of two gray levels. If the red and blue component images were identical, the histogram would have zero value except on the 45° diagonal. Pixels having higher red than blue gray level, or vice versa, contribute to the histogram above or below the diagonal line.

In white light, the microscope field of Figure 5-3 shows considerable color information. The red blood cells appear pinkish, while the white blood cell is gray with a dark blue nucleus due to the staining treatment. Thus the red cells appear dark in blue light, which they absorb, and light in red light, which they transmit. Similarly, the nucleus is much denser in red light. The red/blue histogram therefore has four distinct peaks, one each due to the background (B), the red blood cells (R), and the nucleus (N) and the cytoplasm (C) of the white cell. Further analysis of two-dimensional histograms is discussed in Chapter 17.

Properties of the Histogram

When an image is condensed into a histogram, all spatial information is discarded. The histogram specifies the number of pixels having each gray level but gives no hint as to where those pixels are located within the image. Thus the histogram is unique for any particular image, but the reverse is not true. Vastly different images could have identical histograms. Such operations as moving objects around within an image typically have no effect on the histogram. However, the histogram does possess some useful properties.

If we change variables in Eq. (1) and integrate both sides from D to infinity, we find

$$\int_D^\infty H(P)\,dP = -\Big[A(P)\Big]_D^\infty = A(D) \tag{3}$$

the area function. If we then set $D = 0$, assuming nonnegative gray levels, we obtain

$$\int_0^\infty H(P)\,dP = \text{area of image} \tag{4}$$

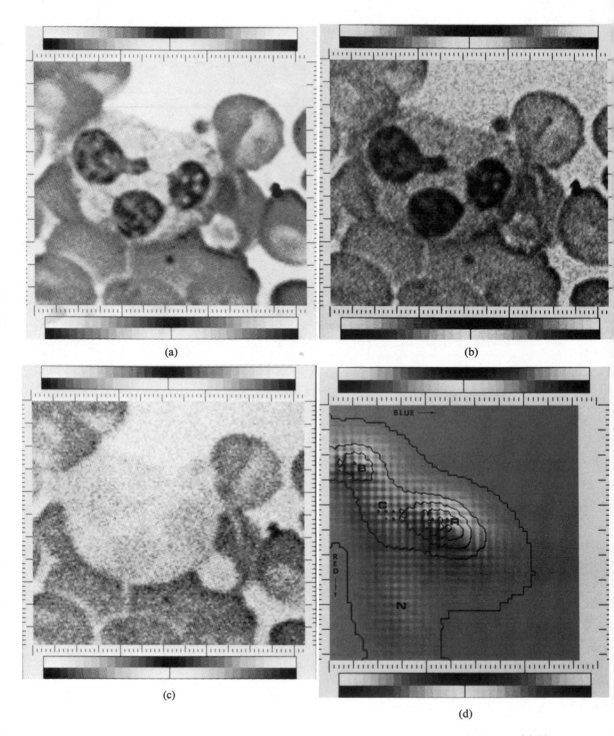

Figure 5-3 A two-dimensional histogram example: (a) white light image; (b) red light image; (c) blue light image; (d) red/blue histogram

or, in the discrete case,

$$\sum_{D=0}^{255} H(D) = NL \times NS \tag{5}$$

If an image contains a single uniformly gray object on a contrasting background and we stipulate that the boundary of that object is the contour line defined by gray level D_1, then

$$\int_{D_1}^{\infty} H(D)\, dD = \text{area of object} \tag{6}$$

If the image contains multiple objects, all of whose boundaries are contour lines at gray level D_1, then Eq. (6) gives the aggregate area of all objects.

Normalizing the gray level histogram by dividing by the area of the image produces the probability density function (pdf) of the image. A similar normalization of the area function produces the cumulative distribution function (CDF) of the image. These functions are useful in statistical treatment of images, as illustrated in Chapter 6.

The histogram has another useful property, which follows directly from its definition as the number of pixels having each gray level. Suppose an image consists of two disjoint regions and the histogram of each region is known. Then the histogram of the entire image is the sum of the two region histograms. Clearly, this can be extended to any number of disjoint regions.

USES OF THE HISTOGRAM

Digitizing Parameters

The histogram indicates whether or not an image is properly scaled within the available range of gray levels. Ordinarily, a digital image should make use of almost all the available gray levels. Failure to do so increases the effective quantizing interval and causes the image to occupy more data storage space than its information content requires. Figure 5-4 shows the histogram of a low-contrast image. Since the histogram is zero for gray levels below 16 and above 200, no pixels have those values. The range between the darkest and lightest pixel in the image is divided into less than the 256 steps available with 8-bit data. The histogram of Figure 5-4 is typical of images produced at a digitizer sensitivity setting that is too low. One could use a point operation (Chapter 6) to spread the gray levels out and fill the entire 8-bit range. Once the image has been digitized to fewer than 256 gray levels, however, the lost information cannot be restored without redigitizing.

The opposite situation is shown in Figure 5-5. This is typical of histograms produced with the digitizing sensitivity set too high. The entire gray scale is used, but there are spikes at gray levels 0 and 255. This suggests that the gray levels in light and dark areas in the image have been "clipped" to fit within the available range. The histogram would extend below 0 and above 255 if clipping were not used. This clipping process destroys the differences between pixels in extremely light and extremely dark areas of the image and causes a loss of detail in those areas. More moderately gray areas of the image are not affected by the clipping.

GUNFIGHTER

MEAN=156.89 σ=31.01

Figure 5-4 Histogram of a low-contrast image

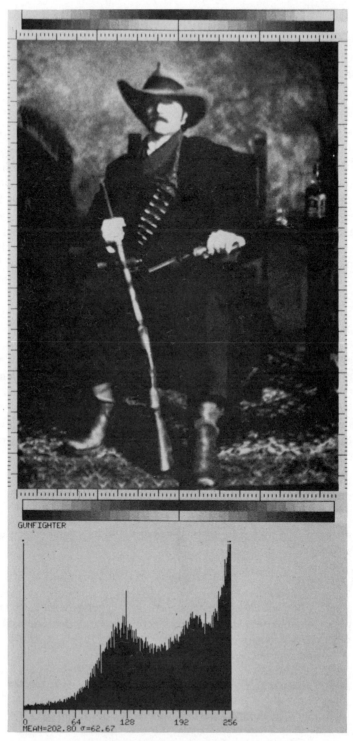

GUNFIGHTER

MEAN=202.80 σ=62.67

Figure 5-5 Histogram showing clipping at 0 and 255

Figure 5-6 shows a histogram in which the peak has been shifted too far toward the high gray levels. This is typical of digitizing with improper offset control settings. The lightest areas within the image have received gray levels down to about 16, but there is severe clipping in the black.

Boundary Threshold Selection

As mentioned earlier, contour lines provide an effective way to establish the boundary of a simple object within an image. The technique of using contour lines as boundaries is called *thresholding*. Optimal techniques for selecting threshold gray levels is a subject of considerable discussion in the literature and is treated in Chapter 15.

Suppose an image contains a dark object on a light background. Figure 5-7 illustrates the appearance of the histogram of such an image. The dark pixels inside the object produce the rightmost peak in the histogram. The leftmost peak is due to the large number of light gray pixels in the background. The relatively few mid-gray pixels around the edge of the object produce the dip between the two peaks. A threshold gray level chosen in the area of the dip will produce a reasonable boundary for the object (Refs. 3, 4).

In one sense, the gray level corresponding to the minimum between the two peaks is optimal for defining the boundary. Recall from Eq. (1) that the histogram is the derivative of the area function. In the vicinity of the dip, the histogram takes on relatively small values. This means that the area function changes slowly with gray level. If we place the threshold gray level at the dip, we minimize its effect upon the boundary of the object. If we are concerned with measuring the object's area, selecting a threshold at the dip in the histogram reduces the sensitivity of the area measurement to variations in threshold gray level (Ref. 1). Given the histogram in Figure 5-7, we could determine an optimal threshold gray level for the object and compute its area [Eq. (6)] without ever seeing the image.

Integrated Optical Density

Another measurement that can be computed directly from the histogram of simple images is the integrated optical density (IOD). A useful measure of the "mass" of an image, it is given by

$$\text{IOD} = \int_0^a \int_0^b D(x, y) \, dx \, dy \tag{7}$$

where a and b delimit the region of the image. When the image consists of a dark object situated on a background of zero gray level, the IOD reflects a combination of the area and density of that object.

For a digital image, the IOD is given by

$$\text{IOD} = \sum_{i=1}^{NL} \sum_{j=1}^{NS} D(i, j) \tag{8}$$

where $D(i, j)$ is the gray level of the pixel at line i, sample j. Let N_k be the number of

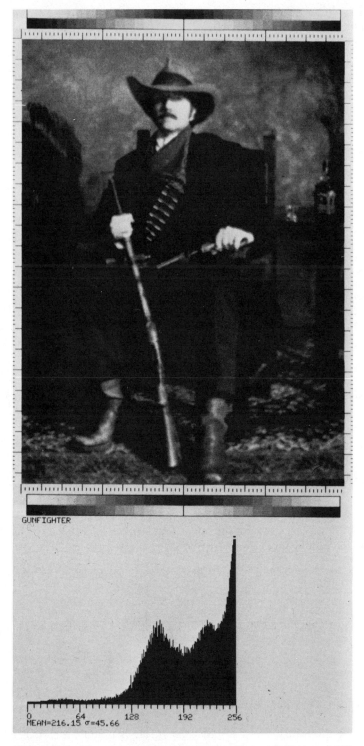

Figure 5-6 Histogram showing gray scale offset and clipping at 255

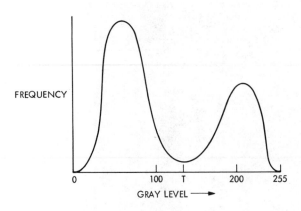

Figure 5-7 A bimodal histogram

pixels in the image with gray level equal to k. Then Eq. (8) can also be written as

$$IOD = \sum_{k=0}^{255} kN_k \qquad (9)$$

since, clearly, this adds up the gray levels of all pixels within the image. However, N_k is merely the histogram evaluated at gray level k. Thus Eq. (9) can be written as

$$IOD = \sum_{k=0}^{255} kH(k) \qquad (10)$$

that is, a gray level weighted summation of the histogram. By equating Eqs. (8) and (10) and taking a limit as the increment between gray levels approaches zero, we derive similar expressions for continuous images:

$$IOD = \int_0^\infty DH(D)\, dD \qquad (11)$$

and

$$\int_0^a \int_0^b D(x, y)\, dx\, dy = \int_0^\infty DH(D)\, dD \qquad (12)$$

If an object within the image is delineated by a threshold boundary at gray level T, the IOD within the object boundary is given by

$$IOD(T) = \int_T^\infty DH(D)\, dD \qquad (13)$$

The mean interior gray level is the ratio of IOD to area:

$$MGL = \frac{IOD(T)}{A(T)} = \frac{\int_t^\infty DH(D)\, dD}{\int_t^\infty H(D)\, dD} \qquad (14)$$

RELATIONSHIP BETWEEN HISTOGRAM AND IMAGE

Since the histogram of a particular image is unique, it should be possible to derive the histogram of simple images whose functional form is known. In this section, we address that problem to obtain insight into the histogram and establish a basis for further study of gray level threshold selection in Chapter 15.

Deriving the Histogram

Suppose we have an image of given functional form, and we desire to compute its histogram. We know that this is the negative of the derivative with respect to gray level of the area function [Eq. (1)]. Thus, we may derive the histogram if we first derive the area function from the expression for the image itself. Sometimes this can be done simply by observation.

Examples

Consider a one-dimensional image consisting of a triangular pulse (Figure 5-8). Assuming that the domain of the image is the interval −1 to 1, it is expressed by

$$D(x) = \begin{cases} 2 - 2x & 0 \leq x \leq 1 \\ 2 + 2x & -1 \leq x < 0 \end{cases} \tag{15}$$

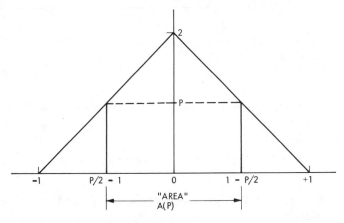

Figure 5-8 The triangular pulse

It is clear from Figure 5-8 that the (one-dimensional) area over which the image equals or exceeds some gray level P is given by

$$A(P) = 2 - P \tag{16}$$

and is shown in Figure 5-9. Now the histogram is given by

$$H(P) = -\frac{d}{dP} A(P) = 1 \tag{17}$$

Thus the histogram of the one-dimensional triangular pulse is flat, as shown in Figure 5-10.

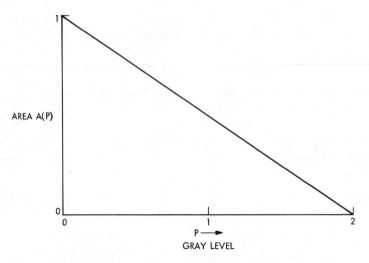

Figure 5-9   Area function for the triangular pulse

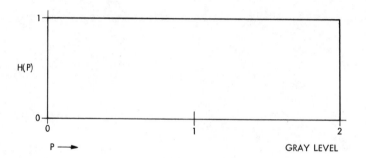

Figure 5-10   Histogram of the triangular pulse

As a second example, consider the one-dimensional Gaussian pulse (Figure 5-11) given by

$$D(x) = e^{-x^2} \qquad -\infty \leq x \leq \infty \tag{18}$$

Notice that for nonnegative x, the function is monotonic (Figure 5-12). Furthermore, the area is merely the inverse of the image function. Thus, for nonnegative values of x, we may solve Eq. (18) for x as a function of gray level to yield

$$x(D) = \sqrt{-\ln(D)} \qquad x \geq 0 \tag{19}$$

which is the area function for the right half of the image. Since the two halves of the images are symmetrical, the overall area function is twice that of Eq. (19). The histogram is given by

$$H(D) = -\frac{d}{dD}[2\sqrt{-\ln(D)}] = \frac{1}{D\sqrt{-\ln(D)}} \tag{20}$$

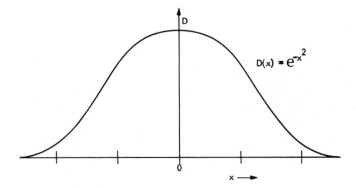

Figure 5-11 The Gaussian pulse

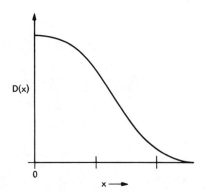

Figure 5-12 The Gaussian pulse for
non-negative *x*

and is shown in Figure 5-13. The histogram builds up to a spike at $D = 0$ because of the vast areas of low gray level at large positive and negative values of *x*. The small spike at $D = 1$ results from the image having zero slope at $x = 0$.

The same procedure may be extended to two-dimensional images by judicious use of symmetry within the image. For example, suppose that the one-dimensional Gaussian pulse of Eq. (18) is actually one line of a two-dimensional image. If all lines are identical, the histogram will have the same shape as that in Figure 5-13, differing only in ordinate scale.

One may take advantage of circular symmetry in the following way. Suppose the image is a circularly symmetric Gaussian pulse centered on the origin (Figure 5-14). The image function in polar coordinates is given by

$$D(r, \theta) = e^{-r^2} \qquad 0 \leq r \leq \infty, 0 \leq \theta < 2\pi \tag{21}$$

A contour of constant gray level *P* is a circle of radius

$$r(P) = \sqrt{-\ln (P)} \tag{22}$$

Such a contour encloses an area given by

$$A(P) = \pi[r(P)]^2 = -\pi \ln (P) \tag{23}$$

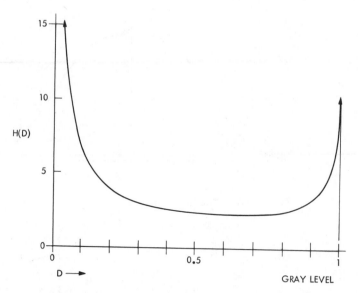

Figure 5-13 Histogram of the Gaussian pulse

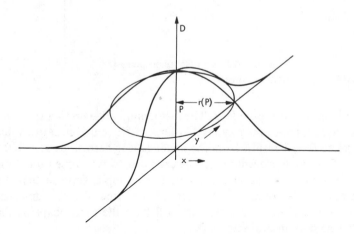

Figure 5-14 The circular Gaussian spot

The area function of Eq. (23) may now be differentiated to yield the histogram (Ref. 1, 2)

$$H(P) = \frac{d}{dP} A(P) = \frac{\pi}{P} \tag{24}$$

shown in Figure 5-15. Notice that the point of zero slope at the origin is not powerful enough to produce a spike at $D = 1$ as it does in the one-dimensional case.

For more complex images, the histogram may be derived by first partitioning

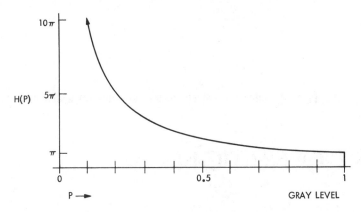

Figure 5-15   Histogram of the circular Gaussian spot

the image into disjoint regions over which the area function may be determined. The histogram of the complete image is then the sum of the histograms of all the disjoint regions.

SUMMARY OF IMPORTANT POINTS

1. The gray level histogram is the negative of the derivative of the threshold area function.

2. The histogram shows how many pixels occur at each gray level.

3. Inspection of the histogram points out improper digitization.

4. The area and IOD of a simple object can be computed from its image histogram.

5. The histogram of an image of specified functional form can be derived with the aid of the area function.

REFERENCES

1. R. J. WALL, "The Gray Level Histogram for Threshold Boundary Determination in Image Processing with Applications to the Scene Segmentation Problem in Human Chromosome Analysis," Ph.D. Thesis, UCLA, 1974.

2. R. J. WALL, A. KLINGER, and K. R. CASTLEMAN, "Analysis of Image Histograms," Proc. Second Int. Conf. on Pattern Recognition, Copenhagen, 1974.

3. J. PREWITT and M. MENDELSOHN, "The Analysis of Cell Images," *Annals of the New York Academy of Sciences*, **128**, 1035–1053, January 1966.

4. M. MENDELSOHN, B. MAYALL, J. PREWITT, R. BOSTROM, and R. HOLCOMB, "Digital Transformation and Computer Analysis of Microscope Images," in *Advances in Optical and Electron Microscopy*, **2**, R. Barer and V. Cosslet, eds., Academic Press, London, 1968.

POINT OPERATIONS

IIIIIIIIIIIIIIIIIIIIIIIIIIIIIIIIIIIIIIIIIIIIIIIIIIIIIIIIIIIIIIIIIIIIIIIIIIIIIIIIIIIIIIIIIIIIIIIIIIIIIIIIIIIIIIIIIIIIIIIIIIIIIIIIIIIIIIIIIIIIIIIIIIIIIIIIIIIIIIIIIIIIII

INTRODUCTION

Point operations constitute a simple but important class of image processing techniques. They allow the user to modify the way in which his data fills the available range of gray levels. A point operation is, by definition, an operation taking a single input image into a single output image in such a way that each output pixel's gray level depends only on the gray level of the corresponding input pixel. This contrasts with local operations, in which a neighborhood of input pixels determines the gray level of each output pixel. Furthermore, in a point operation, each output pixel corresponds directly to the input pixel having the same coordinates. Thus a point operation cannot modify the spatial relationships within an image. Point operations are sometimes called by other names, including *contrast enhancement*, *contrast stretching*, and *gray scale manipulation*.

Point operations modify the gray scale of an image. They may be viewed as pixel-by-pixel copying operations, except that the gray levels are modified in a predetermined way. A point operation which takes an input image $A(x, y)$ into an output image $B(x, y)$ may be expressed mathematically as

$$B(x, y) = f[A(x, y)] \qquad (1)$$

84

The point operation is completely specified by the function f, which specifies the mapping of input gray level to output gray level.

USES FOR POINT OPERATIONS

Point operations may be used to remove the effects of image sensor nonlinearity. For example, suppose an image has been digitized by an instrument with a nonlinear response to light intensity. A point operation can transform the gray scale so that the gray levels represent equal increments in light intensity. This is an example of "photometric decalibration." Another use for the point operation is to transform the units of the gray scale. Suppose an image was digitized by an instrument producing gray level values that are linear with transmittance. A point operation could be used to generate an output image in which the gray levels represent equal steps in optical density.

In the previous examples, point operations are used to overcome digitizer limitations before the actual processing begins. Equally important are point operations used prior to image display. Many display devices do not maintain a linear relationship between the gray level of a pixel in the digital image and the brightness of the corresponding point on the display screen. Similarly, many film recorders are unable to transform gray levels linearly into optical density. These shortcomings may be overcome by a suitably designed point operation prior to display. Taken together, the point operation and the display nonlinearities combine to cancel each other out, preserving linearity in the displayed image.

Frequently, display devices have a preferred range of gray levels over which they make image features most visible. Darker and lighter features, having the same contrast in the digital image, do not show up as well on the display. In this case, the user may employ a point operation to ensure that the features of interest fall into the "maximum visibility" range.

A user might have an image in which the features of interest occupy a relatively narrow range of the gray scale. He might use a point operation to expand the contrast of the features of interest so that they occupy the entire displayed range of gray levels.

In the examples above, point operations are used to match the contrast of interesting portions of the image to the contrast range of the display device. Such operations are sometimes viewed as image processing to bring out detail or increase the contrast of an image. It is usually desirable, however, to have the gray levels of digitized images reflect some physical properties such as light intensity or optical density. Likewise, all gray levels usually should be visible in a displayed image. Thus we can consider photometric decalibration as the software side of digitizing and display decalibration and contrast enhancement as the software side of digital image display.

Other point operation applications include contouring and thresholding. One may use a point operation to add contour lines to an image. One can accomplish thresholding with a point operation that divides an image into disjoint regions on the basis of gray level. This is useful for defining boundaries or for making masks for subsequent operations.

LINEAR POINT OPERATIONS

We first consider point operations in which output gray level is a linear function of input gray level. In this case, the gray scale transformation function of Eq. (1) takes the form

$$D_B = f(D_A) = aD_A + b \tag{2}$$

where D_B is the gray level of the output point corresponding to an input point having gray level D_A (Figure 6-1). Obviously, if $a = 1$ and $b = 0$, we have the identity

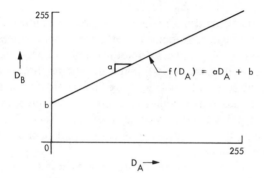

Figure 6-1 The linear point operation

operation that merely copies $A(x, y)$ into $B(x, y)$. If a is greater than 1, the contrast will be increased in the output image. For a less than 1, the contrast is reduced. If $a = 1$ and b is nonzero, the operation merely scales the gray level values of all pixels up or down. The effect of this is to make the entire image appear darker or lighter when displayed. If a is negative, dark areas become light and light areas become dark, and the image is "complemented."

POINT OPERATIONS AND THE HISTOGRAM

The foregoing discussion suggests that a point operation modifies the gray level histogram in a predictable way. We now address the question of predicting the output image histogram given the input image histogram and the functional form of the gray scale transformation. This capability is useful for two reasons. First, one may wish to design a point operation to scale the output gray levels into a predetermined range or to produce an output histogram of a particular form. Second, this exercise yields insight into the effect point operations have on images. This insight proves useful when one is designing a point operation.

The Output Histogram

Suppose a point operation defined by a function f takes an input image $A(x, y)$ into an output image $B(x, y)$. Given $H_A(D)$, the histogram of the input image, we wish to derive the output image histogram. The gray level D_B of an arbitrary output pixel is given by

$$D_B = f(D_A) \tag{3}$$

where D_A is the gray level of the corresponding input pixel. For the present, let us assume that f is a monotonically increasing function with finite slope. Thus, its inverse function exists and we can write

$$D_A = f^{-1}(D_B) \tag{4}$$

We shall later remove this restriction.

Figure 6-2 illustrates the relationship between the input histogram, the gray scale transformation function, and the output histogram. The gray level D_A transforms to D_B; similarly, the gray level $D_A + \Delta D_A$ transforms to $D_B + \Delta D_B$. Furthermore, all pixels with gray level between D_A and $D_A + \Delta D_A$ will transform to gray levels between D_B and $D_B + \Delta D_B$. Thus, the number of output pixels having gray level between D_B and $D_B + \Delta D_B$ equals the number of input pixels with gray level between D_A and $D_A + \Delta D_A$. This implies that the area under $H_B(D)$ between D_B and $D_B + \Delta D_B$ is the same as that under $H_A(D)$ between D_A and $D_A + \Delta D_A$, or

$$\int_{D_B}^{D_B+\Delta D_B} H_B(D)\, dD = \int_{D_A}^{D_A+\Delta D_A} H_A(D)\, dD \tag{5}$$

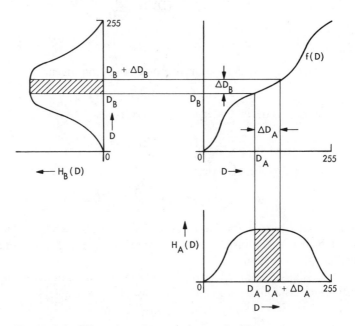

Figure 6-2 Effect of a point operation on the histogram

If we make ΔD_A suitably small, ΔD_B will also be small, and we can write a rectangular approximation to the integral

$$H_B(D_B)\Delta D_B = H_A(D_A)\Delta D_A \tag{6}$$

We now solve for the value of the output histogram to obtain

$$H_B(D_B) = \frac{H_A(D_A)}{\Delta D_B/\Delta D_A} \tag{7}$$

and take the limit as ΔD_A approaches zero. Since f has everywhere nonzero slope, ΔD_B also approaches zero to yield

$$H_B(D_B) = \frac{H_A(D_A)}{dD_B/dD_A} \tag{8}$$

But since D_B is given by Eq. (3), we may substitute to find

$$H_B(D_B) = \frac{H_A(D_A)}{(d/dD_A)f(D_A)} \tag{9}$$

We now have a mixture of independent variables in this equation. The denominator is a simple derivative, and we may change variables freely. In the numerator we use the inverse function of Eq. (4). This yields

$$H_B(D_B) = \frac{H_A[f^{-1}(D_B)]}{df/dD_B} \tag{10}$$

or, after dropping the subscript, the general form

$$H_B(D) = \frac{H_A[f^{-1}(D)]}{df/dD} \tag{11}$$

Examples

As an example, consider the linear point operation given by

$$D_B = f(D_A) = aD_A + b \tag{12}$$

We note that its derivative is a and its inverse is

$$D_A = f^{-1}(D_B) = \frac{D_B - b}{a} \tag{13}$$

Substituting into Eq. (11) yields

$$H_B(D) = \frac{1}{a}H_A\left(\frac{D-b}{a}\right) \tag{14}$$

Notice that $b > 0$ shifts the histogram to the right, while $b < 0$ shifts it left. Also, $a > 1$ broadens the histogram while reducing its amplitude to keep the area under the histogram constant.

To illustrate the effect of a linear point operation, let us assume the input histogram has a Gaussian form, given by

$$H_A(D) = e^{-(D-c)^2} \tag{15}$$

and shown in Figure 6-3. Substituting into Eq. (14) yields

$$H_B(D) = \frac{1}{a} e^{-[D/a - (c+b/a)]^2} \tag{16}$$

as shown in Figure 6-3. The output histogram is also Gaussian, but the peak is moved to $c + b/a$. Also, the width (at the $1/e$ point) goes from unity to a as the height goes from unity to $1/a$.

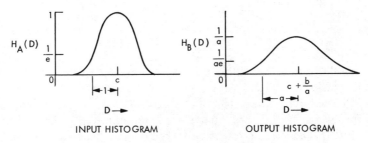

INPUT HISTOGRAM   OUTPUT HISTOGRAM

Figure 6-3 Effect of a linear point operation on a Gaussian histogram

As a second example, consider a square-law point operation given by

$$D_B = f(D_A) = D_A^2 \tag{17}$$

operating upon an image whose histogram

$$H_A(D_A) = e^{-D_A^2} \tag{18}$$

is the right-hand half of a Gaussian pulse. These appear in Figure 6-4.

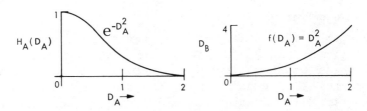

Figure 6-4 Square-law point operation

Using Eq. (11), we obtain the output histogram

$$H_B(D_B) = \frac{e^{-D_B}}{2\sqrt{D_B}} \tag{19}$$

which is shown in Figure 6-5.

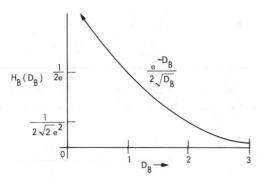

Figure 6-5 Output histogram from square-law point operation

The General Case

In the derivation that led to Eq. (11), we assumed that $f(D)$ has everywhere finite nonzero slope. If $f(D)$ has zero slope over some interval, then the finite area under H_A will be forced into a strip of infinitesimal width in H_B, producing a spike as Eq. (11) suggests. If $f(D)$ has infinite slope, the opposite is the case. An infinitesimally wide strip under H_A is expanded throughout a finite interval in H_B, producing a vanishingly small value for the output histogram there. Thus, the construction of Figure 6-2 is valid in these two extreme cases, and the output histograms behave as Eq. (11) suggests.

If $f(D)$ is not a monotone function, its inverse does not exist, and Eq. (11) cannot be used directly. The input gray level range may be divided into disjoint intervals, however, over which the previously developed technique may be used. This partitions the input image into contiguous disjoint subsets, and the output histogram is the sum of the transformed subset histograms.

APPLICATIONS OF POINT OPERATIONS

Histogram Flattening

Suppose we desire a point operation to take a given input image into an output image with equally many pixels at every gray level (a flat histogram). This can be useful for putting images into a consistent format before comparison. Figure 6-6 shows three images with their normalized histograms and normalized area functions. Recall from Chapter 5 that the probability density function (pdf) of an image is its histogram normalized to unit area

$$p_1(a) = \frac{1}{A_1} H_1(a) \tag{20}$$

where $H(a)$ is the histogram and A_1 is the area of the image. Recall also that the cumulative distribution function (CDF) is the area-normalized threshold area function

$$P_1(a) = \int_0^a p_1(u) \, du = \frac{1}{A_1} \int_0^a H_1(u) \, du \tag{21}$$

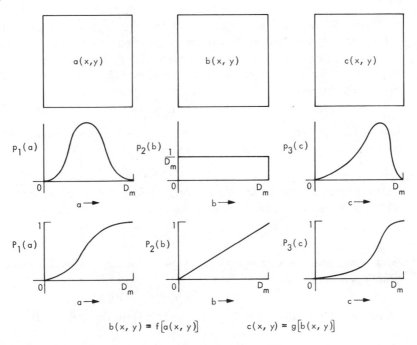

$$b(x, y) = f[a(x, y)] \qquad c(x, y) = g[b(x, y)]$$

Figure 6-6 Histogram flattening and histogram matching

Suppose we wish to select the function f so that, after the point operation given by

$$b(x, y) = f[a(x, y)] \tag{22}$$

$b(x, y)$ will have a flat histogram. As in Eq. (8), we can write

$$\frac{db}{da} = \frac{p_1(a)}{p_2(b)} \tag{23}$$

Substituting Eq. (22) and noting that $p_2(b)$ is constant with value $1/D_m$ produces

$$\frac{df}{da} = D_m p_1(a) \tag{24}$$

where zero to D_m is the gray level range over which the histogram is nonzero.

Equation (24) may be integrated to yield

$$f(a) = D_m \int_0^a p_1(u)\, du = D_m P_1(a) \tag{25}$$

Thus the CDF is the point operation that flattens the histogram. It is a particularly well-behaved function since it is non-negative with non-negative slope.

Histogram Matching

Sometimes it is desirable to transform one image so that its histogram matches that of another image or a specified functional form. This could be used, for example, before subtraction of two similar images that have not been digitized the same way.

In Figure 6-6, suppose we desire to transform $a(x, y)$ into $c(x, y)$ with histogram $p_3(c)$ specified. We shall do this in two steps, first transforming $a(x, y)$ into $b(x, y)$ with a flat histogram as before, and then taking $b(x, y)$ through a second point operation $g(b)$ to produce $c(x, y)$. We know from Eq. (25) what is required to produce $b(x, y)$. Furthermore, we know that the point operation

$$b = D_m P_3(c) \tag{26}$$

will take $c(x, y)$ into $b(x, y)$ and is thus the opposite of what we require. We can express $b(x, y)$ as in Eq. (26) and write the second point operation as

$$c = g[D_m P_3(c)] \tag{27}$$

This says that the sequential application of $g(b)$ and $D_m P(c)$ produces no net effect. Thus $g(b)$ is the inverse function of $DP_3(c)$; that is,

$$g(b) = P_3^{-1}\left(\frac{b}{D_m}\right) \tag{28}$$

Now, if we desire to take $a(x, y)$ to $c(x, y)$ in one step, we may concatenate the two point operations, and

$$c = g[f(a)] = P_3^{-1}[P_1(a)] \tag{29}$$

Notice that when substituting Eqs. (25) and (28) into Eq. (29), the D_m's cancel.

Photometric Decalibration

Historically, one of the most important point operation uses has been the removal of digitizer-induced photometric nonlinearity. Assume that a certain film digitizer has a nonlinear relationship between input film density and output gray level. We may think of this as an ideal digitizer followed by a nonlinear point operation. We wish to design a second point operation that will restore linearity by reproducing the image from the ideal digitizer. This process is shown in Figure 6-7. The gray scale

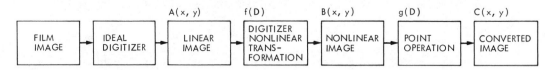

Figure 6-7 Photometric decalibration

transformation of the digitizer either is known in functional form or can be measured. We wish to select $g(D)$ so that the net effect of the two cascaded point operations is zero; that is,

$$C(x, y) = g\{f[A(x, y)]\} = A(x, y) \tag{30}$$

This is satisfied by

$$g(D) = f^{-1}(D) \tag{31}$$

since the effect of a point operation is undone by its inverse function.

The digitizer's gray scale transfer curve $f(D)$ may be measured by digitizing a linear gray wedge test image. This function is easily inverted to produce the required point operation. Difficulty may be encountered when digitizer saturation drives the slope of $f(D)$ to zero.

As an example of photometric decalibration, consider the digitizer transfer curve

$$D_B = f(D_A) = aD_A^2 + b \tag{32}$$

We may solve Eq. (32) for D_A, and the required point operation is

$$g(D_B) = \sqrt{\frac{D_B - b}{a}} \tag{33}$$

Substituting Eq. (32) into Eq. (33) produces

$$D_C = g(D_B) = D_A \tag{34}$$

as expected.

Digitizers which measure each pixel with the same sensing device generally have a gray scale transformation function which is constant throughout the image. Other digitizers, like the vidicon, may have a spatially variant transformation that is different from one pixel to the next. In this case, a simple point operation is not sufficient. An algebraic operation may be required, as discussed in the following chapter. It may be necessary to use a spatially variant point operation, implemented by dividing the image into regions and performing a separate point operation in each region. It might be practical to specify the functional form of a spatially variant gray scale transformation. While a spatially variant point operation does not fit the definition offered at the beginning of this chapter, we may consider it a generalization of the original concept.

In extreme cases, it may be necessary to specify a unique point operation for each picture element. For the Mariner Venus/Mercury 1973 mission, nine different calibration images of flat fields were taken at different illumination levels before the cameras were flown. This defined, for each pixel, a nine-point piecewise linear digitizer transfer curve. These were inverted to form individual piecewise linear gray level transformations for each pixel. While this method required storing considerable data, it produced previously unattained photometric accuracy. Figure 6-8(a) shows a flat field image taken by the B camera of Mariner 10, contrast enhanced by a factor of ten. The shading pattern occupies about 25 of the 256 gray levels. After decalibration the shading pattern occupies only about five gray levels. Figure 6-8(b) shows the decalibrated flat field contrast enhanced by a factor of 50.

Display Decalibration

We may use a similar approach to design a point operation to compensate for display nonlinearity. In this case, we model the imperfect display as an ideal display preceded by a nonlinear point operation, as shown in Figure 6-9. In this case, the transformation g is given and f is to be determined. The desired gray scale transformation is given by

$$f(D) = g^{-1}(D) \tag{35}$$

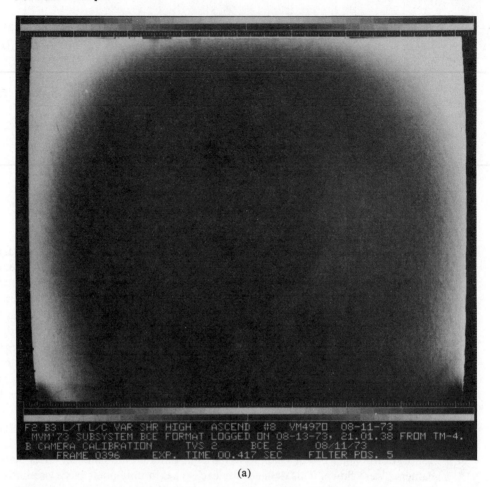

(a)

Figure 6-8   MVM'73 photometric decalibration example: (a) flat field image before decalibration, contrast × 10; (b) after decalibration, contrast × 50

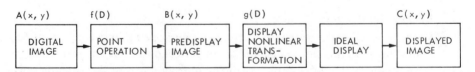

Figure 6-9   Display decalibration

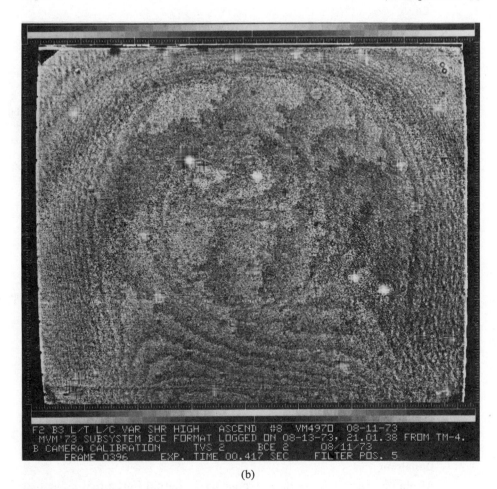

(b)

Figure 6-8   (cont.)

SUMMARY OF IMPORTANT POINTS

1. Point operations transform the gray scale of an image.

2. Point operations are useful for photometric decalibration, display decalibration, and histogram modification.

3. The histogram of an image following a specified point operation can be computed from Eq. (11).

4. A linear point operation can only stretch or compress and shift the histogram.

5. The cumulative distribution function (normalized area function) is the point operation that flattens the histogram.

ALGEBRAIC OPERATIONS

||||||||||||||||||||||||||||||||||||||||||||||||||||||||||||||||||||||||||||||||||||||||||||||||||||||||||||||||||||||||||||||||||||||||||||||||||||||||||||||

INTRODUCTION

Definition

Algebraic operations are those producing an output image that is the pixel-by-pixel sum, difference, product, or quotient of two input images. In the case of sums, more than two input images may be involved. In general, one of the input images may be a constant. However, addition, subtraction, multiplication, and division by a constant can be treated as a linear point operation, as discussed in Chapter 6. The same is true for cases where the input images are identical.

The four algebraic image processing operations are expressed mathematically by

$$C(x, y) = A(x, y) + B(x, y) \tag{1}$$

$$C(x, y) = A(x, y) - B(x, y) \tag{2}$$

$$C(x, y) = A(x, y) \times B(x, y) \tag{3}$$

$$C(x, y) = A(x, y) \div B(x, y) \tag{4}$$

where $A(x, y)$ and $B(x, y)$ are the input images and $C(x, y)$ is the output image. By making suitable combinations, one may form complex algebraic equations involving several images.

Uses of Algebraic Operations

An important application of image addition is averaging together multiple images of the same scene. This is used frequently and successfully to reduce the effects of additive random noise. Image addition may also be used to superimpose the contents of one image upon another.

Image subtraction can be used to remove an undesired additive pattern from an image. This may be a slowly varying background shading pattern, a periodic noise pattern, or any other additive contamination that is known at every point in the image. Subtraction is also useful in detecting changes between two images of the same scene. For example, one could detect motion by subtracting sequential images of a scene. Image subtraction is also required to compute the gradient, a useful function for locating edges.

Multiplication and division find less application in digital image processing; nevertheless, they do have important uses. They can be used to correct for the effects of a digitizer in which the sensitivity of the light sensor varies from point to point within the image. Division can produce ratio images that are important in multi-spectral analysis. Multiplication by a "mask" function can blot out certain portions of an image, leaving only objects of interest.

ALGEBRAIC OPERATIONS AND THE HISTOGRAM

In this section, we examine the output histogram of sum and difference operations. This yields insight into the operations and the scaling necessary to keep the output gray levels within range. We also present a technique for determining the integrated optical density (IOD) of an image contaminated by additive random noise.

Histograms of Sum Images

Suppose, for the operation in Eq. (1), that the input images $A(x, y)$ and $B(x, y)$ have gray level histograms $H_A(D)$ and $H_B(D)$, respectively. We wish to determine the output histogram $H_C(D)$. If the input images are identical, or if one is constant, the process reduces to a point operation and the results of Chapter 6 apply. In this section, we address the case where the images are uncorrelated.

The two input images are uncorrelated if their joint two-dimensional histogram $H_{AB}(D_A, D_B)$ is given by

$$H_{AB}(D_A, D_B) = H_A(D_A)H_B(D_B) \tag{5}$$

the product of the two individual image histograms. In a practical sense, this means that the images are unrelated. Note that Eq. (5) is not satisfied if the input images are identical but is satisfied if at least one image is random and statistically independent of the other.

We may reduce a two-dimensional histogram to a one-dimensional marginal histogram by integrating over one of the independent variables, that is,

$$H(D_A) = \int_{-\infty}^{\infty} H_{AB}(D_A, D_B) \, dD_B \tag{6}$$

Thus, given Eq. (5), we may produce a one-dimensional histogram by

$$H(D) = \int_{-\infty}^{\infty} H_A(D_A) H_B(D_B)\, dD_B \tag{7}$$

Equation (1) implies, however, that at every point,

$$D_A = D_C - D_B \tag{8}$$

Substituting this into the right side of Eq. (7) yields

$$H(D) = \int_{-\infty}^{\infty} H_A(D_C - D_B) H_B(D_B)\, dD_B \tag{9}$$

This one-dimensional histogram is a function of output gray level and thus is the output histogram. We may now write the output histogram of a sum operation of uncorrelated images as

$$H_C(D_C) = H_A(D_A) * H_B(D_B) \tag{10}$$

where the $*$ indicates the convolution operation defined by the integral in Eq. (9).

The convolution integral is discussed in detail in Chapter 9; however, the following development illustrates the operation. Suppose we wish to convolve two identical Gaussian functions, each given by e^{-x^2}. Then

$$e^{-x^2} * e^{-x^2} = \int_{-\infty}^{\infty} e^{-y^2} e^{-(x-y)^2}\, dy \tag{11}$$

Expanding the exponent and collecting terms produces

$$e^{-x^2} * e^{-x^2} = \int_{-\infty}^{\infty} e^{-(x^2 - 2xy - 2y^2)}\, dy \tag{12}$$

We now insert a product that is unity, and

$$e^{-x^2} * e^{-x^2} = \int_{-\infty}^{\infty} e^{-(x^2 - 2xy - 2y^2)} e^{+x^2/2} e^{-x^2/2}\, dy \tag{13}$$

which may be rearranged as

$$e^{-x^2} * e^{-x^2} = \int_{-\infty}^{\infty} e^{-2(x^2/4 - xy + y^2)} e^{-x^2/2}\, dy \tag{14}$$

This may be factored in the exponent to yield

$$e^{-x^2} * e^{-x^2} = \int_{-\infty}^{\infty} e^{-2(y - x/2)^2} e^{-x^2/2}\, dy \tag{15}$$

which may be rearranged to produce

$$e^{-x^2} * e^{-x^2} = e^{-x^2/2} \int_{-\infty}^{\infty} e^{-(y - x/2)^2/[2(1/4)]}\, dy \tag{16}$$

We now use a property of the Gaussian function (Ref. 1)

$$\int_{-\infty}^{\infty} e^{-(x-\mu)^2/2\sigma^2}\, dx = \sqrt{2\pi\sigma^2} \tag{17}$$

and Eq. (16) becomes

$$e^{-x^2} * e^{-x^2} = \sqrt{2\pi(\tfrac{1}{4})}\, e^{-x^2/2} \tag{18}$$

A similar but more general development shows that

$$A_1 e^{-(x-\mu_1)^2/2\sigma_1^2} * A_2 e^{-(x-\mu_2)^2/2\sigma_2^2} = A_1 A_2 \sqrt{2\pi\sigma_1\sigma_2} e^{-(x-\mu_3)^2/2\sigma_3^2} \tag{19}$$

where

$$\mu_3 = \mu_1 + \mu_2 \tag{20}$$

and

$$\sigma_3^2 = \sigma_1^2 + \sigma_2^2 \tag{21}$$

This means that convolving two Gaussians produces a third Gaussian that is shifted and broader, as Eq. (21) indicates. In general, convolution "smears" a function. Since adding uncorrelated images convolves their histograms, we can expect the sum of uncorrelated images to occupy a broader gray level range than that of its component images. Further discussion of the convolution operation is reserved for Chapter 9.

Histograms of Difference Images

For subtraction of uncorrelated images, the result of Eq. (10) holds after redefinition of one image as its negative. Thus addition and subtraction of uncorrelated images behave similarly. There is, however, one case of image subtraction that bears further consideration. This is the subtraction of slightly misaligned but near-identical images. This situation arises when sequential images of a scene are subtracted to detect motion or other change.

Suppose the difference image is given by

$$C(x, y) = A(x, y) - A(x + \Delta x, y) \tag{22}$$

which may be approximated by

$$C(x, y) \approx \frac{\partial}{\partial x} A(x, y) \, \Delta x \tag{23}$$

if Δx is small.

Notice that $\partial A/\partial x$ is itself an image with a histogram we may denote by $H'_A(D)$. Thus, the histogram of the displaced difference image is given by

$$H_c(D) \approx \frac{1}{\Delta x} H'_A(D/\Delta x) \tag{24}$$

recalling from Chapter 6 the effect of a multiplicative constant. Thus subtracting slightly misaligned copies of an image produces a partial derivative image. The direction of the partial derivative is the same as that of the displacement.

IOD of a Noisy Image

Suppose we have an image containing a spot on a uniform contrasting background. Suppose also that the image has been contaminated by additive random noise, and we wish to determine the IOD of the spot. We model the situation as follows: Let $S(x, y)$ represent the noise-free image of the spot and $N(x, y)$, the noise image defined on the same region. Then the observed image is given by

$$M(x, y) = S(x, y) + N(x, y) \tag{25}$$

The histograms of the three images are shown in Figure 7-1. We assume that the noise has a symmetrical histogram centered on the unknown mean value N_0 and that the spot histogram has a sharp spike at the origin due to the uniform background surrounding the spot.

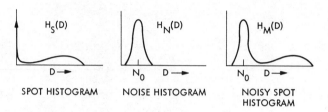

Figure 7-1   Histogram of a noisy spot image

We wish to determine

$$\text{IOD}_S = \int_0^a \int_0^b S(x, y) \, dx \, dy = \int_0^a \int_0^b M(x, y) \, dx \, dy - \int_0^a \int_0^b N(x, y) \, dx \, dy \quad (26)$$

Substituting the property of Chapter 5, Eq. (12) produces

$$\text{IOD}_S = \int_0^\infty D H_M(D) \, dD - N_0 A \quad (27)$$

where A is the area of the region of definition. Now, recalling Eq. (4) of Chapter 5, we can write

$$A = \int_0^\infty H_M(D) \, dD \quad (28)$$

since the total areas of the noise and observed images are the same. Now

$$\text{IOD}_S = \int_0^\infty D H_M(D) \, dD - N_0 \int_0^\infty H_M(D) \, dD \quad (29)$$

and rearrangement yields

$$\text{IOD}_S = \int_0^\infty (D - N_0) H_M(D) \, dD \quad (30)$$

This is a simple expression for IOD provided that N_0 can be determined.

One could estimate N_0 by averaging the gray level of a small area distant from the spot. Under a set of reasonable assumptions, however, we can argue that the leftmost peak of the histogram $H_M(D)$ occurs at N_0. Assume that the noise histogram $H_N(D)$ is symmetrical so that its peak occurs at the mean value N_0. Since $N(x, y)$ is random, the two images are uncorrelated. Equation (10) states that the sum of uncorrelated images has a histogram that is the convolution of the two original histograms. Furthermore, $H_S(D)$ is dominated by the spike at $D = 0$. We show in Chapter 9 that the spike (impulse) is the identity under convolution [Eq. (62)]. Thus, the histogram $H_M(D)$ will be dominated by a peak at N_0, as shown in Figure 7-1. The asymmetry of $H_S(D)$ will skew the peak slightly to the right, but the location of the peak remains a good estimate of N_0 if the spot is surrounded by a reasonable amount of background. Thus, the histogram of a noisy spot image yields an easily computed estimate of the noise-free IOD.

APPLICATIONS OF ALGEBRAIC OPERATIONS

In this section, we illustrate several situations in which algebraic operations are useful.

Averaging for Noise Reduction

In many applications, it is possible to obtain multiple images of a stationary scene. If these images are contaminated by an additive noise source, the multiple images may be averaged to reduce the noise. In the averaging process, the stationary component of the image is unchanged, whereas the noise pattern, different from one image to the next, builds up more slowly in the sum.

Suppose we have a set of M images of the form

$$D_i(x, y) = S(x, y) + N_i(x, y) \tag{31}$$

where $S(x, y)$ is the image of interest and the $N_i(x, y)$ are noise images such as those introduced by film grain or electronic noise in a digitizing system. Each image in the set is degraded by a different noise image. While we do not know the noise images exactly, we assume that each comes from one of several identical uncorrelated ensembles of random noise images with zero mean value. This means that

$$\mathcal{E}\{N_i(x, y)\} = 0 \tag{32}$$

$$\mathcal{E}\{N_i(x, y) + N_j(x, y)\} = \mathcal{E}\{N_i(x, y)\} + \mathcal{E}\{N_j(x, y)\} \qquad i \neq j \tag{33}$$

and

$$\mathcal{E}\{N_i(x, y)N_j(x, y)\} = \mathcal{E}\{N_i(x, y)\}\mathcal{E}\{N_j(x, y)\} \qquad i \neq j \tag{34}$$

where $\mathcal{E}\{\ \}$ indicates the expectation operator; that is, $\mathcal{E}\{N_i(x, y)\}$ is the average of the points at x, y of all the noise images in the ith ensemble. Expectation and random variables are discussed in more detail in Chapter 11.

For any point in the image, we may define the signal-to-noise power ratio as

$$P(x, y) = \frac{S^2(x, y)}{\mathcal{E}\{N^2(x, y)\}} \tag{35}$$

If we average M images to form

$$\bar{D}(x, y) = \frac{1}{M} \sum_{i=1}^{M} [S(x, y) + N_i(x, y)] \tag{36}$$

the signal-to-noise power ratio becomes

$$\bar{P}(x, y) = \frac{S^2(x, y)}{\mathcal{E}\left\{\left[\left(\frac{1}{M}\right) \sum_{i=1}^{M} N_i(x, y)\right]^2\right\}} \tag{37}$$

The numerator is unchanged because averaging does not affect the signal component.

We may factor the $1/M$ out of the denominator to obtain

$$\bar{P}(x, y) = \frac{S^2(x, y)}{\frac{1}{M^2}\mathcal{E}\left\{\sum_{i=1}^{M} N_i(x, y)^2\right\}} \tag{38}$$

or

$$\bar{P}(x, y) = \frac{M^2 S^2(x, y)}{\mathcal{E}\left\{\sum_{i=1}^{M} \sum_{j=1}^{M} N_i(x, y)N_j(x, y)\right\}} \tag{39}$$

Using the property of Eq. (33), we may separate the denominator into two terms, producing

$$\bar{P}(x, y) = \frac{M^2 S^2(x, y)}{\mathcal{E}\left\{\sum_{i=1}^{M} N_i^2(x, y)\right\} + \mathcal{E}\left\{\underbrace{\sum_{i=1}^{M} \sum_{j=1}^{M}}_{i \neq j} N_i(x, y)N_j(x, y)\right\}} \tag{40}$$

The second term may be factored, according to Eq. (34), while the first term may be written as a sum of expectations yielding

$$\bar{P}(x, y) = \frac{M^2 S^2(x, y)}{\sum_{i=1}^{M} \mathcal{E}\{N_i^2(x, y)\} + \underbrace{\sum_{i=1}^{M} \sum_{j=1}^{M}}_{i \neq j} \mathcal{E}\{N_i(x, y)\}\mathcal{E}\{N_j(x, y)\}} \tag{41}$$

Now Eq. (32) implies that the second denominator term is zero. Furthermore, since the M noise samples come from identical ensembles, all terms in the first summation are identical. Therefore,

$$\bar{P}(x, y) = \frac{M^2 S^2(x, y)}{M\mathcal{E}\{N^2(x, y)\}} = MP(x, y) \tag{42}$$

Thus, averaging M images increases the signal-to-noise power ratio by the factor M at all points in the image. The signal-to-noise amplitude ratio is the square root of the power ratio

$$\overline{\text{SNR}} = \sqrt{\bar{P}(x, y)} = \sqrt{M}\sqrt{P(x, y)} \tag{43}$$

and it goes up by the square root of the number of images averaged.

Figure 7-2 illustrates the effect of image averaging. Part (a) shows a telescope photograph of a star cluster, and the image is contaminated by film grain. The images in (b), (c), and (d) are the averages of 2, 4, and 8 consecutive photographs of the star cluster, respectively. The image improvement results because the film grain pattern builds up in the sum more slowly than the image of the stationary star cluster.

Figure 7-3 shows another application of image averaging. It is an electron microscope image of the organic crystal catalase. The crystal is periodic with a unit cell of known size. The lack of periodicity in Figure 7-3(a) results from specimen damage during preparation for the electron microscope. The image in the figure was divided into unit cells, and the cell images were averaged together. Figure 7-3(b) shows the replicated average of all the unit cells. While the edges are somewhat blurred, the basic structure of the catalase crystal unit cell can be seen.

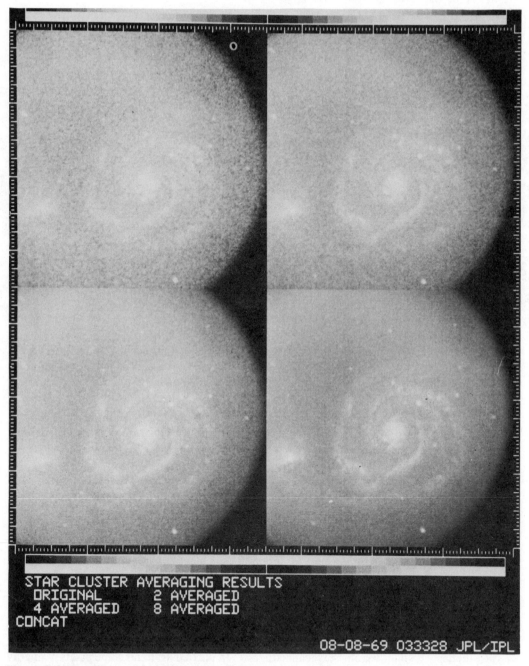

Figure 7-2 Image averaging for film grain reduction

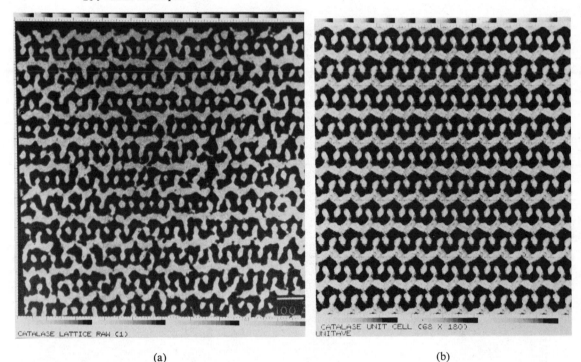

CATALASE LATTICE RAW (1)

CATALASE UNIT CELL (68 X 180)
UNITAVE

(a)   (b)

Figure 7-3 Averaging in a periodic image

Image Subtraction

Background Subtraction. The technique of subtracting a superimposed noise pattern is illustrated in Figure 7-4. Part (a) shows a digitized light microscope image containing two human metaphase chromosome spreads. The image is contaminated by a slowly varying background shading pattern. In part (b), the microscope stage was moved to bring an empty field beneath the objective. Thus (b) contains only the background shading pattern. In part (c), the background has been subtracted from the original image of part (a), thus removing the shading. A constant value 64 was added to each pixel after the subtraction. Below each image appears its gray level histogram. Notice the complexity of the background histogram, how it affects the histogram of part (a), and how the histogram of part (c) resembles the ideal histogram of dark objects on a uniform white background.

The background subtraction technique worked well in Figure 7-4 because an optical density digitizer was used in a case in which the background was superimposed. If some parameter other than optical density had been digitized, the subtraction would have been mathematically invalid and probably less effective. The histogram in part (c) departs slightly from the ideal histogram. In particular, it has some pixels with gray level less than 64, which is the theoretical minimum. This results from noise in the digitizing process. Digitizer noise prevents background pixels in parts (a) and (b) from having identical gray level values.

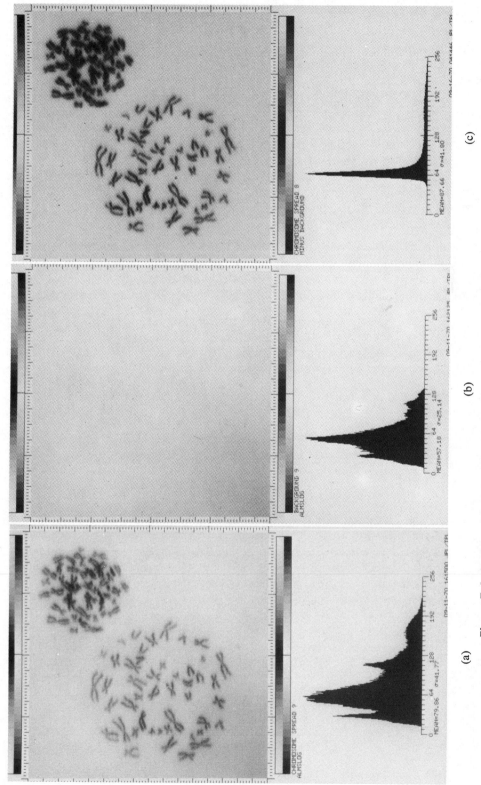

(a)

(b)

(c)

Figure 7-4 Background subtraction: (a) original image, (b) background, (c) difference image

Motion Detection. Figure 7-5 illustrates subtraction for motion detection. Part (a) is one frame from a cinemicrograph of a capillary bed in the mesentery membrane of the rat. This image contains several capillaries and numerous stationary structures. Part (b) is the difference image of two frames taken $\frac{1}{8}$ second apart. The stationary structures cancel out almost completely, while the motion of red blood cells in the capillaries produces considerable contrast. Some of the stationary structures produce residual derivative outlines in the difference image due to imperfect registration between the two frames.

Figure 7-6 shows, in parts (a) and (b), sequential aerial photographs of a freeway. Part (c) is the difference image. The freeway and stationary vehicles subtract out, while vehicle movement is apparent in the difference image. Part (c) is considerably easier to analyze for the detection of moving vehicles than parts (a) and (b).

Gradient Magnitude. Image subtraction can also be used to produce an important derivative of the image, the gradient magnitude function. The gradient is defined as follows: Given a scalar function $f(x, y)$ and a coordinate system with unit vectors $\mathbf{i}$ in the x-direction and $\mathbf{j}$ in the y-direction, the gradient is a vector function defined by

$$\nabla f(x, y) = \mathbf{i}\,\frac{\partial f(x, y)}{\partial x} + \mathbf{j}\,\frac{\partial f(x, y)}{\partial y} \tag{44}$$

where ∇ indicates the vector gradient operator. The vector $\nabla f(x, y)$ points in the direction of maximum upward slope, and its magnitude (length) is equal to the value of the slope. An important scalar function is the gradient magnitude given by

$$|\nabla f(x, y)| = \sqrt{\left(\frac{\partial f}{\partial x}\right)^2 + \left(\frac{\partial f}{\partial y}\right)^2} \tag{45}$$

This represents the steepness of slope at every point, but directional information is lost. Since the square root operation is computationally expensive, Eq. (45) is often approximated by the form

$$|\nabla f(x, y)| \approx \max\left[|f(x, y) - f(x + 1, y)|, |f(x, y) - f(x, y + 1)|\right] \tag{46}$$

that is, the maximum of the absolute vertical and horizontal neighboring pixel differences.

The gradient magnitude takes on large values in areas of steep slope, such as at the edges of objects. Figure 7-7 illustrates the gradient magnitude of a microscope image of a muscle biopsy specimen. The gradient magnitude is high at the edges and low in the interior of the uniformly gray fibers.

Multiplication and Division

The multiplication operation can be used for masking portions of an image. The mask image is unity in areas to be left intact and zero in areas where an image is to be suppressed. Multiplying an image by the mask will blot out or drive to zero the specified area. One may then produce a complement mask for a second image that will blot out those areas retained in the first image. The two masked images may be added to compose the final product.

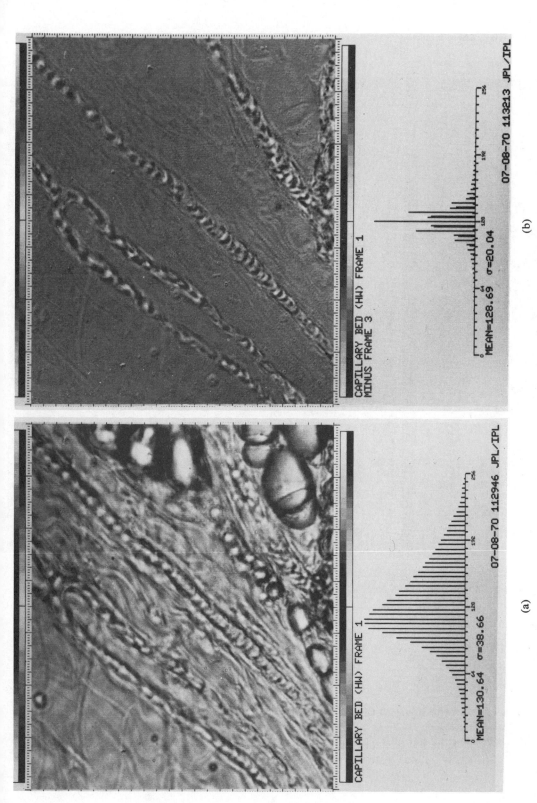

Figure 7-5 Motion detection in a capillary bed: (a) single frame, (b) difference of sequential frames

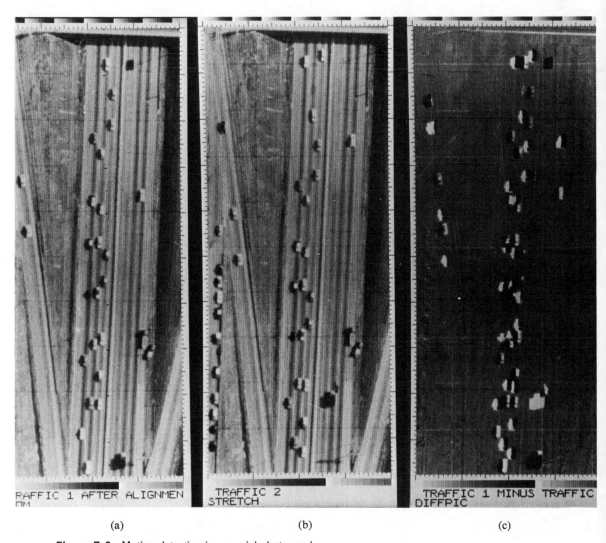

(a) (b) (c)

Figure 7-6 Motion detection in an aerial photograph

Division may be used to remove effects of a spatially varying digitizer sensitivity function. Division also can be used to generate ratio images useful in multispectral analysis.

SUMMARY OF IMPORTANT POINTS

1. The histogram of the sum of two uncorrelated images is given by the convolution of the two input histograms.

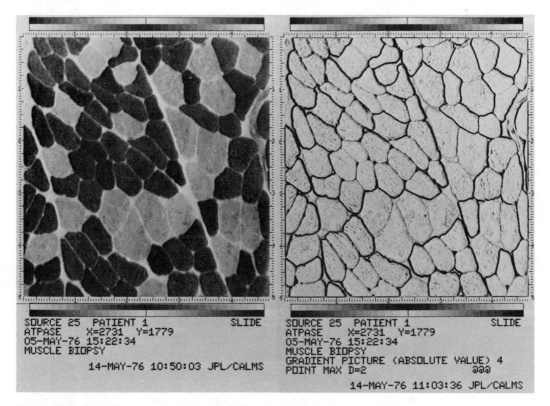

Figure 7-7 A gradient magnitude image: (left) muscle fibers, (right) gradient image

2. Averaging N images of a stationary scene increases the signal-to-noise amplitude ratio by $\sqrt{N}$.

3. Subtracting slightly displaced identical images produces a partial derivative image.

4. The convolution of two Gaussian functions produces another broader Gaussian.

5. The IOD of a noisy image can be computed from the histogram.

6. Image subtraction is useful for background removal and motion detection.

Chapter 8

GEOMETRIC OPERATIONS

INTRODUCTION

Geometric operations are those that change the spatial relationships between objects within an image. Such operations may be thought of as moving things around within the image. The effect is that of printing the image on a rubber sheet, stretching the rubber sheet, and then tacking it down at various points. Actually, a geometric operation is much more general than that, since any point in the image may move to any position. Such an unconstrained geometric operation, however, would constitute a scrambling of the image, so geometric operations are generally constrained to preserve some semblance of order within the image.

The Spatial Transformation

In digital image processing, two separate algorithms are required for a geometric operation. First, there must be an algorithm which defines the spatial transformation itself. This specifies the "motion" of each pixel as it "moves" from its initial to its final position. In most applications, it is desirable to preserve the continuity of curvilinear features and the connectivity of objects within the image. A less constrained algorithm would break up lines and objects and tend to "splatter" the contents of the image. While one could exhaustively specify the motion of each pixel, this would quickly become unwieldy, even for small images. It is more convenient to specify mathematically the spatial relationship between points in the input image and points in the output image. The general definition for a geometric operation is

$$g(x, y) = f(x', y') = f[a(x, y), b(x, y)] \tag{1}$$

where $f(x, y)$ is the input image and $g(x, y)$ is the output image. The functions $a(x, y)$ and $b(x, y)$ uniquely specify the spatial transformation. If they are continuous, connectivity will be preserved within the image.

Gray Level Interpolation

The second requirement for a geometric operation is an algorithm for the interpolation of gray level values. In the input image $f(x, y)$, the gray level values are defined only at integral values of x and y. Equation (1), however, will in general dictate that the gray level value for $g(x, y)$ be taken from $f(x, y)$ at nonintegral coordinates. If the geometric operation is considered as a mapping from f to g, pixels in f can map to positions between pixels in g and vice versa. In this discussion, we are considering pixels to be located exactly at integral coordinates of the sampling grid.

Armed with a spatial transformation and an algorithm for gray level interpolation, we are prepared to perform a geometric operation. Usually, the gray level interpolation algorithm is permanently established in the computer program. The algorithm defining the spatial transformation, however, is specified uniquely for each task. Since the gray level interpolation algorithm is either the same or one of several options, it is the spatial transformation that serves in practice to define each geometric operation.

Pixel Transfer

There are two approaches one may adopt when implementing a geometric transformation algorithm. One may think of the operation as transferring the gray levels from the input image to the output image pixel by pixel. If an input pixel maps to a position between four output pixels, then its gray level is divided among the four output pixels according to the interpolation rule. We may call this the *pixel carryover* approach (see Figure 8-1).

An alternate way to implement the geometric operation is by the *pixel filling* algorithm. In this case, the output pixels are mapped into the input image one at a time to establish their gray level. If an output pixel falls between four input pixels, its gray level is determined by the interpolation algorithm (Figure 8-1).

If one were to program the pixel carryover algorithm, the program would consider the input image pixel by pixel, gradually building up the output image as intensity is carried over. The order in which the output pixels are filled would depend on the spatial transformation. This algorithm could be somewhat wasteful, since many input pixels might map to positions outside the border of the output picture. Furthermore, each output pixel would be addressed several times, since many input

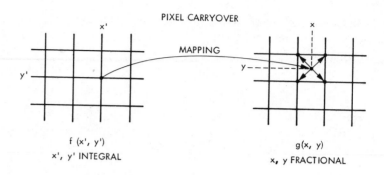

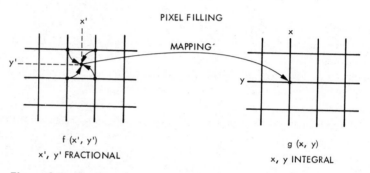

Figure 8-1 Pixel transfer

pixels might contribute to its gray level value. If the spatial transformation involves demagnification, more than four input pixels could contribute. If magnification is involved, certain of the output pixels might be missed when no input pixels mapped to positions near them.

If one uses the pixel filling algorithm, however, the output image may be generated pixel by pixel, line by line. Furthermore, the gray level of each pixel is uniquely determined by one interpolation step between, at most, four input pixels. The input image, of course, must be addressed in a manner defined by the spatial transformation, and this can be quite complex. Nevertheless, the pixel filling algorithm is the more practical for general use, and it is the one programmed in the VICAR system geometric operation programs (see Appendix II).

GRAY LEVEL INTERPOLATION

As mentioned previously, the output pixels map to fractional positions in the input image, generally falling in the space between four input pixels. Interpolation is necessary to determine the gray level of the output pixel.

Nearest Neighbor Interpolation

The simplest interpolation scheme is the so-called zero-order or nearest neighbor interpolation. In this case, the gray level of the output pixel is taken to be that of the input pixel nearest to the position to which it maps. This is computationally simple and produces acceptable results in many cases. However, nearest neighbor interpolation can introduce artifacts in images containing fine structure whose gray level changes significantly over one unit of pixel spacing. Figure 8-2 shows an example of rotating images with nearest neighbor interpolation. The result is a sawtooth effect at some of the edges.

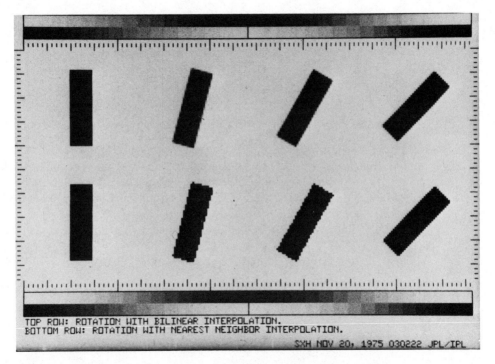

TOP ROW: ROTATION WITH BILINEAR INTERPOLATION.
BOTTOM ROW: ROTATION WITH NEAREST NEIGHBOR INTERPOLATION.
SXH NOV 20, 1975 030222 JPL/IPL

Figure 8-2   Comparison of zero-order and first-order gray level interpolation

Bilinear Interpolation

First-order interpolation produces more desirable results with a slight increase in programming complexity and execution time. Since fitting a plane through four points is an overconstrained problem, first-order interpolation on a rectangular grid requires the bilinear function.

    Let $f(x, y)$ be a function of two variables that is known at $(0, 0)$, $(0, 1)$, $(1, 0)$, and $(1, 1)$, the vertices of the unit square. Suppose we desire to establish by interpolation the value of $f(x, y)$ at an arbitrary point inside the square (Figure 8-3). We shall do

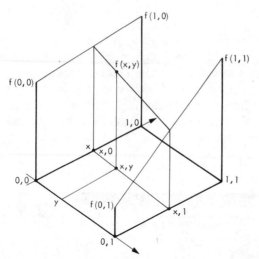

Figure 8-3 Bilinear interpolation

so by fitting a hyperbolic paraboloid, defined by the bilinear equation

$$f(x, y) = ax + by + cxy + d \tag{2}$$

through the four known values.

The four coefficients, a through d, are to be chosen so that $f(x, y)$ fits the known values at the four corners. We shall present an algorithm for determining $f(x, y)$, demonstrate that it produces a bilinear interpolation function, and show that it fits at the corners.

First, we linearly interpolate between the upper two points to establish the value of

$$f(x, 0) = f(0, 0) + x[f(1, 0) - f(0, 0)] \tag{3}$$

and similarly, for the two lower points,

$$f(x, 1) = f(0, 1) + x[f(1, 1) - f(0, 1)] \tag{4}$$

Finally, we linearly interpolate vertically to determine the value of

$$f(x, y) = f(x, 0) + y[f(x, 1) - f(x, 0)] \tag{5}$$

Substituting Eqs. (3) and (4) into Eq. (5), expanding, and collecting terms produces

$$f(x, y) = [f(1, 0) - f(0, 0)]x + [f(0, 1) - f(0, 0)]y$$
$$+ [f(1, 1) + f(0, 0) - f(0, 1) - f(1, 0)]xy + f(0, 0) \tag{6}$$

which is in the form of Eq. (2) and thus bilinear.

Upon inspection, it is clear that Eq. (6) fits the four known values of $f(x, y)$ at the corners of the unit square. Alternatively, Eq. (6) may be derived by interpolating first vertically twice and then horizontally once. This produces the same result because the hyperbolic paraboloid is a two-way ruled surface; that is, it intersects all planes parallel to the xz plane and all planes parallel to the yz plane in straight lines. Notice that if we hold one variable constant, Eq. (2) becomes linear in the other variable.

Bilinear interpolation may be implemented either directly by Eq. (6) or by performing the triple linear interpolation given by Eqs. (3), (4), and (5). Since Eq. (6) involves four multiplications and eight add or subtract operations, the VICAR programs for geometric operations do the latter (Ref. 1). This requires only three multiplications and six addition/subtractions. Although the foregoing development was performed on the unit square, it is easily generalized by a translation, after which x and y represent the fractional pixel position. Figure 8-2 compares bilinear with nearest neighbor interpolation.

When adjacent four-pixel neighborhoods are interpolated with the bilinear equation, the resulting surfaces match in amplitude at the neighborhood boundaries but do not match in slope. Thus a surface generated by piecewise bilinear interpolation is continuous, but its derivatives, in general, have discontinuities at the neighborhood boundaries.

Higher-Order Interpolation

For geometric operations involving magnification, the smoothing effect of bilinear interpolation may degrade fine detail in the image, particularly if magnification is involved. In other applications, the slope discontinuities of bilinear interpolation may produce undesirable effects. In these cases, the extra computational effort of higher-order interpolation may be justified. A function more complex than Eq. (2) and having more than four coefficients is made to fit through a neighborhood of more than four points. If the number of coefficients equals the number of points, the interpolating surface can be made to fit them all. If points outnumber coefficients, an error-minimizing procedure must be used. Examples of higher-order interpolating functions are cubic splines (Ref. 2), Legendre centered functions, and $\sin(x)/x$. The latter is discussed in later chapters. Higher-order interpolation is usually implemented by convolution. This discussion is reserved for Part II of this book.

SPATIAL TRANSFORMATION

Equation (1) gives the general expression for the spatial transformation. It is instructive to consider some special cases before going to more general geometric operations. If we let

$$a(x, y) = x \qquad b(x, y) = y \tag{7}$$

in Eq. (1), we have the identity operation, which merely copies f into g. If we let

$$a(x, y) = x - x_0 \qquad b(x, y) = y - y_0 \tag{8}$$

we have the translation operation, in which the features within f are merely moved by an amount $\sqrt{x_0^2 + y_0^2}$.

We can produce reflection about a vertical line by letting

$$a(x, y) = c - x \qquad b(x, y) = y \tag{9}$$

A similar expression clearly would produce a reflection about a horizontal line. If we let

$$a(x, y) = \frac{x}{c} \qquad b(x, y) = \frac{y}{d} \qquad (10)$$

we can produce magnification of the image. The image will be magnified by the factor c in the x-direction and by d in y. The origin of the image will remain stationary as the image "expands."

Finally, if we let

$$a(x, y) = x \cos \theta - y \sin \theta \qquad b(x, y) = x \sin \theta + y \cos \theta \qquad (11)$$

we produce a rotation about the origin through an angle θ.

Clearly, we can combine translations with magnifications to cause the image to "grow" about a point other than the origin. Likewise, we may combine translation with rotation to produce rotation about an arbitrary point. There is a matrix form which allows geometric operations to be specified quite compactly (see Ref. 1 of Chapter 17).

Control Grids

For relatively simple spatial transformations, it may be practical to use an analytic expression for Eq. (1). In many image processing applications, however, the desired spatial transformation is relatively complex and not amenable to convenient mathematical expression. Furthermore, the desired pixel translations are frequently obtained from actual image measurements, and it is desirable to specify the geometric transformation in these terms. An example is the geometric decalibration of images taken with a camera having geometric distortion. First, a rectangular grid target is digitized and displayed. Because of geometric distortions in the camera, the displayed grid pattern will not be exactly rectangular (see Figure I-1, Appendix I). The desired spatial transformation is that which makes the grid pattern rectangular again, thereby correcting the distortions introduced by the camera (Ref. 3). This same spatial transformation can then be used on subsequent images digitized by the same camera (assuming the distortion is not scene-dependent), thereby producing undistorted images.

It is convenient to specify the spatial transformation as a series of displacement values for specified "control points" in the image. Since only a small percentage of the pixels are actually specified, the displacements of noncontrol points must be determined by interpolation.

The VICAR programs for geometric operations (see Appendix II) allow the user to specify an input "control grid" made up of control points which form the vertices of contiguous quadrilaterals in the input image (Refs. 1, 4, 5). The input control grid maps into a control grid of contiguous, horizontally oriented rectangles in the output image. The quadrilateral vertices (input control points) map directly to the corresponding rectangle vertices (Figure 8-4). Points inside an input quadrilat-

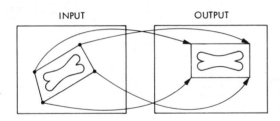

Figure 8-4 Spatial mapping of control points

eral map to points within the corresponding output rectangle. This mapping is accomplished by interpolation based upon the mapping of the vertices.

Control Grid Interpolation

Bilinear interpolation is a good choice for control grid interpolation because it is fast and computationally simple, and it produces a smooth mapping that preserves continuity and connectivity. The general expression for the bilinear spatial transformation is

$$G(x, y) = F(x', y') = F(ax + by + cxy + d, ex + fy + gxy + h) \qquad (12)$$

The bilinear transformation is defined by the values of the eight coefficients a through h. By specifying that the four vertices of a quadrilateral map to the four vertices of the corresponding rectangle, we create two sets of four linear equations in four unknowns. The mapping from x' to x generates four equations in $a, b, c,$ and d and likewise for the mapping from y' to y and the coefficients $e, f, g,$ and h. These sets of equations may be solved for a through h [recall Eq. (6)] to specify the bilinear spatial transformation algorithm that applies to all output points falling inside the rectangle.

While the spatial transformation algorithm could be implemented as Eq. (12), there is a more convenient and computationally more efficient way. By redefining the coefficients a and e, we can write Eq. (12) as

$$G(x, y) = F[x + dx(x, y), y + dy(x, y)] \qquad (13)$$

where dx and dy are pixel displacements that are bilinear functions of x and y. Figure 8-5 shows these displacements with the input quadrilateral superimposed upon the output rectangle to which it maps. The problem now reduces to specifying dx and dy for all points inside the rectangle.

If we implement the pixel filling approach discussed earlier, we can generate the output image line by line. Since $dx(x, y)$ and $dy(x, y)$ are bilinear in x and y, they become linear in x along each output line. Thus we can define an increment Δx so that, assuming unit pixel spacing,

$$dx(x + 1, y) = dx(x, y) + \Delta x \qquad (14)$$

and similarly for dy. The increment Δx changes from line to line but is easily computed from the displacement values at the ends of the output rectangle. These can be interpolated between the given displacements at the vertices. Implementing Eq. (14)

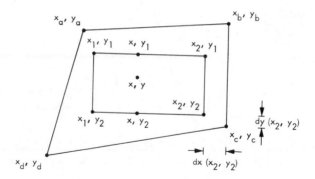

Figure 8-5 Control point displacements

requires only two additions, one for *dx* and one for *dy*, at each output pixel to compute the coordinates of the corresponding input point.

The foregoing procedure specifies the spatial transformation for points falling inside the output rectangle. Frequently, a single quadrilateral-to-rectangle mapping is inadequate to specify the desired spatial transformations, and one can specify contiguous sets of quadrilaterals in the input image that map into contiguous sets of rectangles in the output image (Ref. 1). It is not necessary, however, for the rectangles to cover the output image completely. Figure 8-6 shows an output image in

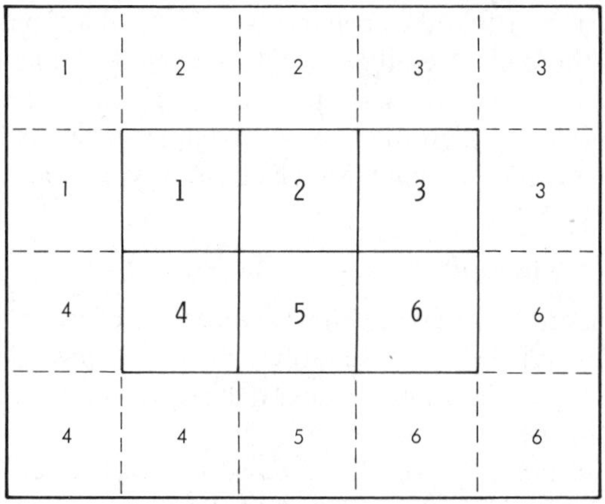

1	2	2	3	3
1	1	2	3	3
4	4	5	6	6
4	4	5	6	6

Figure 8-6 Control grid extrapolation

which are defined six contiguous rectangles. Inside each of the rectangles, the spatial transformation is defined as described above. Figure 8-6 shows how the VICAR programs extrapolate the spatial transformation outside the rectangles by which it is defined. The numbers inside the unspecified (dotted) rectangles indicate the control rectangles from which the bilinear coefficients are used (Ref. 1). For example, the spatial transformation used in the upper left-hand rectangle of the output image uses the bilinear coefficients for rectangle 1.

It is clear from the previous discussion that the bilinear transformation is continuous and unique at the vertices and boundaries of output rectangles. At these boundaries, bilinear interpolation degenerates into linear interpolation between two rectangle vertices. When specifying adjacent rectangles, one must make their vertices coincident. Similarly, adjacent quadrilaterals in the input image must have coincident vertices. Nonadjacent quadrilaterals, however, are not so constrained and may even overlap. Objects inside areas where input quadrilaterals overlap become duplicated in the output image.

APPLICATIONS OF GEOMETRIC OPERATIONS

Geometric Decalibration

An important application of geometric operations is the removal of geometric distortion from digital images (Refs. 4, 5, 6). An example appears in Figure I-1 of Appendix I. Geometric decalibration has proved important in extracting quantitative spatial measurements from a wide variety of digitized images. Certain images, such as those from wide-angle moving mirror scanners and airborne side-looking radar, are subject to rather severe geometric distortions and may require geometric correction before interpretation.

Display Rectification

The Viking Lander camera was designed to digitize Martian panoramas. Its scan lines are spaced at equal angles of elevation, and its pixel spacing represents equal increments of azimuth angle. Figure 8-7(a) shows the distortion this produces on rectangular displays for objects located near the camera.

Rectification of azimuth/elevation images for rectangular display involves the projection of a spherical surface onto a tangent plane. The projection lines emanate from the center of the sphere and carry points on its surface out to the plane. The relationship between input and output pixel location is derived in Ref. 5. In Figure 8-7(b), program GEOTRAN (see Appendix II) was used to project Figure 8-7(a) for rectangular display. Notice that the table edges are straight in Figure 8-7(b).

Program GEOTRAN can project Viking Lander images to other viewpoints. Figure 8-8(a) shows an image from the camera point of view, and Figure 8-8(b) is an orthographic projection designed to show the surface as it would appear from directly overhead. The projection (Ref. 5) assumes that the surface is planar and deviations from this introduce slight distortions. Using three-dimensional scene information from the Viking Lander range finder, GEOTRAN can also transform Lander images to appear as if viewed from other camera positions (Ref. 5).

Image Registration

Another major application of geometric operations is registering similar images for comparison purposes. This is typified by image subtraction for motion or change detection. As pointed out in Chapter 7, if similar images are displaced slightly and

(a)

(b)

Figure 8-7 Viking Lander camera correction: (a) before, (b) after

(a)   (b)

Figure 8-8 Viking orthographic projection: (a) before, (b) after

subtracted, the difference image has a strong partial derivative component. This could easily mask the image differences of interest. If images of a stationary object can be digitized from a fixed camera position, they can be obtained in register. If this is not the case, however, it is likely that the images will have to be registered before subtraction.

While simple translation is easily accomplished, rotation or more complex distortion requires a geometric operation. Registration of film scan images is likely to involve translation and rotation. Serial sections of biological tissue, sliced on a microtome and photographed through a microscope, are subject to rather severe geometric distortions. In cases such as this, simple translation and rotation are inadequate. One such image can be taken as a standard of reference and the others distorted to match it. Small features are located throughout the images and used to define control points. Frequently, techniques such as cross-correlation are helpful in locating these features. Chapter 7 shows several examples of image subtraction in which careful registration is required.

Figure 8-9 shows an example of image subtraction where a geometric operation is required for proper image registration (Ref. 3). The images in (a) and (b) are chest X rays taken seven months apart. In (c), the images are subtracted to bring out changes that might be due to malignancy. Because of differences in rib position, however, the difference image contains considerable contrast due to misregistration. A geometric operation was performed on the second chest film to produce picture (d). In this picture, rib position has been matched to that of picture (a). The result of subtracting (d) from (a) is shown in picture (e).

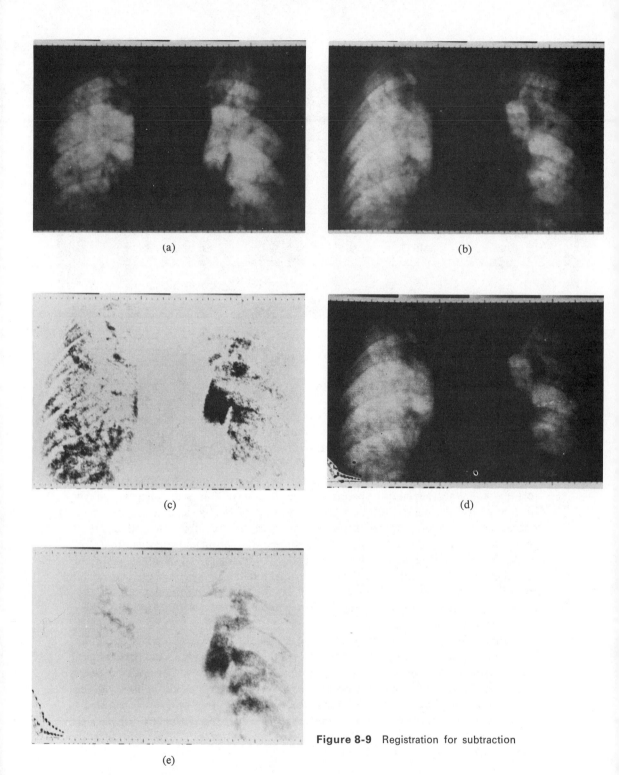

(a)

(b)

(c)

(d)

(e)

Figure 8-9 Registration for subtraction

122

Image Format Conversion

Geometric operations are sometimes useful simply for placing images into a format more convenient for interpretation. Figure 8-10(a) shows a photographic map of the chromosomes of one species of the fruitfly *Drosophila*. The map is made by pasting up photographs of chromosomes taken through a microscope. Geneticists analyze the pattern of bands on the chromosomes to deduce patterns of evolution. The areas are numbered for reference.

Figure 8-10(b) shows the result of using a geometric operation to produce a map in which the chromosomes appear straight. In the input image, each chromosome was overlayed with a control grid of quadrilaterals, each one having two sides parallel to the chromosome axis. These were mapped into horizontal strings of rectangles in the output image. In order to prevent axial distortion of the chromosome, the horizontal length of each rectangle was made equal to the mean of the two axial sides of the corresponding quadrilateral.

The numbers below chromosome 3 suffered less distortion than the others because a second row of quadrilaterals was defined below this chromosome. These were actually parallelograms having vertical ends and sides parallel to the chromosome axis. They mapped into a second row of rectangles falling beneath the straightened chromosome.

Figure 8-11 shows another example of geometric reformatting for interpretive purposes. Figure 8-11(a) is a photomicrograph of the chromosomes from one human white blood cell. In Figure 8-11(b), the chromosomes have been oriented vertically and arranged into groups of similar size and shape. The format of Figure 8-11(b) is called a *karyotype*. Assembled by manual methods, the karyotype is an important diagnostic tool in genetics. Figure 8-11(b) was produced by program GEOM. The input quadrilaterals were actually rectangles fitted around each chromosome, with two sides parallel to the chromosome axis. These were mapped into vertically oriented rectangles. As in Figure 8-10, the control points were determined manually by measurement of the input image. In both cases, this process is quite laborious. For human chromosomes, the orientation and classification can be done automatically. These techniques are discussed in Chapters 15 and 16.

Map Projection

Another major application of geometric operations is projecting images for mapping purposes. For example, it was necessary to produce photomosaic maps of the Moon, Mars, Mercury, and Venus using images transmitted back from spacecraft. The borders of the spacecraft camera image project onto the planet's surface, forming a "footprint" with four curvilinear sides [Figure 8-12(a)]. The spherical surface of the planet is projected onto a flat surface to make a map [Figure 8-12(b)]. The footprint also projects onto the map, producing a further distorted four-sided figure.

A geometric operation may be used to transform the spacecraft camera image into the form it should assume on the map. Multiple images processed in this way may be combined to form a photographic map of the planet. The task of determining

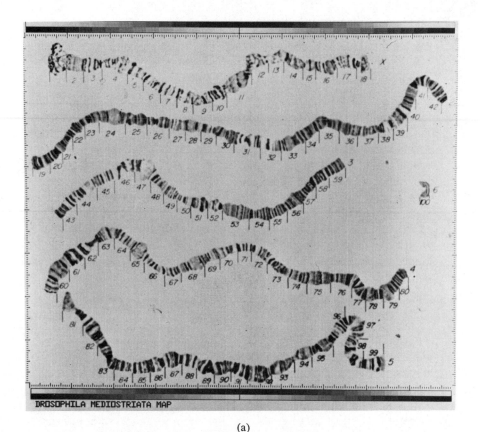

(a)

(b)

Figure 8-10 *Drosophila* chromosome map: (a) original, (b) straightened

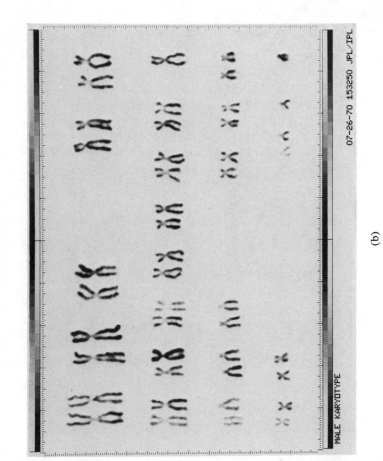

(b)

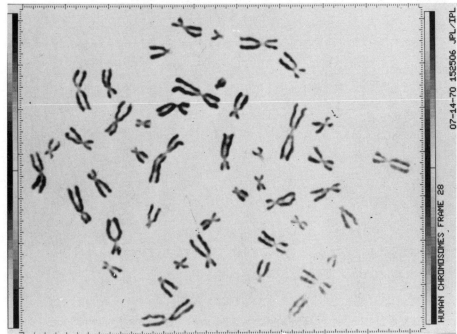

(a)

Figure 8-11 Human karyotype: (a) original, (b) karyotype

125

(a)

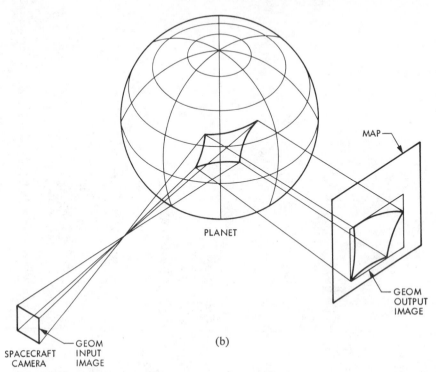

(b)

Figure 8-12 Photographic mapping: (a) spacecraft camera "footprint,"
(b) map projection

the control points for projecting a given image is somewhat involved. A program called MAP2 was developed for this purpose (Ref. 7; see Appendix II). MAP2 takes the spacecraft viewing geometry and the desired cartographic projection parameters and generates input and output control grids.

Determining the spatial transformation between the input and the projected image is a two-step process. MAP2 solves this problem by working backward from the output image to the input image. The specified cartographic projection technique defines the relationship between points in the output image and points on the planet's surface. The spacecraft viewing geometry determines the spatial relationship between points on the planet surface and pixel position in the camera image. MAP2 overlays a rectangular control grid on the output image and maps it back through the cartographic projection and the spacecraft viewing geometry to overlay it on the input image. In the following section, we outline this technique.

Cartographic Projection Techniques

The science of cartography is concerned with producing two-dimensional maps of spherical or ellipsoidal bodies. This is not a simple matter because spherical surfaces cannot be flattened without distortion. Cartographers solve this problem by projecting the spherical surface onto a plane or onto a cylinder or cone that can be "unrolled" to form a flat surface (Refs. 8, 9).

There are two classes of cartographic projection techniques. The first class is called *geometric* or *perspective* projection and includes maps generated purely by geometric projection techniques. The second class is called *semigeometric* or *nonperspective*. These maps are generated by modified or partial geometric projection techniques.

Map Properties. There are three important properties that maps may or may not have depending on their method of generation. A map is said to be *equidistant* if scale is preserved along certain lines. This means distances along those lines are proportional to the distance between corresponding points on the planet. A map has the property of *equivalence* if the area of a region is preserved in the projection. Such maps may be used for comparing the areas of different features. A map is *conformal* or *orthomorphic* if angles are preserved in the projection. Lines on the surface will intersect at the same angle as their projections on the map. A conformal map also preserves shape at a point. This means that the shape of small features is distorted only very slightly. Shape distortion becomes progressively worse as feature size increases.

Cartographic Projections. There are three types of surfaces onto which surface features may be projected to form a two-dimensional map. These are the plane, the cylinder, and the cone. The latter two must be cut along a line parallel to the axis and "unrolled" to form a flat map. The cone may be considered the general case, since the plane can be thought of as a cone with apex angle 180° and the cylinder a cone with apex angle 0°.

While many types of projections have been defined and used throughout cartographic history, four of the most important have been implemented in program MAP2. They are (1) the orthographic, (2) the stereographic, (3) the Mercator, and (4) the Lambert conformal conic projections (Ref. 7). These projection techniques differ in their generation techniques and their properties. They are described below with reference to Figure 8-13.

In the orthographic projection, surface features are projected onto a plane tangent to the sphere at a point called the *center of projection*. Features are projected along parallel lines normal to the plane. When the center of projection is a pole, the

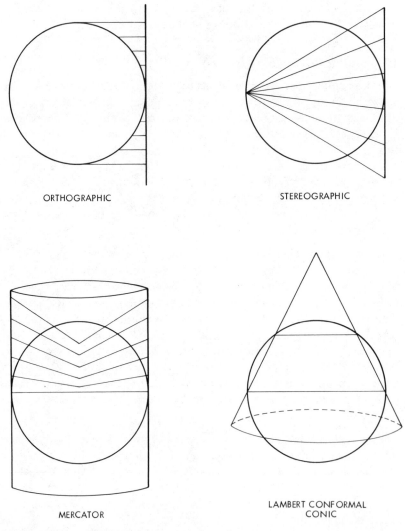

ORTHOGRAPHIC

STEREOGRAPHIC

MERCATOR

LAMBERT CONFORMAL
CONIC

Figure 8-13 Cartographic projections

scale along parallels of latitude is constant. The radial scale, however, decreases away from the center of projection. There is little distortion of features near the center of projection. Parallels of latitude project as concentric circles centered on the pole, and meridians project as straight lines intersecting at the pole. The orthographic projection is useful because it approximates viewing the planet from a large distance. For this reason, the eye is able to visualize the spherical shape in an orthographic projection. Because of scale and shape distortion, however, orthographic maps are of restricted quantitative use except for small features near the center of projection.

The stereographic projection is similar to the orthographic except that the projection rays emanate from a perspective point located directly opposite the center of projection. In the polar case, parallels of latitude project as concentric circles centered on the pole, and meridians project as radial lines intersecting at the pole. The scale along parallels and along meridians increases away from the pole. They increase proportionately, however, so that at any point the longitude and latitude scales are the same. This makes the stereographic projection conformal, and shape is preserved locally. There is little distortion of features near the center of projection. This, coupled with conformality, makes the stereographic projection quite useful.

The Mercator projection maps surface features onto a right-circular cylinder that is tangent at the equator. The cylinder axis is colinear with the polar axis of the sphere. The meridians map to equidistant vertical lines and parallels map to circles on the cylinder, and these open up to form horizontal lines on the map. Scale along latitude lines increases with distance from the equator. The projection is designed so that the perspective point moves up the axis with increasing latitude, keeping the latitude and longitude scales equal and thus making the map conformal. Scale is exaggerated away from the equator and features near the poles become quite large. The poles themselves cannot be mapped.

The vertical position of latitude lines is given by

$$y = R \ln \left[\tan \left(45 + \frac{\phi}{2} \right) \right] \tag{15}$$

where R is the planet radius on the map and ϕ is latitude. Historically the mercator projection has been useful for navigation because a course of constant compass heading projects to a straight line on the map.

In the Lambert conformal conic projection, surface features are projected onto a cone having the same axis as the planet. The cone intersects the sphere at two parallels called the standard parallels. Meridians map to straight lines, and parallels map to circles inside the cone. When the cone is unrolled, the parallels become arcs and the meridians merge at the pole. The spacing of the parallels is adjusted to achieve conformality. The two standard parallels project at true scale. Scale decreases between them and is exaggerated outside the two standard parallels.

Implementation

The steps necessary to project a spacecraft image for mapping purposes are

1. Establish the spacecraft camera viewing geometry.

2. Determine an expression giving camera pixel position as a function of the latitude and longitude of the corresponding point on the planet's surface.

3. Select the map projection parameters (type of projection, center of projection, etc.) and establish the borders of the output picture on the map.

4. Determine an expression giving the latitude and longitude of a point on the planet's surface in terms of the pixel coordinates of the corresponding point on the map.

5. Combine the results of steps 2 and 4 to yield an expression giving camera pixel position as a function of position on the output map.

6. Overlay a rectilinear control grid on the output picture.

7. Use the expression of step 5 to map the output control points into the input image, thus establishing the input control grid.

8. Use the results of step 7 in a geometric operation to effect the projection.

The spacecraft viewing geometry may be established (Ref. 7) with reference to Figure 8-14. In this figure, the spacecraft is located at a distance R_s from the center of the planet directly above the point at latitude ϕ_s and longitude λ_s. Point C is the perspective point that represents the nodal point of the camera lens. Point p is in the camera image and corresponds to point p', which has longitude λ and latitude ϕ on the surface. The distance f represents the focal length of the lens and is exaggerated for clarity in Figure 8-14. The vector $\mathbf{Q}$ extends from C to p'. Notice that the vector

$$\mathbf{P} = \begin{bmatrix} x_p \\ y_p \\ f \end{bmatrix} \tag{16}$$

has components x_p and y_p, which are the camera pixel position coordinates. Since $\mathbf{P}$ and $\mathbf{Q}$ are colinear, they are related by a scale factor

$$\mathbf{P} = \left(\frac{f}{Q_z}\right)\mathbf{Q} \tag{17}$$

From Figure 8-14, we see that

$$\mathbf{Q} = \mathbf{R} - \mathbf{S} \tag{18}$$

which we can write in matrix notation as

$$\begin{bmatrix} Q_x \\ Q_y \\ Q_z \end{bmatrix} = [\mathbf{M}] \begin{bmatrix} R \cos \phi \cos \lambda - R_s \cos \phi_s \cos \lambda_s \\ R \cos \phi \sin \lambda - R_s \cos \phi_s \sin \lambda_s \\ R \sin \phi - R_s \sin \phi_s \end{bmatrix} \tag{19}$$

where $[\mathbf{M}]$ is the 3 by 3 matrix that transforms from planet-centered to spacecraft coordinates.

Finally, Eq. (17) implies

$$x_p = \left(\frac{f}{Q_z}\right)Q_x \quad \text{and} \quad y_p = \left(\frac{f}{Q_z}\right)Q_y \tag{20}$$

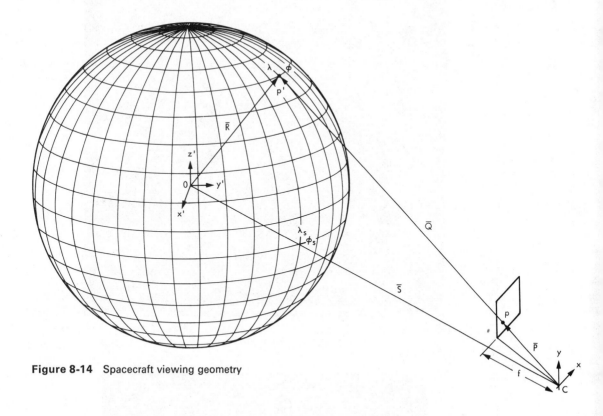

Figure 8-14 Spacecraft viewing geometry

Several cartography texts develop equations that give map position in terms of latitude and longitude on the surface. Since MAP2 works backward from map to planet, however, inverse forms of the equations are required. These are developed in Ref. 7 for the four projections mentioned previously.

Frequently, spacecraft images require both geometric correction and map projection, suggesting two sequential operations. Since pixel interpolation twice might reduce image detail, the two geometric operations can be combined so that one execution corrects and projects the image.

Examples

Figure 8-15 illustrates the steps in producing a photographic map. Picture (a) is a Mariner 10 image of Mercury before correction for photometric and geometric distortion. In picture (b), the image has been subjected to a geometric operation to produce an orthographic projection. In picture (c), several neighboring orthographic projections have been combined to form a mosaic. Finally, in picture (d), a latitude and longitude grid has overlayed the orthographic mosaic.

Figure 8-16 shows a polar stereographic projection of Mariner 6 and 7 images of Mars (Ref. 10). A mosaic of high-resolution narrow-angle images has been inserted

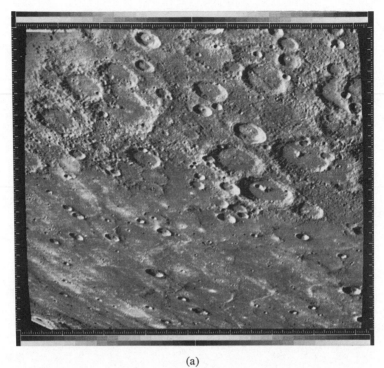

(a)

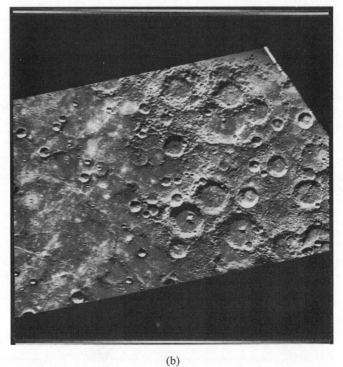

(b)

Figure 8-15 Mariner 10 map projection example: (a) original image, (b) orthographic projection, (c) mosaic of several projections, (d) map grid overlay

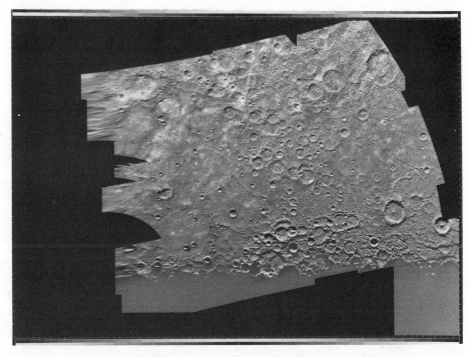

(c)

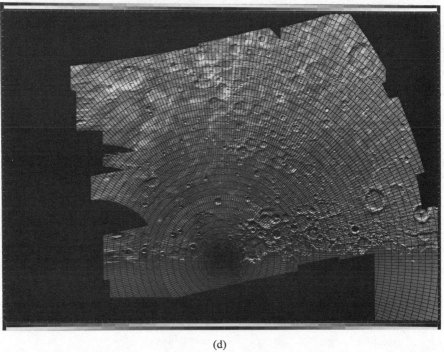

(d)

Figure 8-15 (cont.)

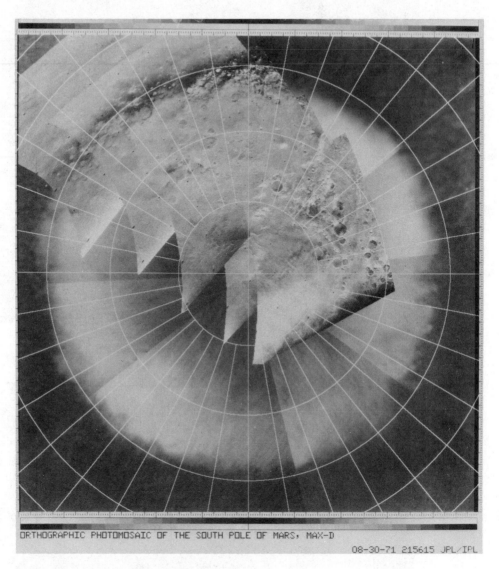

ORTHOGRAPHIC PHOTOMOSAIC OF THE SOUTH POLE OF MARS, MAX-D

08-30-71 215615 JPL/IPL

Figure 8-16 Polar orthographic map of Mars

into a mosaic of wide-angle low-resolution pictures of the entire polar area. Figure I-5 in Appendix I shows a 4-ft diameter globe covered with orthographic projections of 2000 Mariner 9 images of Mars (Ref. 6).

SUMMARY OF IMPORTANT POINTS

1. A geometric operation requires a means for specifying the spatial transformation and an algorithm for gray level interpolation.

2. A geometric operation can be thought of as mapping each output image pixel into the input image, where its gray level value is determined by interpolation.

3. Bilinear gray level interpolation is generally superior to nearest neighbor interpolation, and it produces only a modest increase in program complexity and execution time.

4. The spatial transformation can be specified by a pair of control grids, one defined in the input and one in the output image.

5. The input control points map to the corresponding output control points.

6. Between control points, the spatial transformation is obtained by interpolation.

7. Bilinear interpolation is useful for control grid mapping.

8. Geometric operations are useful for digitizer decalibration, display rectification, image registration, reformatting images for display, and for map projection.

REFERENCES

1. H. Frieden, "GEOM and LGEOM," in *Image Processing Laboratory Users Documentation of Applications Programs*, Jet Propulsion Laboratory Internal Document No. 900-670, October 11, 1975.

2. S. S. Rifman and D. M. McKinnon, *Evaluation of Digital Correction Techniques for ERTS Images*, TRW Report No. 20634-6003-TV-00, Clearinghouse for Federal Scientific and Technical Information, Springfield, Virginia 22151.

3. R. Nathan, *Digital Video Data Handling*, JPL Tech. Report 32-877, January 5, 1966.

4. J. E. Kreznar, *User and Programmer Guide to the MM'71 Geometric Calibration and Decalibration Programs*, Jet Propulsion Laboratory Internal Document No. 900-575, March 1, 1973.

5. M. Girard, *Viking 75 Project Software Requirements Document for the GEOTRAN Program*, Jet Propulsion Laboratory Internal Document No. 620-52, January 12, 1975.

6. W. B. Green, P. L. Jepsen, J. E. Kreznar, R. M. Ruiz, A. A. Schwartz, and J. B. Seidman, "Removal of Instrument Signature From Mariner 9 Television Images of Mars," *Applied Optics*, **14**, 105, January 1975.

7. C. A. Rofer, A. A. Schwartz, J. B. Seidman, and W. B. Green, *Computer Cartographic Projections for Planetary Mosaics—Final Report* (3 Vol.), Jet Propulsion Laboratory Internal Document No. 900-636, November 15, 1973.

8. G. P. Keloway, *Map Projections*, Methuen and Co., London, 1946.

9. P. Richardis and R. K. Adler, *Map Projections*, American Elsevier, New York, 1972.

10. A. R. Gillespie and J. M. Soha, "An Orthographic Photomap of the South Pole of Mars," *Icarus*, **16**, 522, 1972.

Part II

LINEAR FILTERING

LINEAR SYSTEM THEORY

■■

INTRODUCTION

In preceding chapters, we have examined the effects that certain image processing operations have on images. The effects studied have been those that can be explained by relatively simple mathematics. We have thus far avoided discussions of sampling effects, spatial resolution, and operations commonly referred to as *enhancement*. In Part II, we address questions of sampling, resolution, and enhancement, a term commonly taken to mean *linear filtering*. In this chapter and the next, we develop the analytical tools required to approach these questions.

Linear system theory is a well-developed field commonly used to describe the behavior of electrical circuits and optical systems. It provides a firm mathematical basis upon which to examine the effects of sampling, filtering, and spatial resolution. Linear system theory is also helpful in a variety of other applications, and it makes a useful addition to one's technical background.

Definitions

In the context of this book, we consider a system to be anything that accepts an input and produces an output. Since we are concerned only with the relationship between input and output, we have no interest in what is inside the system. The input and output can be one-dimensional, two-dimensional, or higher-dimensional. In this development, however, we shall restrict our examples to two cases: one-dimensional functions of time, and two-dimensional functions of two spatial variables. This keeps the notation simpler, makes the analysis somewhat less abstract since the development is tied to real physical processes, and avoids a burden of excess generality. The analysis can be easily generalized to higher dimensions when necessary. In most cases, the development will be done for one-dimensional functions of time and generalized to two-dimensional images.

Figures 9-1 and 9-2 show the conventional notation for one- and two-dimensional linear systems. In each case, an input to the system produces a response that is another function of the same variables as the input.

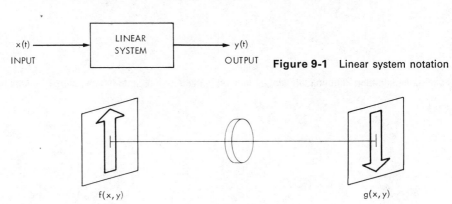

Figure 9-1 Linear system notation

Figure 9-2 Two-dimensional linear system

Linearity. Linear systems have a special property that gives rise to their name. Suppose that, for a particular system, an input $x_1(t)$ gives rise to an output $y_1(t)$:

$$x_1(t) \longrightarrow y_1(t) \tag{1}$$

(where the arrow may be read as *produces*) and a second input $x_2(t)$ gives rise to an output $y_2(t)$:

$$x_2(t) \longrightarrow y_2(t) \tag{2}$$

The system is linear if and only if it has the property

$$x_1(t) + x_2(t) \longrightarrow y_1(t) + y_2(t) \tag{3}$$

A third input signal, which is the sum of the first two, produces an output signal that is the sum of the original two output signals. Any system that does not obey this constraint is nonlinear. Nonlinear system analysis has produced many useful results in a variety of areas. However, the analysis of nonlinear systems is considerably more complex than that of linear systems, and that additional complexity is not required for our purposes. Therefore we shall restrict ourselves to the analysis of linear systems.

The definition of a linear system states that an input which is the sum of two signals produces an output which is the sum of the outputs produced by the individual signals acting alone. From this it follows that, if an input signal is multiplied by a rational number, the output is increased or decreased by the same factor; that is,

$$ax_1(t) \longrightarrow ay_1(t) \tag{4}$$

We take it as an axiom that Eq. (4) also holds for irrational numbers.

The property defined in Eqs. (1), (2), and (3) and its corollary in Eq. (4) serve to define a linear system. When using linear system theory to analyze a process, it is imperative that the process being modeled is linear. If the system under study does not satisfy these criteria, then it is nonlinear, and linear system theory will produce inaccurate and possibly misleading results. If the system is only slightly nonlinear, it may be assumed linear for purposes of analysis, but the results of the analysis will be only as good as the assumption. Frequently, systems known to be slightly nonlinear are studied with linear system theory because this approach is mathematically tractable. One must be cautious when dealing with nonlinear systems because the protective canopy of linear system theory disintegrates as the assumption of linearity breaks down. The analyst has the responsibility not only for the mathematics but also for the validity of the linearity assumption.

Shift Invariance. A useful property of certain systems is called *shift invariance*. It is illustrated by the following. Assume, for a particular linear system, that

$$x(t) \longrightarrow y(t) \tag{5}$$

Suppose we now shift the input signal in time by an amount T. The system is shift invariant if

$$x(t - T) \longrightarrow y(t - T) \tag{6}$$

that is, the output is shifted by the same amount but otherwise unchanged. Thus, for a shift invariant system, shifting the input merely shifts the output by the same amount. The important point is that the nature of the output is not changed by a time shift at the input. Spatial shift invariance is the two-dimensional analog of time shift invariance. If the input image is shifted relative to its origin, the output image is the same as before except for an identical shift.

Most of the analysis in the next few chapters is directed toward shift invariant linear systems. These assumptions are valid to a very good approximation for electrical networks, well-designed linear electronic networks, and optical systems, which are the principal components of image processing systems.

HARMONIC SIGNALS AND COMPLEX SIGNAL ANALYSIS

In ordinary usage, signals and images can be represented by real-valued functions of one or two variables, respectively. The value of the function represents the magnitude of some physical parameter such as voltage as a function of time or light intensity as a function of two spatial coordinates. The development of linear system properties proceeds much more smoothly, however, if we allow the inputs and outputs to be complex-valued functions. Since real-valued functions can be considered a special case of complex-valued functions, we lose nothing by this generalization. The advantages will become clear during the course of the development.

Harmonic Signals

Consider a complex-valued signal of the form

$$x(t) = e^{j\omega t} = \cos(\omega t) + j\sin(\omega t) \tag{7}$$

where $j^2 = -1$. This is called a *harmonic signal*. It is a complex-valued function of time that can be viewed as a unit length vector rotating in the complex plane with an angular velocity ω (Figure 9-3). The angular frequency ω, in radians per second, is related by $\omega = 2\pi f$ to f, the frequency in revolutions or cycles per second (hertz).

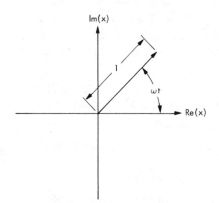

Figure 9-3 Harmonic signal generating vector

Response to a Harmonic Input

Suppose a shift invariant linear system is subjected to a harmonic input given by

$$x_1(t) = e^{j\omega t} \tag{8}$$

We may express the response of the system as

$$y_1(t) = K(\omega, t)e^{j\omega t} \tag{9}$$

where

$$K(\omega, t) = \frac{y_1(t)}{e^{j\omega t}} \tag{10}$$

is a complex function of ω and t selected so that, when multiplied by $e^{j\omega t}$, it yields $y_1(t)$. For any $y_1(t)$, there is a $K(\omega, t)$ that works.

    Now suppose we generate a second input signal by time shifting $x_1(t)$. This becomes

$$x_2(t) = e^{j\omega(t-T)} = e^{-j\omega T}e^{j\omega t} = e^{-j\omega T}x_1(t) \tag{11}$$

Notice that $x_2(t)$ is merely $x_1(t)$ multiplied by a complex constant. This results because $x_1(t)$ is a harmonic signal.

The linear system's response to $x_2(t)$ is now given by

$$y_2(t) = K(\omega, t - T)e^{j\omega(t-T)} \tag{12}$$

which is

$$y_2(t) = K(\omega, t - T)e^{-j\omega T}e^{j\omega t} \tag{13}$$

or

$$y_2(t) = K(\omega, t - T)e^{-j\omega T}x_1(t) \tag{14}$$

Recalling Eq. (4), we can write

$$x_2(t) = e^{-j\omega T}x_1(t) \longrightarrow e^{-j\omega T}y_1(t) = e^{-j\omega T}K(\omega, t)e^{j\omega t} \tag{15}$$

From Eq. (8), we recognize the exponential factor on the right as $x_1(t)$. Also, we know that the response side of Eq. (15) has to be $y_2(t)$ since it is the system's response to $x_2(t)$. Therefore, we can write

$$y_2(t) = e^{-j\omega T}K(\omega, t)x_1(t) \tag{16}$$

which is a second expression for the system's response to the shifted harmonic input.

Equation (14) was obtained by inserting a time shift into Eq. (9). Equation (16) resulted from the linearity property of Eq. (4). Both equations, however, are expressions for the linear system's response to the time shifted harmonic input, and thus they must be equal. Combining Eqs. (14) and (16) produces

$$K(\omega, t - T)e^{-j\omega T}x_1(t) = K(\omega, t)e^{-j\omega T}x_1(t) \tag{17}$$

and it is clear that

$$K(\omega, t - T) = K(\omega, t) \tag{18}$$

must be true for any amount of shift T. Equation (18) can only be true, however, if $K(\omega, t)$ is independent of t. Now Eq. (9) can be rewritten in general form as

$$y(t) = K(\omega)x(t) \tag{19}$$

The general function whose form was assumed in Eq. (10) turns out to be a function only of frequency ω. Equation (19) states the important property that the response of a shift invariant linear system to a harmonic input is simply that input multiplied by a frequency-dependent complex constant. Notice that a harmonic input always produces a harmonic output at the same frequency.

Harmonic Signals and Sinusoids

When we use a linear system to model the behavior of a physical (electronic or optical) system, the inputs and outputs are conveniently represented by real-valued functions. Thus we can add another restriction to shift invariant linear systems, namely that they preserve realness. By definition, this means that a real-valued input can only produce a real-valued output. From this, it can be shown that such a system also preserves imaginariness and that removing the imaginary part of a complex input merely removes the imaginary part of the corresponding complex output; that is,

$$x(t) \longrightarrow y(t) \Longrightarrow \mathcal{R}e\{x(t)\} \longrightarrow \mathcal{R}e\{y(t)\} \tag{20}$$

In a sense, the real and imaginary parts of a harmonic input go through the system independently of each other.

The real preserving restriction on linear systems allows us to simplify this analysis. For example, if the input is a cosine, we can add an imaginary sine component to form a harmonic signal [recall Eq. (7)], determine the system's response to that harmonic input, and then discard the imaginary part of the complex output. This indirect approach is justified by a significant simplification of the analysis.

Any sinusoidal signal can be thought of as the real part of a (unique) harmonic signal. This allows us to derive a linear system's response to a sinusoid by (1) representing the input sinusoid by a harmonic signal, (2) deriving the linear system's response to the harmonic input, and (3) taking the real part of the harmonic output to yield the output. In so doing, we are using a transform method of solution; that is, we transform from sinusoids to harmonic signals, solve the problem in terms of harmonics, and then transform the harmonic output back into a sinusoid. The technique is analogous to using logarithms for multiplication where one transforms the multiplier and multiplicand into logarithms, adds them to effect multiplication, and transforms the result back from logarithms to ordinary numbers to obtain the desired product. As with logarithms, the transformation to harmonic signals considerably simplifies the analysis of linear systems.

The Transfer Function

The function $K(\omega)$ is called the transfer function of the linear system and is sufficient to specify the system completely. For a shift invariant linear system, the transfer function contains all the information that exists about the system.

We can convert $K(\omega)$ to polar form to obtain

$$K(\omega) = A(\omega)e^{j\phi(\omega)} \qquad (21)$$

where $A(\omega)$ is a real-valued function of frequency and the complex exponential is a unit vector in the complex plane, that is, a complex number having unit magnitude.

The effect of the transfer function is illustrated by the following. Suppose the input is a cosine, taken to be the real part of a harmonic signal:

$$x(t) = \cos(\omega t) = \mathcal{R}e\{e^{j\omega t}\} \qquad (22)$$

The system's response to the harmonic input is

$$K(\omega)e^{j\omega t} = A(\omega)e^{j\phi}e^{j\omega t} = A(\omega)e^{j(\omega t + \phi)} \qquad (23)$$

Finally, the actual output signal is given by

$$y(t) = \mathcal{R}e\{A(\omega)e^{j(\omega t + \phi)}\} = \mathcal{R}e\{A(\omega)[\cos(\omega t + \phi) + j\sin(\omega t + \phi)]\}$$
$$= A(\omega)\cos(\omega t + \phi) \qquad (24)$$

$A(\omega)$ is a multiplicative gain factor and represents the degree to which the system amplifies or attenuates the input signal. The $\phi(\omega)$ is the phase shift angle. Its only effect is to shift the time origin of the harmonic input function.

In the remainder of this book, the analysis will be done in terms of harmonic signals, with the conversion to sinusoids left as a step of interpretation.

In summary, we have developed three important properties of shift invariant linear systems: (1) A harmonic input always produces a harmonic output at the same frequency. (2) The system is completely specified by its transfer function, a complex-valued function of frequency alone. (3) The transfer function produces only two effects upon a harmonic input, a change in amplitude and a phase shift (a shift of the time origin).

THE CONVOLUTION OPERATION

Consider the linear system shown in Figure 9-4. It would be useful to have a general expression that relates the output signal to the input. We can obtain such a relation in the following way. The linear functional expression (superposition integral)

$$y(t) = \int_{-\infty}^{\infty} f(t, \tau)x(\tau)\, d\tau \tag{25}$$

is general enough to express the relationship between $x(t)$ and $y(t)$. A function $f(t, \tau)$ of two variables can be chosen to make Eq. (25) hold for any linear system; however, we would prefer to characterize a linear system with a function of only one variable.

Figure 9-4 A linear system $x(t)$ ⟶ [$f(t,\tau)$] ⟶ $y(t)$

We shall now impose the shift invariance constraint in an effort to simplify Eq. (25). Substituting Eq. (6) into Eq. (25) produces

$$y(t - T) = \int_{-\infty}^{\infty} f(t, \tau)x(t - T)\, d\tau \tag{26}$$

We now make a change of variables by adding T to both t and τ. This produces

$$y(t) = \int_{-\infty}^{\infty} f(t + T, \tau + T)x(\tau)\, d\tau \tag{27}$$

If we compare Eqs. (25) and (27), we see that

$$f(t, \tau) = f(t + T, \tau + T) \tag{28}$$

must be true for all values of T. This means that $f(t, \tau)$ does not change if we add the same constant to both of its variables. In other words, $f(t, \tau)$ is constant as long as the difference between t and τ is constant. Thus we can define a new function of only this difference

$$g(t - \tau) = f(t, \tau) \tag{29}$$

and Eq. (25) becomes

$$y(t) = \int_{-\infty}^{\infty} g(t - \tau)x(\tau)\, d\tau \tag{30}$$

This is the familiar convolution integral. Equation (30) states that the output of a shift invariant linear system is given by the convolution of the input signal with a function $g(t)$ that is characteristic of that system. This characteristic function is called the *impulse response* of the system for reasons pointed out later. Notice that the system preserves realness if and only if $g(t - \tau)$ is a real function.

We now have two ways to specify the relationship between the input and output of a shift invariant linear system: (1) Every such system has a complex transfer function which, when multiplied by a harmonic input, yields the harmonic output; and (2) every such system has a real impulse response which, when convolved with the input signal, yields the output signal.

Since the transfer function and the impulse response of a shift invariant linear system are each adequate to specify the system completely, we suspect that the two functions are related. This relationship is developed in the next chapter.

Convolution in One Dimension

The convolution integral in Eq. (30) may be abbreviated by the shorthand notation

$$y = g * x \tag{31}$$

where $*$ is used to indicate the convolution of two functions. Figure 9-5 presents a graphic illustration of the convolution operation. One point on the curve $y(t)$ is obtained in the following way. One function g is reflected about its origin and shifted by an amount t to the right. The point-by-point product of x and the reflected, shifted g is formed, and that product is integrated to produce the value of the output at point t. This process is repeated for all values of t to produce other points on the output curve. As t is varied, the reflected function is shifted past the stationary function, and the value of $y(t)$ depends on the amount of overlap of the two functions.

The convolution operation has several useful properties. First, convolution is commutative; that is,

$$f * g = g * f \tag{32}$$

and we may reflect either function and obtain the same result. This can be shown by writing

$$f * g = \int_{-\infty}^{\infty} f(\tau)g(t - \tau)\, d\tau \tag{33}$$

making the change of variables

$$x = t - \tau \qquad \tau = t - x \qquad dx = -d\tau \tag{34}$$

and rearranging to produce

$$f * g = \int_{-\infty}^{\infty} f(t - x)g(x)\, dx = g * f \tag{35}$$

In Eq. (35), the limits had to be interchanged, and this compensated for the minus sign on $d\tau$.

The convolution operation is also distributive over addition, so that

$$f * (g + h) = f * g + f * h \tag{36}$$

This can be shown by writing

$$f * (g + h) = \int_{-\infty}^{\infty} f(t - \tau)\,[g(\tau) + h(\tau)]\, d\tau \tag{37}$$

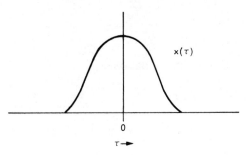

INPUT FUNCTION

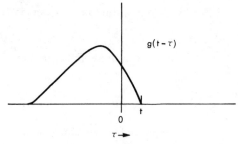

$g(\tau)$ REFLECTED AND SHIFTED

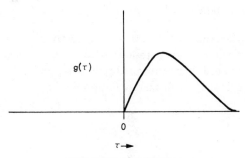

CONVOLVING FUNCTION

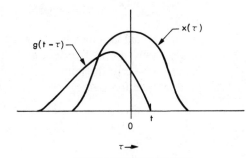

FUNCTIONS SUPERIMPOSED

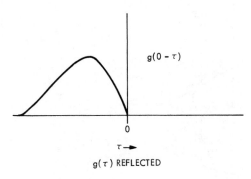

$g(\tau)$ REFLECTED

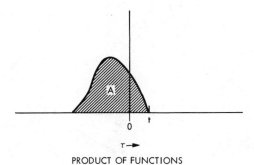

PRODUCT OF FUNCTIONS

$$y(t) = \int_{-\infty}^{\infty} x(\tau) g(t-\tau) \, d\tau = A$$

Figure 9-5   Convolution

and rearranging to yield

$$f * (g + h) = \int_{-\infty}^{\infty} f(t - \tau) g(\tau) \, d\tau$$

$$+ \int_{\infty-}^{\infty} f(t - \tau) h(\tau) \, d\tau = f * g + f * h \tag{38}$$

Convolution is also associative, which means that

$$f * (g * h) = (f * g) * h \tag{39}$$

This equation may be verified by the reader. Under differentiation

$$\frac{d}{dt}[f * g] = f' * g = f * g' \tag{40}$$

Convolution in Two Dimensions

The convolution of two-dimensional functions is similar to one-dimensional convolution. The equation is given by

$$h(x, y) = f * g = \int_{-\infty}^{\infty} \int_{-\infty}^{\infty} f(u, v)g(x - u, y - v) \, du \, dv \tag{41}$$

and illustrated graphically in Figure 9-6. Notice that $g(0 - u, 0 - v)$ is merely $g(u, v)$ rotated 180° about its origin and that $g(x - u, y - v)$ is translated so as to move the origin of g to the point x, y. The functions are then multiplied pointwise, and the product function is integrated over two dimensions. As an example, suppose that

$$f(x, y) = Ae^{-(x^2 + y^2)/2\sigma^2} \tag{42}$$

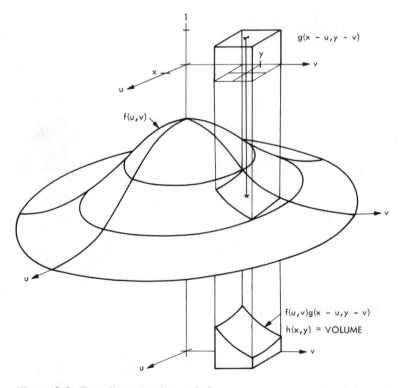

Figure 9-6 Two-dimensional convolution

and

$$g(x, y) = \begin{cases} 1 & -1 \leq x \leq 1, -1 \leq y \leq 1 \\ 0 & \text{elsewhere} \end{cases} \tag{43}$$

as shown in Figure 9-6. In this case, a two-dimensional rectangular pulse is convolved with a two-dimensional Gaussian. Since $g(x, y)$ is symmetric about the origin, the 180° rotation has no effect. The value of $h(x, y)$ is merely the volume of the product function when the rectangular pulse is shifted to the position x, y.

Sampling with a Finite Spot

Suppose a particular image digitizer samples an image with a square sampling spot. At each pixel location, the digitized gray level is the local average of a small square section of the image. In Figure 9-6, $f(x, y)$ could represent the image and $g(x, y)$ could be the spatial sensitivity function of the sampling spot. Then $h(x, y)$, the convolution of $f(x, y)$ with $g(x, y)$, is the same local average that the digitizer sees. Thus convolution is a valid way to model the action of a finite sampling spot on an image. The function $g(x, y)$ can be chosen to model the spatial sensitivity function of whatever sampling aperture is used.

Digital Convolution

Convolution of digital images is similar to that for continuous images except that the variables take on integral values and the double integral becomes a double summation. Thus, for a digital image,

$$H(i, j) = F * G = \sum_m \sum_n F(m, n)G(i - m, j - n) \tag{44}$$

Since both F and G are nonzero only over a finite domain, the summations are taken only over the area of nonzero overlap.

Digital convolution is illustrated in Figure 9-7. The function g is rotated 180°

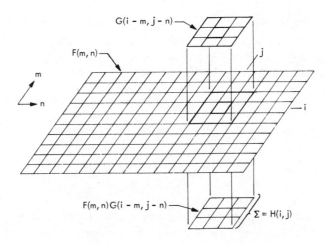

Figure 9-7 Digital convolution

and its origin shifted to the coordinates i, j. The two functions are multiplied together point by point and the resulting products summed to give the output value.

In Figure 9-7, a 3×3 array g is convolved with a larger digital image. Clearly, the number of multiply and add operations is equal to the number of pixels in g times the number of pixels in f (ignoring edge effects). Unless at least one of the functions has relatively small nonzero domain, convolution becomes a computationally expensive operation.

APPLICATIONS OF DIGITAL FILTERING

Digitally implemented linear filtering is useful for three major classes of image processing applications:

1. Deconvolution—removing the effects of previously applied linear systems that have operated on the image beyond the user's control. An example of this is using convolution to restore the detail lost by a lens system or by motion blur, both of which may be assumed linear operations.

2. Noise removal—reducing the effects of undesirable contaminating signals that have been linearly added to the image. Examples are
   (a) estimating what the signal was before the noise was added,
   (b) detecting the presence of known features embedded in a noisy background, and
   (c) coherent (periodic) noise removal.

3. Feature enhancement—increasing the contrast of specific features (edges, spots, etc.) at the expense of other objects in the scene.

SOME USEFUL FUNCTIONS

In the development of linear system theory and its application to image processing, we shall make particular use of five functions. At this point, we shall introduce these five functions and derive some of their properties. This will greatly simplify the development and examples in the following chapters.

The Rectangular Pulse

Following the notation of Ref. 1, we shall denote the rectangular pulse by

$$\Pi(x) = \begin{cases} 1 & -\tfrac{1}{2} < x < \tfrac{1}{2} \\ \tfrac{1}{2} & x = \pm\tfrac{1}{2} \\ 0 & \text{elsewhere} \end{cases} \tag{45}$$

The rectangular pulse of height A and width a is shown in Figure 9-8. This function is useful for modeling rectangular sampling windows and smoothing functions.

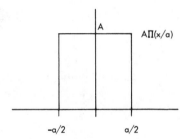

Figure 9-8 The rectangular pulse

The Triangular Pulse

We denote the triangular pulse by

$$\Lambda(x) = \begin{cases} 1 - |x| & |x| \leq 1 \\ 0 & |x| > 1 \end{cases} \tag{46}$$

This function is shown in Figure 9-9. Its applications are similar to those of the rectangular pulse. Convolving two identical rectangular pulses produces a triangular pulse.

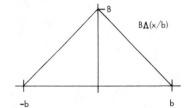

Figure 9-9 The triangular pulse

The Gaussian Function

The Gaussian function is given by

$$e^{-x^2/2\sigma^2} \tag{47}$$

and shown in Figure 9-10. It can be shown that the area under the Gaussian function

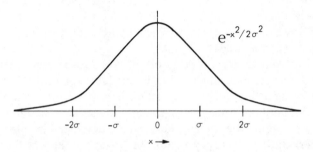

Figure 9-10 The Gaussian function

is given by

$$\int_{-\infty}^{\infty} e^{-x^2/2\sigma^2} \, dx = \frac{1}{\sqrt{2\pi\sigma^2}} \tag{48}$$

In probability theory, the *normal distribution* is given by

$$p(x) = \frac{1}{\sqrt{2\pi\sigma^2}} e^{-(x-x_0)^2/2\sigma^2} \tag{49}$$

which is a Gaussian function adjusted to unit area. The term σ^2 is called the *variance* and σ, the *standard deviation*. Table 9-1 lists the values of the Gaussian at several points.

Table 9-1. Values of the Gaussian function

x	$e^{-x^2/2\sigma^2}$
0	1
0.5σ	0.8825
1.0σ	0.6065
1.177σ	0.5000
1.5σ	0.3247
$1.177\sigma\sqrt{2}$	0.2500
2.0σ	0.1353
3.0σ	0.0111

The Gaussian has a very useful property, mentioned in Chapter 7. The convolution of two Gaussian functions always produces another Gaussian. In particular,

$$Ae^{-(x-a)^2/2\sigma_1^2} * Be^{-(x-b)^2/2\sigma_2^2} = ABe^{-(x-c)^2/2\sigma_3^2} \tag{50}$$

where

$$c = a + b \quad \text{and} \quad \sigma_3^2 = \sigma_1^2 + \sigma_2^2 \tag{51}$$

Thus the resulting Gaussian is broadened—its standard deviation is the root-mean-square of the two original standard deviations—and its offset from the origin is the sum of the two original offsets. The amplitude of the peak is the product of the two original peak amplitudes.

This convolution property of the Gaussian is quite useful for studying linear systems. Furthermore, the smooth unimodal shape of the Gaussian makes it appropriate for modeling sampling spots, display spots, and a variety of other entities encountered in digital image processing and the analysis of optical systems. Several more useful properties of the Gaussian are developed in subsequent chapters. Taken together, these properties explain the frequent usage of the Gaussian function in linear system analysis.

The Impulse

The impulse, or Dirac delta function $\delta(x)$, is not a function by the traditional definition. Instead, it is a symbolic function defined by its integral property,

$$\int_{-\infty}^{\infty} \delta(x)\, dx = \int_{-\epsilon}^{\epsilon} \delta(x)\, dx = 1 \tag{52}$$

where ϵ is an arbitrarily small number greater than 0. Notice that $\delta(x) = 0$ for $x \neq 0$. The impulse is undefined at the origin.

Since $\delta(x)$ is not a function, its use as such somewhat undermines our level of mathematical rigor. There is a mathematically rigorous approach treating the impulse as a concept in the theory of distributions (Refs. 2, 3, 4), but its use here would only complicate the notation while producing the same results. We shall adhere to common engineering practice and treat $\delta(x)$ as if it were a function.

The impulse can be modeled as the limit of a narrow rectangular pulse

$$\delta(x) = \lim_{a \to 0} \frac{1}{a} \Pi\left(\frac{x}{a}\right) \tag{53}$$

as shown in Figure 9-11. As a becomes smaller, the pulse becomes narrower but taller to maintain unit area. In the limit, the pulse becomes infinitely tall with infinitesimal width. The symbol for the shifted impulse of nonunit area is shown in Figure 9-12.

From Eq. (52), we can write

$$\int_{-\infty}^{\infty} A\delta(x)\, dx = A \tag{54}$$

and furthermore,

$$\int_{-\infty}^{\infty} f(x)\delta(x)\, dx = f(0) \tag{55}$$

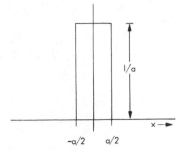

Figure 9-11 Rectangular pulse model of the impulse

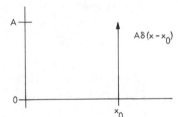

Figure 9-12 Notation for the shifted impulse

since the impulse is zero for nonzero x. This more general integral property is commonly taken as the definition of the impulse.

Properties of the Impulse. The impulse has a "sifting property" because of its ability to isolate a single point on a curve. This is expressed by

$$\int_{-\infty}^{\infty} f(x)\delta(x - x_0)\, dx = \int_{-\infty}^{\infty} f(x + x_0)\delta(x)\, dx = f(x_0) \qquad (56)$$

When we multiply a function by a shifted impulse and integrate the product, we are left with only the value of the function at the location of the impulse. We can prove Eq. (56) by substituting $x - x_0 = \tau$, which implies $dx = d\tau$. Substituting into Eq. (56) produces

$$\int_{-\infty}^{\infty} f(x)\delta(x - x_0)\, dx = \int_{-\infty}^{\infty} f(\tau + x_0)\delta(\tau)\, d\tau \qquad (57)$$

which, from Eq. (55), becomes

$$\int_{-\infty}^{\infty} \delta(\tau)f(\tau + x_0)\, d\tau = f(\tau + x_0)\Big|_{\tau=0} = f(x_0) \qquad (58)$$

to complete the proof.

The delta function exhibits rather curious behavior under changes of abscissa scale; namely,

$$\delta(ax) = \frac{1}{|a|}\delta(x) \qquad (59)$$

This says that a change in abscissa scale actually produces a change in ordinate scale. This property must be kept in mind while performing algebraic manipulations with the impulse. We can prove Eq. (59) by letting $f(x)$ be an arbitrary function and writing

$$\int_{-\infty}^{\infty} \delta(ax)f(x)\, dx = \frac{1}{a}\int_{-\infty}^{\infty} \delta(\tau)f\left(\frac{\tau}{a}\right) d\tau = \frac{1}{|a|}f(0) \qquad (60)$$

where $ax = \tau$, $x = \tau/a$, $dx = (1/a)d\tau$. For $a < 0$, the required limit interchange counteracts the minus sign and hence requires the absolute value bars. Now we can write

$$\int_{-\infty}^{\infty} \delta(ax)f(x)\, dx = \frac{1}{|a|}f(0) = \frac{1}{a}\int_{-\infty}^{\infty} \delta(x)f(x)\, dx = \int_{-\infty}^{\infty}\left[\frac{1}{|a|}\delta(x)\right]f(x)\, dx \qquad (61)$$

which, since $f(x)$ is an arbitrary function, can only be true if Eq. (59) is true. Notice that setting $a = -1$ proves that the delta function is symmetric about the origin.

Impulse Response of a Linear System. Notice that in convolution

$$\delta(x) * f(x) = \int_{-\infty}^{\infty} \delta(\tau)f(x - \tau)\, d\tau = f(x - \tau)\Big|_{\tau=0} = f(x) \qquad (62)$$

which means that the impulse is the identity function under convolution. For this reason, the characteristic function of the linear system [recall the discussion surrounding Eq. (30)] is called the *impulse response* of the system. The impulse response is the system's output produced in response to an impulse at the input.

The Step Function

The step function is a symbolic function discontinuous at $x = 0$. It is given by

$$u(x) = \begin{cases} 1 & x > 0 \\ \frac{1}{2} & x = 0 \\ 0 & x < 0 \end{cases} \tag{63}$$

and its integral property by

$$\int_{-\infty}^{\infty} u(x)f(x)\, dx = \int_{0}^{\infty} f(x)\, dx \tag{64}$$

where $f(x)$ is an arbitrary function. The shifted step function $u(x - x_0)$ is shown in Figure 9-13. Notice that the step function is the integral of the impulse:

$$u(x - x_0) = \int_{-\infty}^{x} \delta(\tau - x_0)\, d\tau = \begin{cases} 1 & x > x_0 \\ 0 & x < x_0 \end{cases} \tag{65}$$

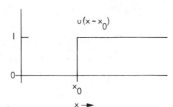

Figure 9-13   The step function

As one might expect, the impulse is also the derivative of the step function:

$$u'(x) = \frac{du(x)}{dx} = \delta(x) \tag{66}$$

We can prove Eq. (66) in the following way. First we integrate by parts the expression

$$\int_{-\infty}^{\infty} u'(x)f(x)\, dx = u(x)f(x)\Big|_{-\infty}^{\infty} - \int_{-\infty}^{\infty} u(x)f'(x)\, dx \tag{67}$$

where $f(x)$ is an arbitrary function that goes to zero at $x = \pm\infty$. With this restriction, Eq. (67) reduces to

$$\int_{-\infty}^{\infty} u'(x)f(x)\, dx = - \int_{-\infty}^{\infty} u(x)f'(x)\, dx \tag{68}$$

Making use of the definition of the step function [Eq. (64)], we can write

$$\int_{-\infty}^{\infty} u(x)f'(x)\, dx = - \int_{0}^{\infty} f'(x)\, dx = [f(\infty) - f(0)] = f(0) \tag{69}$$

since $f(\infty) = 0$. Using the definition of the impulse [Eq. (55)], we can write

$$\int_{-\infty}^{\infty} u'(x)f(x)\, dx = f(0) = \int_{-\infty}^{\infty} \delta(x)f(x)\, dx \tag{70}$$

which must be true for arbitrarily selected $f(x)$. This can be the case only if Eq. (66) is true.

Convolution is commonly used to implement linear operations on signals and images. This section illustrates this with a few examples.

Smoothing

Figure 9-14 illustrates the use of convolution for smoothing a noisy function $f(x)$. A rectangular pulse $g(x)$ is the impulse response of the smoothing filter. As the convolution proceeds, the rectangular pulse moves from left to right, producing $h(x)$, which

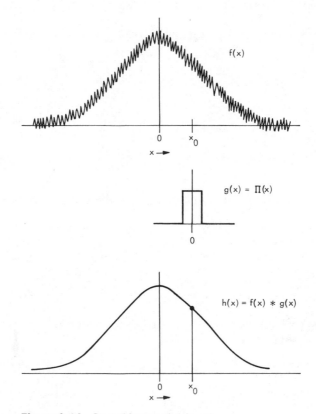

Figure 9-14 Smoothing a noisy function

is, at every point, a local average of $f(x)$ over a unit width interval. The local averaging has the effect of suppressing the high-frequency variations while preserving the basic shape of the input function. This application is typical of using filters with nonnegative impulse responses to smooth noisy data. We could equally well use the triangular pulse or the Gaussian pulse as the smoothing function.

Edge Enhancement

Figure 9-15 illustrates another type of filtering, this time for edge enhancement. The edge function $f(x)$ is a rather slowly varying transition from low to high amplitude. The impulse response $g(x)$ is a positive peak with negative side lobes. As the convolution proceeds, $g(x)$ moves from left to right with the side lobes and main lobe progressively encountering the transition of the edge. The filter output is shown as $h(x)$.

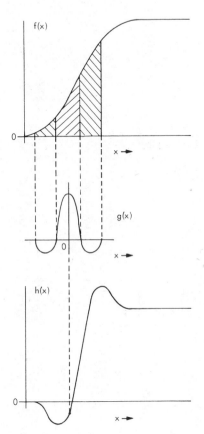

Figure 9-15 Edge enhancement, Example 1

The edge enhancement filter in Figure 9-15 has two effects. First, it tends to increase the slope of the transition at the edge. It also produces overshoot or "ringing" on either side of the edge. This behavior is typical of common edge enhancement filters.

As a second edge enhancement example, consider the impulse response given by

$$g(x) = 2\delta(x) - e^{-x^2/2\sigma^2} \qquad (71)$$

and shown in Figure 9-16. Notice that

$$h = f * g = f(x) * 2\delta(x) - f(x) * e^{-x^2/2\sigma^2} = 2f(x) - f(x) * e^{-x^2/2\sigma^2} \qquad (72)$$

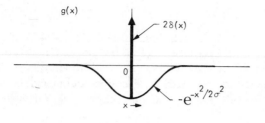

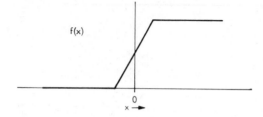

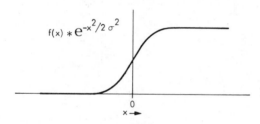

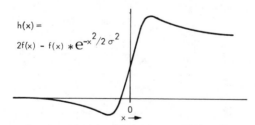

Figure 9-16 Edge enhancement, Example 2

So the output is merely twice the input minus the input convolved with a Gaussian. Convolving with the Gaussian blurs the edge as illustrated in Figure 9-16. The output has the enhanced edge form shown in the figure. Again there is sharpening of the edge with ringing. This exercise points out that subtracting a blurred image from the original image has the effect of edge enhancement.

CONCLUSION

In this chapter, we have established a framework within which to analyze the behavior of electronic circuits, optical systems, and digital filtering operations. This almost completely covers the components encountered in image processing systems. In Chapter 10, we develop another powerful tool for linear system analysis, the Fourier transform. In the remainder of Part II, we apply these tools to develop concise methods for expressing the effects of digitizing systems, display systems, and image processing operations.

SUMMARY OF IMPORTANT POINTS

1. When the input to a linear system is the sum of two signals, the output is the sum of the outputs produced by each of those signals acting alone.

2. Changing the time (or spatial) origin of the input to a shift invariant system merely shifts the output by the same amount.

3. Harmonic signals are used to represent sinusoidal signals because they simplify the analysis of linear systems.

4. Harmonic (sinusoidal) inputs to a shift invariant linear system produce harmonic outputs.

5. A linear, shift invariant system is completely specified by its transfer function.

6. The transfer function is a complex-valued function of frequency that relates the amplitude and phase of harmonic inputs and outputs.

7. The harmonic input, multiplied by the value of the transfer function at the input frequency, yields the output of a shift invariant linear system.

8. The convolution of two functions consists of reflecting and shifting one function and then integrating their product. The output is the value of the integral as a function of shift.

9. The output of a shift invariant linear system is given by the convolution of the input with a function called the *impulse response* of that system.

10. The impulse response of a particular shift invariant linear system is unique, and it completely specifies the system.

11. The convolution operation models the effect a sampling spot has on an image.

12. Convolution may be implemented digitally to perform linear filtering on digitized signals and images.

13. Digitally implemented linear filtering may be used for deconvolution, noise reduction, and feature enhancement.

14. Convolving two Gaussian functions produces another Gaussian, broader than either of the inputs.

15. The impulse $\delta(x)$ is the identity function under convolution [Eq. (62)].

16. Abscissa scale changes affect the strength of the impulse [Eq. (59)].

17. The impulse is the derivative of the step function.

18. The impulse response of an edge enhancement filter typically has a positive peak at the origin surrounded by negative side lobes.

19. Edge enhancement filters can produce an artifact called *ringing*.

REFERENCES

1. BRACEWELL, R., *The Fourier Transform and Its Applications*, McGraw-Hill Book Company, New York, 1965.

2. E. O. BRIGHAM, *The Fast Fourier Transform*, Prentice-Hall, Inc., Englewood Cliffs, New Jersey, 1974.

3. A. PAPOULIS, *The Fourier Integral and Its Applications*, McGraw-Hill Book Company, New York, 1962.

4. S. C. GUPTA, *Transform and State Variable Methods in Linear Systems*, John Wiley & Sons, New York, 1966.

THE FOURIER TRANSFORM

INTRODUCTION

The Fourier transform is a powerful tool in linear system analysis. It allows us to quantify the effects of digitizing systems, sampling spots, electronic amplifiers, convolution filters, noise, and display spots. Those who combine a theoretical knowledge of Fourier transform properties with a practical knowledge of their physical interpretation are well prepared to approach most image processing problems. It is usually only students of electrical engineering and physical optics who develop this combination in the course of their studies. For anyone who intends to use digital image processing seriously in his work, however, the time spent becoming familiar with the Fourier transform is well invested.

In a sense, the Fourier transform is like a second language for describing functions. Bilingual persons frequently find one language better than another for expressing their ideas. The image processing analyst may similarly move back and forth between the spatial domain and the frequency domain while proceeding through an image processing problem. When first learning a new language, a person tends to think in his or her native tongue and mentally translate before speaking. After becoming fluent, however, one can think in either language. Similarly, once familiar with the Fourier transform, the analyst can think in either the spatial or the frequency domain.

In the first part of this chapter, we develop the properties of the Fourier transform using one-dimensional functions for simplicity of notation. Later we generalize the results to two dimensions. The convention in Part II of this book is to consider one-dimensional functions of time as simple examples and functions of two spatial variables as image processing examples.

In our study of linear system analysis, we shall restrict our discussions to only one part of this well-developed field. For example, we shall use only the Fourier transform and not the LaPlace transform or the Z-transform because they are not required for our purposes. This restriction allows us to develop the techniques we need for the analysis of digital image processing systems with a minimum of mathematical complexity.

One reason we do not require the generality of the LaPlace transform and other techniques from the field of linear system analysis is that we are working with recorded data. This relieves us of the burden of physical realizability (causality) and its implications for the analysis. Linear systems implemented with electronic hardware are referred to as *causal* because the input signal causes the output signal to occur. In particular, this means that if the input is zero for all negative time, then the output must likewise be zero for $t < 0$. While this is intuitively obvious, consider the constraint it places upon the impulse response of a linear system. If the input is an impulse at $t = 0$, the impulse response must be 0 for all negative t. Thus, with physically realizable systems, the impulse response is always one-sided. This means it can be neither even nor odd except in the trivial case. This restriction considerably complicates linear system analysis of physically realizable systems.

Working with recorded data leaves us not so constrained. Digitally implemented convolution can easily deal with even and odd functions as well as those that are zero for negative time. Furthermore, in the spatial domain of image processing, the coordinate origin is arbitrary, and negative values of x and y have no special significance. Readers who find the mathematics in the following chapters burdensome should be thankful that we are working with recorded data and do not have to impose the causality restriction upon the analysis.

Definition

The Fourier transform of a one-dimensional function $f(t)$ is defined as (Ref. 1)

$$\mathcal{F}\{f(t)\} = F(s) = \int_{-\infty}^{\infty} f(t)e^{-j2\pi st}\, dt \qquad (1)$$

The Fourier transform is a linear integral transformation that, in the general case, takes a complex function of n real variables into another complex function of n real variables. The inverse Fourier transform of $F(s)$ is defined as

$$\mathcal{F}^{-1}\{F(s)\} = \int_{-\infty}^{\infty} F(s)e^{j2\pi st}\, ds \qquad (2)$$

The only difference between the direct and inverse Fourier transformation is the sign of the exponent.

Fourier's integral theorem states that

$$f(t) = \int_{-\infty}^{\infty} \left[\int_{-\infty}^{\infty} f(t) e^{-j2\pi st} \, dt \right] e^{j2\pi st} \, ds \tag{3}$$

This means that the transformation is reciprocal, and

$$\mathcal{F}\{f(t)\} = F(s) \Longrightarrow \mathcal{F}^{-1}\{F(s)\} = f(t) \tag{4}$$

The functions $f(t)$ and $F(s)$ are called a *Fourier transform pair*. For any $f(t)$, the Fourier transform $F(s)$ is unique and vice versa.

There are alternate ways to write Eqs. (1), (2), and (3), depending on where the factor 2π is placed in the equations. The convention used here corresponds to System 1 of Ref. 1. In this convention, the frequency variable is measured in whole cycles per unit of t.

Example: The Fourier Transform of a Gaussian

As an illustrative exercise, we now derive the Fourier transform of the Gaussian function

$$f(t) = e^{-\pi t^2} \tag{5}$$

From Eq. (1), we can write

$$F(s) = \int_{-\infty}^{\infty} e^{-\pi t^2} e^{-j2\pi st} \, dt \tag{6}$$

or

$$F(s) = \int_{-\infty}^{\infty} e^{-\pi(t^2 + j2st)} \, dt \tag{7}$$

We multiply the right-hand side by

$$e^{-\pi s^2} e^{+\pi s^2} = 1 \tag{8}$$

which yields

$$F(s) = e^{-\pi s^2} \int_{-\infty}^{\infty} e^{-\pi(t + js)^2} \, dt \tag{9}$$

We now make the variable substitution

$$u = t + js \qquad du = dt \tag{10}$$

and Eq. (9) becomes

$$F(s) = e^{-\pi s} \int_{-\infty}^{\infty} e^{-\pi u^2} \, du \tag{11}$$

The integral in Eq. (11) is known to equal unity, so Eq. (11) reduces to

$$F(s) = e^{-\pi s^2} \tag{12}$$

Thus Eqs. (5) and (12) are a Fourier transform pair, and the Fourier transform of a Gaussian is also a Gaussian. This property makes the Gaussian function quite useful in later analysis.

Existence of the Fourier Transform

Since the Fourier transform is an integral transformation, we must address the question of the existence of the integrals in Eqs. (1) and (2).

Transient Functions. Some functions go to zero for large positive and negative arguments rapidly enough that the integrals in Eqs. (1) and (2) exist. For our purposes, if the integral of the absolute value of a function exists,

$$\int_{-\infty}^{\infty} |f(t)| \, dt \neq \infty \tag{13}$$

and it is either continuous or has only finite discontinuities, then its Fourier transform exists for all values of s. We call these *transient functions* since the useful ones characteristically die out at large $|t|$.

In a sense, these are the only functions we shall ever process. Any digitized signal or image is necessarily truncated to finite duration and bounded. Therefore the transform exists for any function we shall ever use. Nevertheless, it is convenient to be able to discuss other functions whose transforms do not exist in the strict sense.

Periodic and Constant Functions. Clearly, the Fourier transform does not exist for all values of s if $f(t) = \cos(2\pi t)$ or if $f(t) = 1$. However, the impulse $\delta(t)$, introduced in Chapter 9, allows us to handle these cases conveniently. Consider the inverse transform of a pair of impulses

$$f(t) = \mathcal{F}^{-1}\{\delta(s-f) + \delta(s+f)\} = \int_{-\infty}^{\infty} [\delta(s-f) + \delta(s+f)]e^{j2\pi st} \, ds \tag{14}$$

which, by the sifting property of the impulse, is

$$\begin{aligned} f(t) &= \int_{-\infty}^{\infty} \delta(s-f)e^{j2\pi st} \, ds + \int_{-\infty}^{\infty} \delta(s+f)e^{j2\pi st} \, ds \\ &= e^{-j2\pi ft} + e^{+j2\pi ft} = 2\cos(2\pi ft) \end{aligned} \tag{15}$$

where we have used the Euler relation (Chapter 9, Eq. 7). Dividing by 2, we can write

$$\mathcal{F}\{\cos(2\pi ft)\} = \tfrac{1}{2}[\delta(s-f) + \delta(s+f)] \tag{16}$$

This means that the Fourier transform of a cosine of frequency f is a pair of impulses located at $s = \pm f$ in the frequency domain. A similar development yields

$$\mathcal{F}\{\sin(2\pi ft)\} = \frac{j}{2}[\delta(s+f) - \delta(s-f)] \tag{17}$$

If we let $f = 0$ in Eq. (16), we can show that

$$\mathcal{F}\{1\} = \delta(s) \tag{18}$$

the Fourier transform of a constant is an impulse at the origin.

We now have usable expressions for the Fourier transform of constant and sinusoidal functions. It is well known in the theory of Fourier series that any periodic function of frequency f can be expressed as a summation of sinusoids having frequencies nf, where n takes on integer values. By the addition theorem [see Eq. (34)], this means that the Fourier transform of a periodic function is a series of equally spaced impulses in the frequency domain.

Random Functions. We lump nonconstant aperiodic functions of infinite extent whose absolute integral [(Eq. (13)] does not exist into a class called *random functions*. In later chapters, we shall use these to model the output of a random process. In most cases, we shall require only the autocorrelation function of a random function. It is given by

$$R_f(\tau) = \lim_{T \to \infty} \frac{1}{2T} \int_{-T}^{T} f(t)f(t + \tau)\, dt \tag{19}$$

and exists for the functions of interest to us. It is real and even, and its Fourier transform is the power spectrum of $f(t)$ as is shown later.

If it becomes necessary to transform a random function, we can redefine the Fourier transform of Eq. (1) as

$$F(s) = \lim_{T \to \infty} \frac{1}{2T} \int_{-T}^{T} f(t)e^{-j2\pi st}\, dt \tag{20}$$

and similarly for the inverse transform. We can then work with a class of functions for which these redefined transforms exist. In this book, however, we shall stay with the definitions in Eqs. (1) and (2) since they are appropriate for bounded signals of finite duration. Any development carried out with this convention could be redone with the convention suggested by Eq. (20), thereby extending the result to random functions for which $R_f(\tau)$ exists.

We conclude this discussion by taking the position that, for our purposes, the existence of the Fourier transform is not a major problem.

Fourier Transforms of Some Useful Functions

Table 10-1 lists the Fourier transforms of some common functions we will find useful. A more comprehensive set of Fourier transforms appears in Appendix III.

Table 10-1. Fourier transforms of some common functions

Function	$f(t)$	$F(s)$
Gaussian	$e^{-\pi t^2}$	$e^{-\pi s^2}$
Rectangular pulse	$\Pi(t)$	$\dfrac{\sin(\pi s)}{\pi s}$
Triangular pulse	$\Lambda(t)$	$\dfrac{\sin^2(\pi s)}{(\pi s)^2}$
Impulse	$\delta(t)$	1
Unit step	$u(t)$	$\dfrac{1}{2}\left[\delta(s) - \dfrac{j}{\pi s}\right]$
Cosine	$\cos(2\pi ft)$	$\frac{1}{2}[\delta(s + f) + \delta(s - f)]$
Sine	$\sin(2\pi ft)$	$j\frac{1}{2}[\delta(s + f) - \delta(s - f)]$
Complex exponential	$e^{j2\pi ft}$	$\delta(s - f)$

PROPERTIES OF THE FOURIER TRANSFORM

Symmetry Properties

In the general case, a complex function of a single real variable has a Fourier transform that is also a complex function of a real variable. However, there are several restricted classes of functions that are of interest because of their symmetry properties under the Fourier transformation.

Evenness and Oddness. An even function $f_e(t)$ has the property

$$f_e(t) = f_e(-t) \tag{21}$$

and a function $f_o(t)$ is odd if and only if

$$f_o(t) = -f_o(-t) \tag{22}$$

A function $f(t)$, which is neither even nor odd, can be broken into even and odd components given by

$$f_e(t) = \tfrac{1}{2}[f(t) + f(-t)] \tag{23}$$

and

$$f_o(t) = \tfrac{1}{2}[f(t) - f(-t)] \tag{24}$$

where

$$f(t) = f_e(t) + f_o(t) \tag{25}$$

We now investigate the effect of evenness and oddness on the Fourier transformation. Recall the Euler relation

$$e^{jx} = \cos(x) + j\sin(x) \tag{26}$$

We can rewrite the Fourier transform [Eq. (1)] as

$$F(s) = \int_{-\infty}^{\infty} f(t) e^{-j2\pi st}\, dt = \int_{-\infty}^{\infty} f(t) \cos(2\pi st)\, dt - j \int_{-\infty}^{\infty} f(t) \sin(2\pi st)\, dt \tag{27}$$

Expressing $f(t)$ as a sum of even and odd components [Eq. (25)] produces

$$\begin{aligned} F(s) = &\int_{-\infty}^{\infty} f_e(t) \cos(2\pi st)\, dt + \int_{-\infty}^{\infty} f_o(t) \cos(2\pi st)\, dt \\ &- j \int_{-\infty}^{\infty} f_e(t) \sin(2\pi st)\, dt - j \int_{-\infty}^{\infty} f_o(t) \sin(2\pi st)\, dt \end{aligned} \tag{28}$$

Notice that the second and third terms are infinite integrals of the product of an even and an odd function. These terms are zero. The Fourier transform reduces to

$$F(s) = \int_{-\infty}^{\infty} f_e(t) \cos(2\pi st)\, dt - j \int_{-\infty}^{\infty} f_o \sin(2\pi st)\, dt = F_e(s) + jF_o(s) \tag{29}$$

Now we can list the symmetry properties of the Fourier transform:

1. An even component function produces an even component function in the transform.

2. An odd component function produces an odd component function in the transform.

166

3. An odd component function introduces the coefficient $-j$.

4. An even component function does not introduce a coefficient.

Real and Imaginary Components. We can use the four rules stated above to deduce the effect of the Fourier transformation on complex functions. If we express a general complex function as a sum of four components, an even and an odd real part and an even and an odd imaginary part, we can write the following four rules for the Fourier transformation:

1. The real even part produces a real even part.

2. The real odd part produces an imaginary odd part.

3. The imaginary even part produces an imaginary even part.

4. The imaginary odd part produces a real odd part.

Of particular interest is the case of input functions that are real, since we ordinarily use real functions to represent input images. Notice that a real function produces a transform that has an even real part and an odd imaginary part. This is referred to as a *Hermite* function, and it has the property

$$F(s) = F^*(-s) \tag{30}$$

where * denotes the complex conjugate.

Table 10-2 lists the full expansion of Fourier transform symmetry properties. Notice that the inverse transformation [Eq. (2)] differs from the direct transformation [Eq. (1)] only in the sign of the odd component.

Table 10-2. Symmetry properties of the Fourier transform

$f(t)$	$F(s)$
Even	Even
Odd	Odd
Real and even	Real and even
Real and odd	Imaginary and odd
Imaginary and even	Imaginary and even
Complex and even	Complex and even
Complex and odd	Complex and odd
Real	Hermite
Imaginary	Anti-Hermite
Real even plus imaginary odd	Real
Real odd plus imaginary even	Imaginary

The Addition Theorem

Suppose we have two Fourier transform pairs

$$\mathcal{F}\{f(t)\} = F(s) \tag{31}$$

and

$$\mathcal{F}\{g(t)\} = G(s) \tag{32}$$

If the two time functions are added, the Fourier transform of their sum is given by

$$\mathcal{F}\{f(t) + g(t)\} = \int_{-\infty}^{\infty} [f(t) + g(t)]e^{-j2\pi st}\, dt \tag{33}$$

This may be rearranged to yield

$$\mathcal{F}\{f(t) + g(t)\} = \int_{-\infty}^{\infty} f(t)e^{-j2\pi st}\, dt + \int_{-\infty}^{\infty} f(t)e^{-j2\pi st}\, dt = F(s) + G(s) \tag{34}$$

Thus addition in the time or spatial domain corresponds to addition in the frequency domain as illustrated in Figure 10-1. This fits well with the concept of linearity in a system. It follows from the addition theorem that

$$\mathcal{F}\{cf(t)\} = cF(s) \tag{35}$$

where c is a rational constant. We take this as an axiom to be true for any constant.

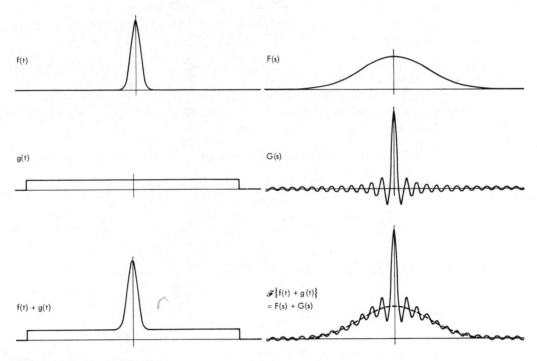

Figure 10-1 The addition theorem

The Shift Theorem

The shift theorem describes the effect that moving the origin of (shifting) a function has upon its transform. Using the function $f(t)$ as before, we can write

$$\mathcal{F}\{f(t - a)\} = \int_{-\infty}^{\infty} f(t - a)e^{-j2\pi st}\, dt \tag{36}$$

where a is the amount of shift. Multiplying the right-hand side of the equation by

$$e^{j2\pi as}e^{-j2\pi as} = 1 \tag{37}$$

produces

$$\mathscr{F}\{f(t-a)\} = \int_{-\infty}^{\infty} f(t-a)e^{-j2\pi s(t-a)}e^{-j2\pi as}\, dt \tag{38}$$

We make the variable substitution

$$u = t - a \qquad du = dt \tag{39}$$

and move the second exponential outside the integral, leaving

$$\mathscr{F}\{f(t-a)\} = e^{-j2\pi as}\int_{-\infty}^{\infty} f(u)e^{-j2\pi su}\, du = e^{-j2\pi as}F(s) \tag{40}$$

Thus shifting a function introduces a complex exponential coefficient into its Fourier transform. Notice that if $a = 0$, this coefficient is unity. The complex coefficient

$$e^{-j2\pi as} = \cos(2\pi as) - j\sin(2\pi as) \tag{41}$$

has unit magnitude and revolves in the complex plane with increasing s. This means that shifting a function does not change the amplitude (modulus) of the Fourier transform but does alter the distribution of energy between its real and imaginary parts. This phase shift is proportional to both frequency and the amount of shift.

The Convolution Theorem

Perhaps the most important theorem for linear system analysis is the convolution theorem. For the functions given in Eqs. (31) and (32), we can express the Fourier transform of their convolution as

$$\mathscr{F}\{f(t) * g(t)\} = \int_{-\infty}^{\infty}\int_{-\infty}^{\infty} f(u)g(t-u)\, du\; e^{-j2\pi st}\, dt \tag{42}$$

which, after rearrangement, becomes

$$\mathscr{F}\{f(t) * g(t)\} = \int_{-\infty}^{\infty} f(u) \int_{-\infty}^{\infty} g(t-u)e^{-j2\pi st}\, dt\, du \tag{43}$$

By the shift theorem, we can write

$$\mathscr{F}\{f(t) * g(t)\} = \int_{-\infty}^{\infty} f(u)e^{-j2\pi su}G(s)\, du = G(s)\int_{-\infty}^{\infty} f(u)e^{-j2\pi su}\, du \tag{44}$$

This means that

$$\mathscr{F}\{f(t) * g(t)\} = F(s)G(s) \tag{45}$$

and convolution in one domain corresponds to multiplication in the other domain. It follows that

$$\mathscr{F}^{-1}\{F(s)G(s)\} = f(t) * g(t) \tag{46}$$

The convolution theorem points out a major benefit of the Fourier transform. Rather than performing convolution in one domain, which is complicated to visualize and expensive to implement, we can perform multiplication in the other domain for the same effect.

We can use the convolution therorem to derive the Fourier transform of the impulse. Recall that

$$f(t) * \delta(t) = f(t) \tag{47}$$

that is, the impulse is the identity under convolution. By the convolution theorem,

$$F(s)\mathfrak{F}\{\delta(t)\} = F(s) \tag{48}$$

Since this is true for any $f(t)$, we can choose one such that $F(s)$ has no zeros, for example, the Gaussian. Then we can divide by $F(s)$ to show that

$$\mathfrak{F}\{\delta(t)\} = 1 \tag{49}$$

proving that the Fourier transform of the impulse is unity.

The Similarity Theorem

The similarity theorem describes the effect that a change of abscissa scale has on the Fourier transform of a function. Changing the abscissa scale broadens or narrows a function. For the function given in Eq. (31), we can stretch or compress it by placing a coefficient in front of its argument. Its Fourier transform becomes

$$\mathfrak{F}\{f(at)\} = \int_{-\infty}^{\infty} f(at)e^{-j2\pi st}\, dt \tag{50}$$

We can multiply both the integral and the exponent by a/a to produce

$$\mathfrak{F}\{f(at)\} = \frac{1}{a} \int_{-\infty}^{\infty} f(at)e^{-j2\pi at(s/a)} a\, dt \tag{51}$$

We now make the variable substitution

$$u = at \qquad du = a\, dt \tag{52}$$

and write

$$\mathfrak{F}\{f(at)\} = \frac{1}{|a|} \int_{-\infty}^{\infty} f(u)e^{-j2\pi u(s/a)}\, du \tag{53}$$

which we recognize as

$$\mathfrak{F}\{f(at)\} = \frac{1}{|a|} F\left(\frac{s}{a}\right) \tag{54}$$

If the coefficient a is greater than unity, it contracts the function $f(t)$ horizontally, which, by Eq. (54), reduces the amplitude of the Fourier transform and expands it horizontally by the factor a. If a is less than unity, it has the opposite effect. This is illustrated in Figure 10-2. The similarity theorem implies that a narrow function has a broad Fourier transform and vice versa.

We can use the similarity theorem to derive a general expression for the Fourier transform of a Gaussian. Recall from Eqs. (5) and (12) that the Fourier transform of a Gaussian is a Gaussian:

$$\mathfrak{F}\{e^{-\pi t^2}\} = e^{-\pi s^2} \tag{55}$$

By the similarity theorem

$$\mathfrak{F}\{e^{-\pi(at)^2}\} = \frac{1}{a} e^{-\pi(s/a)^2} \tag{56}$$

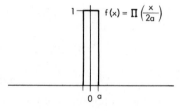

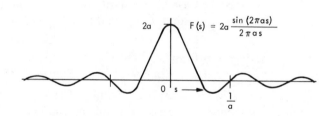

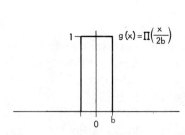

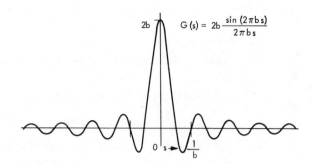

Figure 10-2 The similarity theorem

We now let

$$e^{-\pi(at)^2} = e^{-t^2/2\sigma^2} \tag{57}$$

and solve for

$$a = \frac{1}{\sqrt{2\pi\sigma^2}} \tag{58}$$

Now the transform is given by

$$\mathcal{F}\{e^{-\pi(at)^2}\} = \sqrt{2\pi\sigma^2}\,e^{-2\pi^2\sigma^2 s^2} \tag{59}$$

but since it, too, is a Gaussian, we can define a standard deviation α such that

$$e^{-2\pi^2\sigma^2 s^2} = e^{-s^2/2\alpha^2} \tag{60}$$

This means that

$$2\pi^2\sigma^2 s^2 = \frac{s^2}{2\alpha^2} \tag{61}$$

or

$$\alpha = \frac{1}{2\pi\sigma} \tag{62}$$

Thus, for a Gaussian of arbitrary standard deviation σ, the Fourier transform is given by

$$\mathcal{F}\{e^{-t^2/2\sigma^2}\} = \sqrt{2\pi\sigma^2}\,e^{-s^2/2\alpha^2} \qquad \alpha = \frac{1}{2\pi\sigma} \tag{63}$$

Thus the Fourier transform of a unit-amplitude Gaussian with standard deviation σ is another Gaussian with amplitude $\sqrt{2\pi\sigma^2}$ and standard deviation $1/(2\pi\sigma)$.

We can use the similarity theorem to illustrate again that the transform of the impulse is constant. Suppose that

$$f(t) = ae^{-\pi(at)^2} \tag{64}$$

and its transform is

$$F(s) = e^{-\pi(s/a)^2} \tag{65}$$

If we let a approach infinity, $f(t)$ narrows and grows in amplitude to approach an impulse, while $F(s)$ expands to approach constant unit amplitude. Thus, in the limiting case, the Gaussian approaches an impulse and its transform approaches unity.

Rayleigh's Theorem

An important class of functions are those that are nonzero only over a finite portion of their domain. For such functions, we can discuss the total energy content. The energy of a function is given by

$$\text{energy} = \int_{-\infty}^{\infty} |f(t)|^2 \, dt \tag{66}$$

provided the integral exists. For transient functions, the integral in Eq. (66) exists, and the energy is a convenient parameter regarding the total "size" of the function. Rayleigh's theorem states that

$$\int_{-\infty}^{\infty} |f(t)|^2 \, dt = \int_{-\infty}^{\infty} |F(s)|^2 \, ds \tag{67}$$

which means that the transform has the same energy as the original function.

The proof of Rayleigh's theorem is as follows. First we write

$$\int_{-\infty}^{\infty} |f(t)|^2 \, dt = \int_{-\infty}^{\infty} f(t)f^*(t) \, dt = \int_{-\infty}^{\infty} f(t)f^*(t)e^{j2\pi ut} \, dt \qquad u = 0 \tag{68}$$

where the second equality holds for $u = 0$. Again the superscript asterisk indicates the complex conjugate, since $f(t)$ is in general complex. We recognize Eq. (68) as the inverse Fourier transform of a product of two functions evaluated at frequency $u = 0$. Since

$$\mathcal{F}^{-1}\{f(t)f^*(t)\} = F(u) * F^*(-u) \qquad u = 0 \tag{69}$$

we can write the convolution integral as

$$\mathcal{F}^{-1}\{f(t)f^*(t)\} = \int_{-\infty}^{\infty} F(s)F^*(s-u) \, ds \qquad u = 0 \tag{70}$$

Substituting $u = 0$ produces

$$\mathcal{F}^{-1}\{f(t)f^*(t)\} = \int_{-\infty}^{\infty} F(s)F^*(s) \, ds \qquad u = 0 \tag{71}$$

which proves Eq. (67) and states that the energy is the same in both domains. If $f(t)$ is real and even, $F(s)$ is also real and even and

$$\int_{-\infty}^{\infty} f^2(t) \, dt = \int_{-\infty}^{\infty} F^2(s) \, ds \tag{72}$$

Notice how Rayleigh's theorem agrees with the similarity theorem. If we narrow a function at constant amplitude, we clearly reduce its energy. The similarity theorem states that narrowing a function broadens its transform but also reduces its amplitude to keep the energy equal in both domains.

LINEAR SYSTEMS AND THE FOURIER TRANSFORM

As stated previously, the Fourier transform is an important tool in linear system analysis. In this section, we examine the role of the Fourier transform in linear system analysis.

Linear System Terminology

Figure 10-3 shows, in both domains, the terminology commonly used for a linear system. In general, the Fourier transform of a signal is its spectrum, and the inverse Fourier transform of a spectrum is a signal. Similarly, the impulse response and the transfer function are a Fourier transform pair.

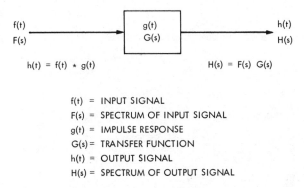

f(t)　=　INPUT SIGNAL
F(s)　=　SPECTRUM OF INPUT SIGNAL
g(t)　=　IMPULSE RESPONSE
G(s)　=　TRANSFER FUNCTION
h(t)　=　OUTPUT SIGNAL
H(s)　=　SPECTRUM OF OUTPUT SIGNAL

Figure 10-3　Linear system terminology

Linear System Identification

Frequently the impulse response and transfer function of a system are unknown and must be determined. This process is called *system identification*. For the linear system shown in Figure 10-3, the convolution theorem implies that

$$H(s) = F(s)G(s) \tag{73}$$

We can now write

$$G(s) = \frac{H(s)}{F(s)} \qquad F(s) \neq 0 \tag{74}$$

and, therefore,

$$g(t) = \mathscr{F}^{-1}\left\{\frac{\mathscr{F}\{h(t)\}}{\mathscr{F}\{f(t)\}}\right\} \tag{75}$$

This means that we can input a known $f(t)$, measure $h(t)$, and compute $g(t)$ by numerical integration. For instance, suppose $f(t)$ is an impulse. In this case, $h(t)$ is merely the impulse response and no further action is necessary to identify the system.

As a more interesting example, assume that

$$f(t) = \Pi(t) \tag{76}$$

is the input, and

$$h(t) = \Lambda(t) \tag{77}$$

is measured at the output, as shown in Figure 10-4. Now

$$g(t) = \mathcal{F}^{-1}\left\{\frac{\dfrac{\sin^2(\pi s)}{(\pi s)^2}}{\dfrac{\sin(\pi s)}{(\pi s)}}\right\} = \Pi(t) \tag{78}$$

is the impulse response.

As a second example, consider the case shown in Figure 10-5. Suppose as an input we choose

$$f(t) = u(t) - \tfrac{1}{2} = \begin{cases} -\tfrac{1}{2} & t < 0 \\ 0 & t = 0 \\ +\tfrac{1}{2} & t > 0 \end{cases} \tag{79}$$

which is an edge function having the spectrum

$$F(s) = \frac{-j}{2\pi s} \tag{80}$$

If the system's response is given by

$$h(t) = \begin{cases} -\tfrac{1}{2} & t < -1 \\ t & -1 \le t \le 1 \\ +\tfrac{1}{2} & t > 1 \end{cases} \tag{81}$$

which has the transform

$$H(s) = -j\frac{\sin(\pi s)}{2(\pi s)^2} \tag{82}$$

we can write

$$G(s) = \frac{H(s)}{F(s)} = \frac{\sin(\pi s)}{\pi s} \tag{83}$$

which implies that the impulse response is

$$g(t) = \Pi(t) \tag{84}$$

In the preceding examples, the system output was expressed analytically and the problem solved directly. In the usual case, however, the output is digitized, both input and output are transformed by numerical integration, the ratio in Eq. (74) is computed directly, and the inverse Fourier transformation of Eq. (75) is performed by numerical integration. Computationally efficient algorithms for computing the Fourier transform exist (Refs. 2–8).

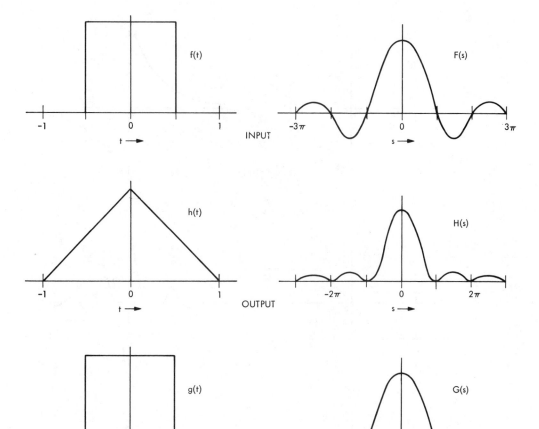

Figure 10-4   System identification, Example 1

Notice that it is prudent to choose an input function whose spectrum does not have zeros. In the second example, we violated this constraint but were fortunate enough to find an impulse response that also had zeros at the same points in the frequency domain. If $F(s)$ has zero crossings, $H(s)$ will also, and $G(s)$ can be interpolated from surrounding values before inverse transformation by numerical means.

Sinusoidal Decomposition

The Fourier transform is a linear integral transformation that uses the imaginary exponential as its kernel function. As shown in Eq. (27), the Fourier transform can be expressed as a sum of two transformations using the sine and cosine functions as

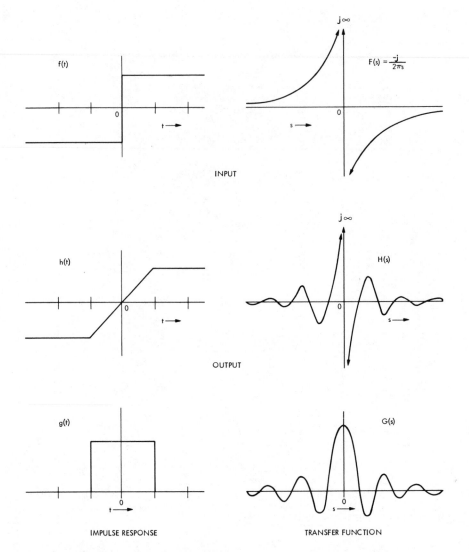

Figure 10-5 System identification, Example 2

kernels. Thus it should come as no surprise that sine and cosine functions exhibit specialized behavior in the Fourier transform.

The following exercise yields insight into the relationship between the impulse response and the transfer function of a linear system. Consider again the linear system shown in Figure 10-3 and assume, for graphic convenience, that $f(t)$ and $g(t)$ are real and even. In Figure 10-6, the input and the impulse response are graphed in both domains.

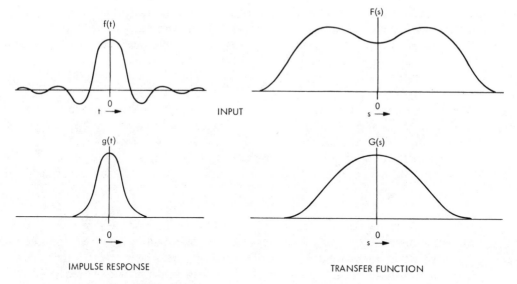

Figure 10-6   Linear system example

For the input spectrum, let us divide the s-axis into small equal intervals Δs and divide $F(s)$ into narrow strips Δs wide. If Δs is sufficiently small, $F(s)$ is well approximated by a sum of rectangular pulses, as shown in Figure 10-7. Note that an approximation to $F(s)$ is given by an infinite summation of such pulse pairs.

$$F(s) \approx \sum_{i=1}^{\infty} F(i\Delta s)\left[\Pi\left(\frac{s - i\Delta s}{\Delta s}\right) + \Pi\left(\frac{s + i\Delta s}{\Delta s}\right)\right] \tag{85}$$

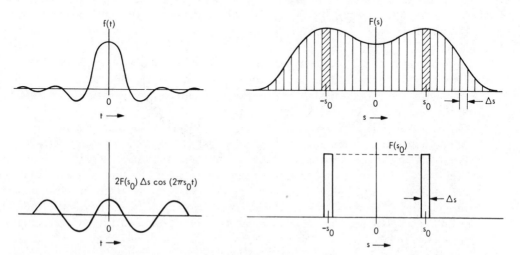

Figure 10-7   Sinusoidal decomposition

Consider a particular pair of pulses, namely those situated at $s = \pm s_0$ having amplitude $F(s_0)$, width Δs and area $F(s_0)\Delta s$. As Δs approaches zero, the pulse pair approaches an even impulse pair at $s = \pm s_0$, with infinitesimal strength $F(s_0)\Delta s$. The inverse transform of the even pulse pair approaches

$$2F(s_0)\Delta s \cos (2\pi s_0 t) \qquad s_0 = i\Delta s \tag{86}$$

Since $F(s)$ approaches a sum of even pulse pairs [Eq. (85)], $f(t)$ approaches a sum of cosines of the form of Eq. (86). This means that any even function can be decomposed into a sum of infinitely many cosines of infinitesimal amplitude.

Since the output spectrum is the product of the input spectrum and the transfer function, the output signal $h(t)$ can be expressed as a sum of cosines of the form

$$2G(s_0)F(s_0)\Delta s \cos (2\pi s_0 t) \tag{87}$$

We may now view the action of a linear filter as follows. The input signal $f(t)$ is first decomposed into a sum of cosines of all different frequencies. The amplitudes of the individual cosines are uniquely determined by $F(s)$, which in turn is the (unique) Fourier transform of $f(t)$. Inside the linear system, each cosine of frequency s_0 is multiplied by $G(s_0)$, the amplitude of the transfer function evaluated at frequency s_0. Finally, all the cosines of modified amplitude are summed at the output of the filter to form the output signal $h(t)$.

Notice that this interpretation is consistent with two previously discussed linear system properties: (1) A sinusoidal input always produces a sinusoidal output at the same frequency, and (2) the transfer function at frequency f is the factor by which the amplitude of an input sinusoid of frequency f is multiplied.

If we had made $f(t)$ odd, $F(s)$ would have been imaginary and odd, the pulse pairs would have been imaginary and odd, and $f(t)$ would decompose into sine functions. The remainder of the process would be identical except that; at the output, the sine functions of modified amplitude would be summed to produce the odd output signal $h(t)$.

Similarly, if $f(t)$ were neither even nor odd, it could be decomposed first into even and odd component functions, which would then be decomposed as before into sines and cosines, respectively. The modified sines and cosines again would be summed at the output to produce the output signal $h(t)$.

The foregoing discussion assumes that the transfer function is real and even. Suppose instead that the input is real and even but the impulse response $g(t)$ is real and odd. This makes the transfer function imaginary and odd. When the incoming even pulse pairs are multiplied by the imaginary odd transfer function, they are converted into imaginary odd impulse pairs. This converts the incoming cosines into output sines. The output then becomes a summation of sine functions. Therefore $h(t)$ is odd. This illustrates that convolving an even input function with an odd impulse response produces an odd function. From the graphical interpretation of the convolution integral, one can satisfy himself that this is correct.

Consider now the case where the input function is a cosine:

$$f(t) = \cos (2\pi f t) \qquad f \geq 0 \tag{88}$$

and the impulse response is real, consisting of even and odd components

$$g(t) = g_e(t) + g_o(t) \qquad (89)$$

The transfer function

$$G(s) = G_e(s) + jG_o(s) \qquad (90)$$

is Hermite, which means that

$$G(f) = G_e(f) + jG_o(f) \qquad f \geq 0 \qquad (91)$$

and

$$G(-f) = G^*(f) = G_e(f) - jG_o(f) \qquad f \geq 0 \qquad (92)$$

Recall that the spectrum of the cosine is

$$F(s) = \tfrac{1}{2}[\delta(s - f) + \delta(s + f)] \qquad (93)$$

We can now write the output spectrum as

$$H(s) = \tfrac{1}{2}G_e(f)[\delta(s - f) + \delta(s + f)] + j\tfrac{1}{2}G_o(f)[\delta(s - f) - \delta(s + f)] \qquad (94)$$

which means that the output signal is

$$h(t) = G_e(f) \cos (2\pi ft) + G_o(f) \sin (2\pi ft) \qquad (95)$$

This can be written as

$$h(t) = A \cos (2\pi ft + \phi) \qquad (96)$$

where

$$A = \sqrt{G_e^2(f) + G_o^2(f)} \quad \text{and} \quad \phi = \arctan\left[\frac{G_o(f)}{G_e(f)}\right] \qquad (97)$$

This is an expected result in view of the property that a linear system can change the amplitude and phase of a sinusoidal input but cannot change its frequency or functional form.

The foregoing exercise illustrates the relationship between the even and odd components of a real impulse response and the real and imaginary components of the transfer function. It shows how an odd component in the impulse response introduces an imaginary odd component into the transfer function. This produces a sine component output from a cosine component of the input and reflects itself in phase shift at the output. Finally, it illustrates that the amplitude of the output depends on the root-mean-square amplitude (modulus) of the complex transfer function.

Notice that we have now accumulated two equivalent ways of viewing the operation of a linear system. We may visualize convolution, with functions being reflected, shifted, multiplied, and integrated, or we may visualize sinusoidal decomposition followed by multiplication and resummation. We also understand the restrictions that evenness and oddness in one domain place on the other domain. These equivalences afford us considerable variety when approaching a problem with linear system analysis, and they illustrate the bilingual analogy mentioned at the beginning of this chapter.

Negative Frequency

Often students are disturbed by the concept of frequencies less than zero. Usually these are persons having experience with a waveform analyzer or a spectrum analyzer.

These devices incorporate narrow bandpass filters that allow energy to pass only in a narrow range about certain sinusoidal frequencies. One can derive the spectrum of a signal by tuning the narrow band filter across the frequency range and plotting the amplitude of the output. Students with experience of this type are frequently uncomfortable with negative frequency.

Recall that the Fourier transform of the cosine is an even impulse pair and the transform of the sine is an imaginary odd impulse pair. Since the cosine is an even function, it must have an even spectrum, and similarly for oddness and the sine function. For any real function, the spectrum is Hermite, and the left half is merely an image of the right half. For real functions, we are using a double-sided mathematical technique somewhat redundantly.

Since the left half of the spectrum is redundant for real functions, it could be ignored, as it implicitly is ignored in the use of a spectrum analyzer. However, we are using an abstract mathematical approach to model a physical process. The analysis is much simpler if we allow the negative frequency part of the spectrum to tag along.

Throughout Part II of this book, we graph double-sided spectra, although spectra are frequently plotted elsewhere only for positive frequency. We should keep in mind that, as long as we are using double-sided mathematics to model linear system operation, the left half of the function, redundant though it may be, does exist.

As far as our analysis is concerned, when the spectrum analyzer operator sets the frequency dial at 2000 Hz, he is adjusting a bandpass filter whose response is approximately an even pulse pair. Thus, even though the frequencies are marked positive on the dial, he is actually passing energy at $+2000$ Hz and -2000 Hz simultaneously. If not, the output signal would have to be complex, not real.

THE FOURIER TRANSFORM IN TWO DIMENSIONS

So far we have considered the Fourier transform of one-dimensional functions of time. In digital image processing and in optical system analysis, the inputs and outputs are commonly two-dimensional and, in some cases, higher-dimensional. Our investment in the one-dimensional Fourier transform will not prove to be wasted effort, however, since the transform generalizes easily to higher dimensions.

Definition

For functions of two dimensions, the direct and inverse Fourier transforms are defined by

$$F(u, v) = \int_{-\infty}^{\infty} \int_{-\infty}^{\infty} f(x, y)e^{-j2\pi(ux+vy)} \, dx \, dy \qquad (98)$$

and

$$f(x, y) = \int_{-\infty}^{\infty} \int_{-\infty}^{\infty} F(u, v)e^{j2\pi(ux+vy)} \, du \, dv \qquad (99)$$

where $f(x, y)$ is an image and $F(u, v)$ is its spectrum. Both u and v are frequency variables. The variable u corresponds to frequency along the x-axis, and the same is true for v and the y-axis. Figure 10-8 shows an image and its two-dimensional amplitude spectrum. The origin is located at the center of the transform image. Periodic noise in the image produces the "spikes" in the transform.

Properties of the Two-Dimensional Fourier Transform

The theorems of the two-dimensional Fourier transform are summarized in Table 10-3. Notice that the generalization from one dimension is quite direct.

Table 10-3. Properties of the two-dimensional Fourier transform

Property	Spatial Domain	Frequency Domain
Addition theorem	$f(x, y) + g(x, y)$	$F(u, v) + G(u, v)$
Similarity theorem	$f(ax, by)$	$\frac{1}{\|ab\|}F\left(\frac{u}{a}, \frac{v}{b}\right)$
Shift theorem	$f(x - a, y - b)$	$e^{-j2\pi(au+bv)}F(u, v)$
Convolution theorem	$f(x, y) * g(x, y)$	$F(u, v)G(u, v)$
Separable product	$f(x)g(y)$	$F(u)G(v)$
Differentiation	$\left(\frac{\partial}{\partial x}\right)^m\left(\frac{\partial}{\partial y}\right)^n f(x, y)$	$(j2\pi u)^m(j2\pi v)^n F(u, v)$
Rotation	$f(x\cos\theta + y\sin\theta,$ $-x\sin\theta + y\cos\theta)$	$F(u\cos\theta + v\sin\theta,$ $-u\sin\theta + v\cos\theta)$
LaPlacian	$\nabla^2(x, y) = \left(\frac{\partial^2}{\partial x^2} + \frac{\partial^2}{\partial y^2}\right)f(x, y)$	$-4\pi^2(u^2 + v^2)F(u, v)$
Rayleigh's theorem	$\int_{-\infty}^{\infty}\int_{-\infty}^{\infty}\|f(x, y)\|^2\, dx\, dy = \int_{+\infty}^{\infty}\int_{-\infty}^{\infty}\|F(u, v)\|^2\, du\, dv$	

The two-dimensional Fourier transform has several properties with no one-dimensional counterpart. One is the property that if a two-dimensional image factors into a product of one-dimensional components, the same is true for its two-dimensional spectrum.

Separability. Suppose that

$$f(x, y) = f_1(x)f_2(y) \tag{100}$$

In this case,

$$F(u, v) = \int_{-\infty}^{\infty}\int_{-\infty}^{\infty} f_1(x)f_2(y)e^{-j2\pi(ux+vy)}\, dx\, dy \tag{101}$$

can be rearranged to yield

$$F(u, v) = \int_{-\infty}^{\infty} f_1(x)e^{-j2\pi ux}\, dx \int_{-\infty}^{\infty} f_2(y)e^{-j2\pi vy}\, dy = F_1(u)F_2(v) \tag{102}$$

Thus if a two-dimensional image factors into one-dimensional components, so does its spectrum.

Consider as an example the elliptical two-dimensional Gaussian

$$e^{-(x^2/2\sigma_x^2 + y^2/2\sigma_y^2)} = e^{-x^2/2\sigma_x^2}e^{-y^2/2\sigma_y^2} \tag{103}$$

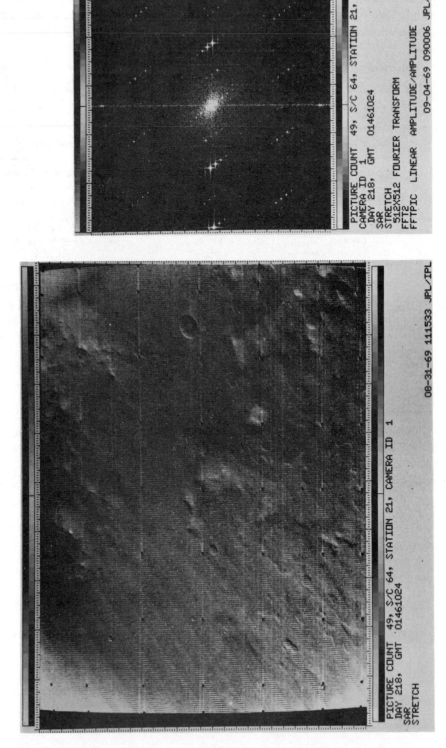

PICTURE COUNT   49, S/C 64, STATION 21,
CAMERA ID 1
DAY 218, GMT 01461024
SAR
STRETCH
512X512 FOURIER TRANSFORM
FFT2
FFTPIC   LINEAR AMPLITUDE/AMPLITUDE
                    09-04-69 090006 JPL/IPL

PICTURE COUNT   49, S/C 64, STATION 21, CAMERA ID 1
DAY 218, GMT 01461024
SAR
STRETCH
                                08-31-69 111533 JPL/IPL

Figure 10-8 A two-dimensional Fourier transform

which factors into the product of two one-dimensional Gaussians. If the standard deviations are equal, we have

$$e^{-(x^2+y^2)/2\sigma^2} = e^{-x^2/2\sigma^2}e^{-y^2/2\sigma^2} \tag{104}$$

which is the circular Gaussian. This function is extremely useful in the analysis of optical systems because it has circular symmetry and yet can be factored into one-dimensional components.

Similarity. For two-dimensional transforms, the similarity theorem may be generalized. We can write

$$\mathscr{F}\{f(a_1x + b_1y, a_2x + b_2y)\}$$
$$= \iint_{-\infty}^{\infty} f(a_1x + b_1y, a_2x + b_2y)e^{-j2\pi(ux+vy)}\,dx\,dy \tag{105}$$

We make the substitutions

$$w = a_1x + b_1y \qquad z = a_2x + b_2y \tag{106}$$

in which case

$$x = A_1w + B_1z \qquad y = A_2w + B_2z$$
$$dx = A_1\,dw + B_1\,dz \qquad dy = A_2\,dw + B_2\,dz \tag{107}$$

where

$$A_1 = \frac{b_2}{a_1b_2 - a_2b_1} \qquad B_1 = \frac{-b_1}{a_1b_2 - a_2b_1}$$
$$A_2 = \frac{-a_2}{a_1b_2 - a_2b_1} \qquad B_2 = \frac{a_1}{a_1b_2 - a_2b_1} \tag{108}$$

Then the Fourier transform becomes

$$\mathscr{F}\{f(a_1x + b_1y, a_2x + b_2y)\}$$
$$= \int_{-\infty}^{\infty}\int_{-\infty}^{\infty} f(w, z)e^{-j2\pi[(A_1u+A_2v)w+(B_1u+B_2v)z]}\,dz\,dw\,(A_1B_2 + A_2B_1) \tag{109}$$
$$= (A_1B_2 + A_2B_1)F(A_1u + A_2v, B_1u + B_2v)$$

The Rotation Property. From the two-dimensional similarity theorem it follows that a rotation of $f(x, y)$ through an angle θ also rotates its spectrum by the same amount. We let

$$a_1 = \cos\theta \qquad b_1 = \sin\theta \qquad a_2 = -\sin\theta \qquad b_2 = \cos\theta \tag{110}$$

so that

$$A_1 = \cos\theta \qquad A_2 = \sin\theta \qquad B_1 = -\sin\theta \qquad B_2 = \cos\theta \tag{111}$$

and

$$\mathscr{F}\{f(x\cos\theta + y\sin\theta, -x\sin\theta + y\cos\theta)\}$$
$$= F(u\cos\theta + v\sin\theta, -u\sin\theta + v\cos\theta) \tag{112}$$

The Projection Property. Suppose we collapse a two-dimensional function $f(x, y)$ into a one-dimensional function by projection onto the x-axis to form

$$p(x) = \int_{-\infty}^{\infty} f(x, y) \, dy \tag{113}$$

Then its Fourier transform is

$$P(u) = \int_{-\infty}^{\infty} \int_{-\infty}^{\infty} f(x, y) \, dy \, e^{-j2\pi ux} \, dx \tag{114}$$

But $P(u)$ can be written

$$P(u) = \int_{-\infty}^{\infty} \int_{-\infty}^{\infty} f(x, y) e^{-j2\pi(ux+0y)} \, dx \, dy = F(u, 0) \tag{115}$$

so the transform of the projection of $f(x, y)$ onto the x-axis is $F(u, v)$ evaluated along the u-axis. This combines with the rotation property to imply that the one-dimensional Fourier transform of $f(x, y)$ projected onto a line at an angle θ with the x-axis is just $F(u, v)$ evaluated along a line at an angle θ with the u-axis. This property forms the basis for system identification by line spread functions (Chapter 13) and for computerized axial tomography (Chapter 17).

Circular Symmetry. Many important two-dimensional functions exhibit the property of circular symmetry. This means that the function can be expressed as a profile function of a single radial variable

$$f(x, y) = f_r(r) \tag{116}$$

where

$$r^2 = x^2 + y^2 \tag{117}$$

We now investigate the effect this has upon the two-dimensional Fourier transform. We can write the Fourier transform of $f(x, y)$ as

$$\int_{-\infty}^{\infty} \int_{-\infty}^{\infty} f(x, y) e^{-j2\pi(ux+vy)} \, dx \, dy = \int_{0}^{\infty} \int_{0}^{2\pi} f_r(r) e^{-j2\pi qr \cos(\theta-\phi)} r \, dr \, d\theta \tag{118}$$

where we have converted the integration from rectangular to annular and made the variable substitution

$$x + jy = re^{j\theta} \quad \text{and} \quad u + jv = qe^{j\phi} \tag{119}$$

We can now rearrange Eq. (118), dropping ϕ because the integral is taken over a full cycle of the cosine, to yield

$$\mathcal{F}\{f(x, y)\} = \int_{0}^{\infty} f_r(r) \left[\int_{0}^{2\pi} e^{-j2\pi qr \cos(\theta)} \, d\theta \right] r \, dr \tag{120}$$

Consider the integral in brackets and recall the definition of the zero-order Bessel function of the first kind

$$J_0(z) = \frac{1}{2\pi} \int_{0}^{2\pi} e^{-jz \cos(\theta)} \, d\theta \tag{121}$$

Recognizing Eq. (121) in Eq. (120) allows us to write

$$\mathcal{F}\{f(x, y)\} = 2\pi \int_0^\infty f_r(r) J_0(2\pi q r) r \, dr \tag{122}$$

Notice that the Fourier transform of a circularly symmetric function is a function only of a single radial frequency variable q. This means that

$$F(u, v) = F_r(q) \tag{123}$$

where

$$q^2 = u^2 + v^2 \tag{124}$$

For circularly symmetric functions, the direct transform is given by

$$F_r(q) = 2\pi \int_0^\infty f_r(r) J_0(2\pi q r) r \, dr \tag{125}$$

and the inverse transformation by

$$f_r(r) = 2\pi \int_0^\infty F_r(q) J_0(2\pi q r) q \, dq \tag{126}$$

These equations define a special case of the two-dimensional Fourier transform that is called the *Hankel transform of zero order*. It is a one-dimensional linear integral transformation similar to the Fourier transform except that the kernel is a Bessel function. Hence two-dimensional functions with circular symmetry may be treated as one-dimensional functions of a single radial variable.

Hankel transforms of some familiar functions are listed in Table 10-4. Table 10-5 illustrates the theorems of the Hankel transform.

Table 10-4. Hankel transforms of certain functions

Function	$f(r)$	$F(q)$
Reciprocal	$\dfrac{1}{r}$	$\dfrac{1}{q}$
Gaussian	$e^{-\pi r^2}$	$e^{-\pi q^2}$
Impulse	$\dfrac{\delta(r)}{\pi r}$	1
Rectangular pulse	$\Pi\left(\dfrac{r}{2a}\right)$	$\dfrac{a J_1(2\pi a q)}{q}$
Triangular pulse	$\Lambda\left(\dfrac{r}{2a}\right)$	$\dfrac{2\pi}{ax^3}\displaystyle\int_0^x J_0(x) \, dx - \dfrac{2\pi}{ax^2} J_0(x)$
Shifted impulse	$\delta(r - a)$	$2\pi a J_0(2\pi a q)$
Exponential decay	e^{-ar}	$\dfrac{2\pi a}{[(2\pi q)^2 + a^2]^{3/2}}$
	$\dfrac{e^{-ar}}{r}$	$\dfrac{2\pi}{[(2\pi q)^2 + a^2]^{1/2}}$
	$\pi r^2 e^{-\pi r^2}$	$\left(\dfrac{1}{\pi} - q^2\right) e^{-\pi q^2}$
	$\dfrac{\sin(2\pi a r)}{r}$	$\dfrac{\Pi(q/2a)}{\sqrt{a^2 - q^2}}$

Table 10-5. Properties of the Hankel transform

Property	Spatial Domain	Frequency Domain
Addition theorem	$f(r) + g(r)$	$F(q) + G(q)$
Similarity theorem	$f(ar)$	$\frac{1}{a^2}F\left(\frac{q}{a}\right)$
Convolution theorem	$\int_0^\infty \int_0^{2\pi} f(\rho)g(r^2 + \rho^2 - 2r\rho \cos\theta)\rho \, d\rho \, d\theta$	$F(q)G(q)$
LaPlacian	$\nabla^2 f(r) = \frac{d^2 f}{dr^2} + \frac{1}{r}\frac{df}{dr}$	$-4\pi^2 q^2 F(q)$
Rayleigh's theorem	$\int_0^\infty |f(r)|^2 r \, dr = E$	$\int_0^\infty |F(q)|^2 q \, dq = E$
Power theorem	$\int_0^\infty f(r)q^*(r)r \, dr = P$	$\int_0^\infty F(q)G^*(q)q \, dq = P$

CORRELATION AND THE POWER SPECTRUM

In this section, we develop a series of analytical tools useful for studying the effects of noise in a linear system.

Autocorrelation

Recall that the self-convolution of a function is given by

$$f(t) * f(t) = \int_{-\infty}^{\infty} f(t)f(\tau - t) \, dt \tag{127}$$

If we do not reflect one term in the product, we form instead the autocorrelation function given by

$$R_f(\tau) = f(t) * f(-t) = \int_{-\infty}^{\infty} f(t)f(t + \tau) \, dt \tag{128}$$

The autocorrelation function is always even and has a maximum at $\tau = 0$. The autocorrelation function has the property

$$\int_{-\infty}^{\infty} R_f(\tau) \, d\tau = \left[\int_{-\infty}^{\infty} f(t) \, dt\right]^2 \tag{129}$$

Every function has a unique autocorrelation function, but the converse is not true.

The Power Spectrum

The Fourier transform of the autocorrelation function is given by

$$P_f(s) = \mathfrak{F}\{R_f(\tau)\} = \mathfrak{F}\{f(t) * f(-t)\} = F(s)F(-s) = F(s)F^*(s) = |F(s)|^2 \tag{130}$$

and is called the *power spectrum of $f(t)$*. If $f(t)$ is real, its autocorrelation function is real and even, and therefore its power spectrum is real and even. Again, for any $f(t)$, the power spectrum is unique, but the converse is not the case.

Cross-correlation

Given two functions $f(t)$ and $g(t)$, their cross-correlation function is given by

$$R_{fg}(\tau) = f(t) * g(-t) = \int_{-\infty}^{\infty} f(t)g(t + \tau)\, dt \qquad (131)$$

In a sense, the cross-correlation function indicates the relative amount of agreement between two functions for various degrees of misalignment (shifting).

The Fourier transform of the cross-correlation function is the cross power spectral density function

$$P_{fg}(s) = \mathfrak{F}\{R_{fg}(\tau)\} \qquad (132)$$

SUMMARY OF FOURIER TRANSFORM PROPERTIES

In this chapter, we have developed a number of properties of the Fourier transform that will prove useful in subsequent analyses of image processing systems. For convenience of reference, these properties are summarized in Table 10-6. Appendix III contains graphs of a number of one-dimensional Fourier transform pairs.

Table 10-6.   Summary of Fourier transform properties

Property	Time (or Spatial) Domain	Frequency Domain
Terminology	Signal	Spectrum
	Impulse response	Transfer function
	Autocorrelation function	Power spectrum
	Cross-correlation function	Cross power spectrum
	$f(x)$	$F(s)$
Definition	$\int_{-\infty}^{\infty} F(s)e^{j2\pi xt}\, ds$	$\int_{-\infty}^{\infty} f(x)e^{-j2\pi xs}\, dx$
Addition theorem	$af(x) + bg(x)$	$aF(s) + bG(s)$
Similarity theorem	$f(ax)$	$\dfrac{1}{\|a\|}F\left(\dfrac{s}{a}\right)$
Shift theorem	$f(x - a)$	$e^{-j2\pi as}F(s)$
Convolution theorem	$f(x) * g(x)$	$F(s)G(s)$
Differentiation	$\dfrac{d}{dx}f(x)$	$j2\pi sF(s)$
Autocorrelation theorem	$R_f(\tau) = f(x) * f^*(-x)$	$\|F(s)\|^2 = P_f(s)$
Rayleigh's theorem	$\int_{-\infty}^{\infty} \|f(x)\|^2\, dx = E$	$\int_{-\infty}^{\infty} \|F(s)\|^2\, ds = E$
Power theorem	$\int_{-\infty}^{\infty} f(x)g^*(x)\, dx = P$	$\int_{-\infty}^{\infty} F(s)G^*(s)\, ds = P$

SUMMARY OF IMPORTANT POINTS

1. The Fourier transform is a linear integral transformation that establishes a unique correspondence between a complex-valued function of time (or a spatial variable) and a complex-valued function of frequency.

2. The Fourier transform of a Gaussian function is another Gaussian.

3. Evenness and oddness are preserved by the Fourier transform.

4. The Fourier transform of a real function is a Hermite function.

5. The Fourier transform of a sum of functions is the sum of their individual transforms (addition theorem).

6. Shifting the origin of a function introduces into its spectrum phase shift, which is linear with frequency and which alters the distribution of energy between real and imaginary parts without changing total energy (shift theorem).

7. Convolution of two functions corresponds to multiplication of their Fourier transforms (convolution theorem).

8. Narrowing a function broadens its Fourier transform and vice versa (similarity theorem).

9. The energy of a function is the same as that of its Fourier transform.

10. The transfer function of a linear system can be determined as the ratio of its output spectrum to its known input spectrum.

11. The Fourier transform of a sinusoidal function is an equally spaced impulse pair.

12. An input signal can be decomposed into an infinite sum of infinitesimal sinusoids.

13. A linear system can be thought of as operating separately on the sinusoidal components, which are summed at the output to form the output signal.

14. The Fourier transform generalizes readily to two-dimensional functions.

15. If a function of two variables separates as a product of two single variable functions, then so does its Fourier transform.

16. Rotating a function of two dimensions rotates its Fourier transform by the same amount.

17. Projecting (collapsing) a two-dimensional function onto a line at an angle θ to the x-axis and transforming the resulting one-dimensional function yields a profile of the two-dimensional spectrum taken along a line at an angle θ to the u-axis.

18. Circularly symmetric two-dimensional functions have circularly symmetric spectra.

19. The Hankel transform relates the profile function of a circularly symmetric function to that of its spectrum.

20. Autocorrelation is self-convolution without reflection of either function.

21. Cross-correlation is like convolution, except that neither function is reflected.

22. The Fourier transform of the autocorrelation function is the power spectrum.

REFERENCES

1. R. BRACEWELL, *The Fourier Transform and Its Applications*, McGraw-Hill Book Company, New York, 1965.

2. J. W. COOLEY and J. W. TUKEY, "An Algorithm for the Machine Computation of Complex Fourier Series," *Math. Computation*, **19** 297–301, April 1965.

3. E. O. BRIGHAM and R. E. MORROW, "The Fast Fourier Transform," *IEEE Spectrum*, **4** 63–70, December 1967.

4. G. D. BERGLAND, "A Guided Tour of the Fast Fourier Transform," *IEEE Spectrum*, **6** 41–52, July 1969.

5. E. O. BRIGHAM, *The Fast Fourier Transform*, Prentice-Hall, Englewood Cliffs, N.J., 1974.

6. B. GOLD and C. M. RADER, *Digital Processing of Signals*, McGraw-Hill Book Company, New York, 1969.

7. L. R. RABINER and C. M. RADER, eds., *Digital Signal Processing*, IEEE Press, New York, 1972.

8. H. D. HELMS and L. R. RABINER, eds., *Literature in Digital Signal Processing*, IEEE Press, New York, 1973.

FILTER DESIGN

llllllllllllllllllllllllllllllllllllllllllllllllllllllllllllllllllllllllllllllllllllllllllllllllllllllllllllllllllllllll

INTRODUCTION

In Chapters 9 and 10, we laid the groundwork for the analysis of linear filtering operations. Linear filters can be implemented either by convolution in the time (or spatial) domain or by multiplication in the frequency domain. In later chapters, we deal with the techniques for and limitations of implementing linear filtering digitally. In this chapter, however, we discuss techniques for designing filters to accomplish particular goals. To gain insight, we shall first examine the time domain and frequency domain behavior of certain simple but useful filters. Later in this chapter, we approach the problem of designing filters that are optimal for doing a specific job.

As in Chapters 9 and 10, we shall perform the analysis with one-dimensional (time) signals for simplicity of mathematics. The generalization to two dimensions is straightforward.

EXAMPLES OF COMMON FILTERS

In this section, we consider some conceptually simple filters in order to gain insight into the time domain and frequency domain characteristics of filters and their effect upon input signals.

190

The Ideal Bandpass Filter

Suppose we desire to implement, by convolution, a filter that passes energy only at frequencies between f_1 and f_2, where $f_2 > f_1$. The desired transfer function is given by

$$G(s) = \begin{cases} 1 & f_1 \leq |s| \leq f_2 \\ 0 & \text{elsewhere} \end{cases} \tag{1}$$

and shown in Figure 11-1. Since $G(s)$ is an even rectangular pulse pair, it can be thought of as a rectangular pulse convolved with an even impulse pair. If we let

$$s_0 = \tfrac{1}{2}(f_1 + f_2) \quad \text{and} \quad \Delta s = f_2 - f_1 \tag{2}$$

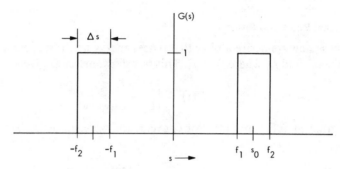

Figure 11-1   Ideal bandpass transfer function

we can write the transfer function of the ideal bandpass filter as

$$G(s) = \Pi\left(\frac{s}{\Delta s}\right) * [\delta(s - s_0) + \delta(s + s_0)] \tag{3}$$

With the transfer function expressed in this form, we can easily write the impulse response

$$g(t) = \Delta s \frac{\sin(\pi \Delta s t)}{\pi \Delta s t} 2 \cos(2\pi s_0 t) = 2\Delta s \frac{\sin(\pi \Delta s t)}{\pi \Delta s t} \cos(2\pi s_0 t) \tag{4}$$

Since $\Delta s < s_0$, Eq. (4) describes a cosine of frequency s_0 enclosed in a $\sin(x)/x$ envelope having frequency $\Delta s/2$. This impulse response is graphed in Figure 11-2. The number of cosine cycles between envelope zero crossings depends on the relationship between s_0 and Δs. Notice that if s_0 is held constant and Δs becomes small, the envelope expands to include more and more cosine cycles between zero crossings. As Δs approaches zero, the impulse response approaches a cosine. In the limiting case, the convolution actually becomes a cross-correlation of the input with the cosine at frequency s_0.

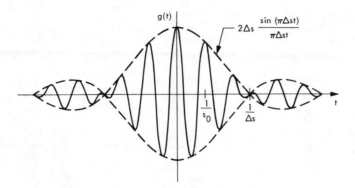

Figure 11-2 Ideal bandpass impulse response

The Ideal Bandstop Filter

Suppose now we desire a filter that passes energy at all frequencies except for a band between f_1 and f_2, where $f_2 > f_1$. This transfer function is given by

$$H(s) = \begin{cases} 0 & f_1 \leq |s| \leq f_2 \\ 1 & \text{elsewhere} \end{cases} \tag{5}$$

and graphed in Figure 11-3. For convenience, we again let s_0 be the center frequency

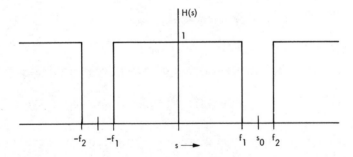

Figure 11-3 Ideal bandstop transfer function

and Δs the bandwidth (Eq. 2). Now we can write the transfer function as one minus a bandpass filter

$$H(s) = 1 - \Pi\left(\frac{s}{\Delta s}\right) * [\delta(s - s_0) + \delta(s + s_0)] \tag{6}$$

from which the impulse response is

$$h(t) = \delta(t) - 2\Delta s \frac{\sin(\pi \Delta s t)}{\pi \Delta s t} \cos(2\pi s_0 t) \tag{7}$$

The impulse response is graphed in Figure 11-4. Its behavior with changing bandwidth

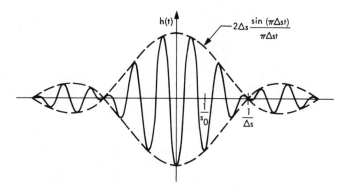

Figure 11-4 Ideal bandstop impulse response

and center frequency is similar to that of the bandpass filter, which it resembles. If Δs is small, this filter is referred to as a "notch filter."

The General Bandpass Filter

We now consider a class of bandpass filters constructed in the following way. We select a nonnegative unimodal function $F(s)$ and convolve it with an even impulse pair at frequency s_0. This yields a bandpass transfer function, as shown in Figure 11-5.

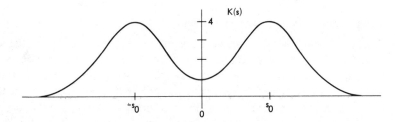

Figure 11-5 The general bandpass filter

The transfer function is given by

$$K(s) = F(s) * [\delta(s - s_0) + \delta(s + s_0)] \tag{8}$$

and the impulse response by

$$k(t) = 2f(t) \cos (2\pi s_0 t) \tag{9}$$

This impulse response is a cosine of frequency s_0 in an envelope that is the inverse Fourier transform of $F(s)$.

Suppose, for example, that $F(s)$ is a Gaussian

$$K(s) = Ae^{-s^2/2\alpha^2} * [\delta(s - s_0) + \delta(s + s_0)] \tag{10}$$

Then the impulse response becomes

$$k(t) = \frac{2A}{\sqrt{2\pi\sigma^2}} e^{-t^2/2\sigma^2} \cos(2\pi s_0 t) \tag{11}$$

This impulse response is a cosine in a Gaussian envelope. It is graphed in Figure 11-6. Notice that we could easily generate a class of bandstop filters as before.

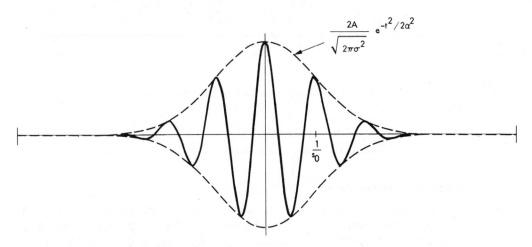

Figure 11-6 The Gaussian bandpass filter

The Gaussian High-Frequency Enhancement Filter

The term *high-frequency enhancement filter* is generally taken to describe a transfer function that is unity at zero frequency and increases away from the origin. A high-frequency enhancement filter may either level off at some value greater than one or, more commonly, may fall back toward zero at high frequencies. In the latter case, the high-frequency enhancement filter is merely a type of bandpass filter with the restriction of unity gain at zero frequency. In practice, it is often desired to have less than unity gain at zero frequency to reduce the contrast of large, slowly varying components in the image.

We can produce a high-frequency enhancement transfer function by expressing it as the difference of two Gaussians of different widths.

$$H(s) = Ae^{-s^2/2\alpha_1^2} - Be^{-s^2/2\alpha_2^2} \qquad A \geq B, \alpha_1 > \alpha_2$$

This is shown in Figure 11-7. The impulse response of this filter is given by

$$h(t) = \frac{A}{\sqrt{2\pi\sigma_1^2}} e^{-t^2/2\sigma_1^2} - \frac{B}{\sqrt{2\pi\sigma_2^2}} e^{-t^2/2\sigma_2^2} \qquad \sigma_i = \frac{1}{2\pi\alpha_i} \tag{12}$$

and graphed in Figure 11-8. Notice that the broad Gaussian in the frequency domain produces a narrow Gaussian in the time domain and vice versa. The impulse response

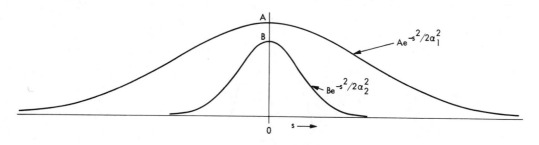

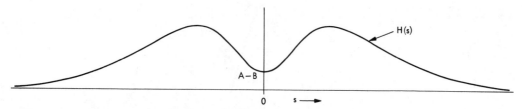

Figure 11-7   The Gaussian high-frequency enhancement transfer function

shown in Figure 11-8 is typical of bandpass and high-frequency enhancement filters, having a positive pulse situated in a negative dish.

If we let α_1 approach infinity, the narrow Gaussian in the time domain narrows down to an impulse and the filter has the form shown in Figure 11-9. Notice that the difference between a filter that rolls off (returns toward zero) at high frequencies and one that does not is the width of the central pulse in the time domain.

Rules of Thumb for High-Frequency Enhancement Filter Design

In this section, we develop two approximate rules to estimate the behavior of high-frequency enhancement filters. Suppose the impulse response of the filter is expressed as a narrow pulse minus a broad pulse,

$$h(t) = h_1(t) - h_2(t) \tag{13}$$

as illustrated in Figure 11-10. We know that the transfer function $H(s)$ will have the general shape of a high-frequency enhancement filter. We would like to estimate the transfer function at zero frequency to determine its effect on the contrast of large objects within the image. We also would like to estimate the maximum value the transfer function takes on at any frequency.

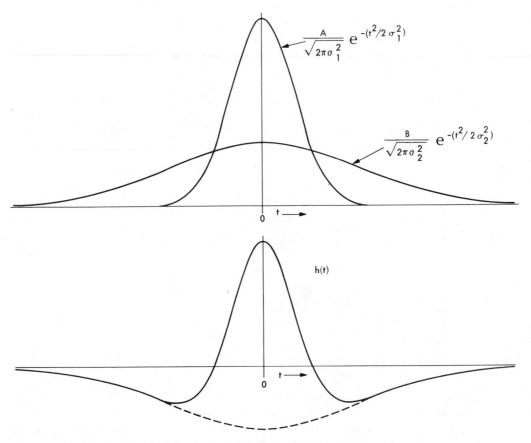

Figure 11-8 The Gaussian high-frequency enhancement impulse response

If we write the Fourier transform of Eq. (13) and substitute the value $s = 0$, we obtain

$$H(0) = \int_{-\infty}^{\infty} h(t)\, dt = \int_{-\infty}^{\infty} h_1(t)\, dt - \int_{-\infty}^{\infty} h_2(t)\, dt = A_1 - A_2 \tag{14}$$

where A_1 and A_2 represent the areas under the two component functions.

We can place an upper bound on the magnitude of the transfer function if we assume that $H_2(s)$ goes to zero (dies out) before $H_1(s)$ decreases from its maximum value; that is,

$$H_{\max} \leq H_1(0) = \int_{-\infty}^{\infty} h_1(t)\, dt = A_1 \tag{15}$$

We now have two rules of thumb for high-frequency enhancement filters composed of the difference of two pulses:

$$H(0) = A_1 - A_2 \quad \text{and} \quad H_{\max} \leq A_1 \tag{16}$$

If $h_1(t)$ is an impulse (recall Figure 11-9), then equality holds in Eq. (16).

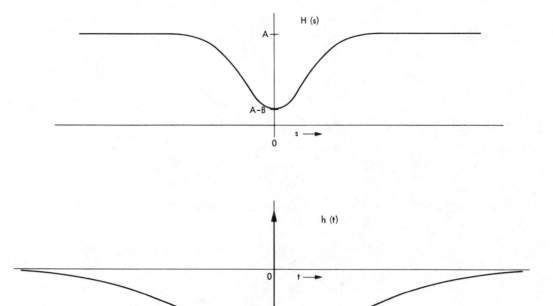

Figure 11-9 The Gaussian highpass filter

Low-Frequency Response

In this section, we examine the effect a filter has upon large constant areas within an image.

Assume the impulse response $g(t)$ is duration-limited, that is, zero outside a finite interval. Assume also that the input signal $f(t)$ is constant over an interval larger than the duration of $g(t)$. This situation is shown in Figure 11-11. The output of the system is given by the convolution integral

$$h(x) = \int_{-\infty}^{\infty} f(\tau)g(x - \tau)\, d\tau \tag{17}$$

Over the interval of interest, however, the input signal is constant and Eq. (17) becomes

$$h(x) = \int_{-\infty}^{\infty} cg(x - \tau)\, d\tau = c \int_{-\infty}^{\infty} g(\tau)\, d\tau \tag{18}$$

Notice that if we substitute $s = 0$ into the definition of the Fourier transform, we have

$$G(0) = \int_{-\infty}^{\infty} g(t)\, dt$$

which means

$$h(x) = cG(0) \tag{19}$$

Thus if $G(0) = 1$, the filter will not change the amplitude of large constant areas of

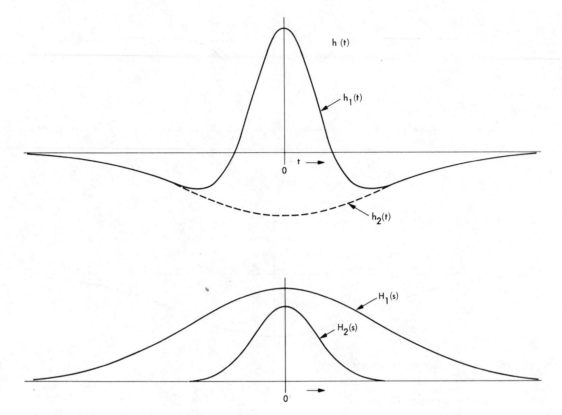

Figure 11-10 The general highpass filter

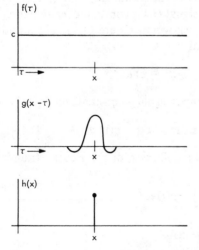

Figure 11-11 Low-frequency response

$f(x)$. Generalizing to two dimensions, this means that the filter does not change the contrast of large flat areas within the input image. If $G(0)$ does not equal one, it becomes a gain factor controlling the overall amplitude relationship between large components of $h(t)$ and $f(t)$.

OPTIMAL FILTER DESIGN

In this section, we develop techniques to design filters that are, in some sense, optimal for the job for which they are intended. We shall accomplish this by first establishing a criterion of "goodness" and then maximizing that criterion by proper selection of the impulse response of the filter.

The history of digital image processing has seen considerable filter design done, the way flying was done in World War I, "by the seat of the pants." Filters have been chosen for reasons of computational simplicity, past success, convenience, aesthetic appeal, and whim. Such filter design bears the unwholesome label "suboptimal," with all its negative connotations. It almost never produces the best filter for the job, and it can be dangerous.

Suboptimal filters, particularly those easy to implement by computer, can introduce artifacts into an image, frequently without warning. Filters involving the rectangular pulse in the spatial domain, a favorite of computer programmers, have a most unsavory spectrum due to the infinite undulations of the sin $(x)/x$ function. Users of such filters are often plagued by "ringing" and other artifactual phenomena in the opposite domain. They frequently regard these undesirable characteristics as indigenous to digital processing, or lament the lack of computer power necessary to do the job correctly.

In this section, we develop design techniques for optimal filters and show that they are, in general, quite well behaved. It is hoped that the reader, armed with this knowledge, can intelligently trade off optimality for computational simplicity without courting disastrous artifacts.

In this section, we review the concept of an ergodic random variable and develop design techniques for two optimal filters. They are the Wiener estimator (Refs. 1, 2, 3, 4), which is optimal for recovering an unknown signal from additive noise, and the matched detector (Refs. 4, 5, 6), which is optimal for finding a known signal buried in additive noise. Even if the reader never designs an optimal filter, these two developments will sharpen his insight considerably.

Random Variables

In previous chapters, we have referred to the concept of a random variable, particularly for describing system noise. Since random variables play a major role in the following development, we consider them now in some detail.

We use the term *random noise* to describe an unknown contaminating signal. The word *random* is a euphemism for our incomplete knowledge. This ignorance results from dealing with a process, the physics of which is not well understood, or with a process too complicated to analyze in detail.

When we record a signal, we know that, during the recording process, an undesired contaminating signal will appear superimposed upon (added to) the desired signal. Though we might know the origin of the noise, we cannot express its functional form mathematically. After observing the noise for a period of time, we may develop a partial knowledge of it and be able to characterize some aspects of its behavior, though we can never predict it in detail. Thus the concept of a random variable becomes a useful tool in dealing with noise.

We may think of a random variable as follows. Consider an ensemble of functions consisting of infinitely many member functions. When we make our recording, one of those member functions emerges to contaminate our record. We have no way of knowing which member function will appear, but we can make general statements about the ensemble as a group. In this way, we can express our partial knowledge of the contaminating signal.

Ergodic Random Variables. In the remainder of this book it is sufficient to concern ourselves only with random variables that are *ergodic*. The definition of the ergodicity property can be approached as follows. There are two ways by which we can compute averages of a random variable. We can compute a "time average" by integrating a particular member function over all time, or we can average together the values of all member functions evaluated at some particular time. The latter technique produces an "ensemble average" at some point in time. A random variable is ergodic if and only if (1) the time averages of all member functions are equal, (2) the ensemble average is constant with time, and (3) the time average and the ensemble average are numerically equal. Thus, for ergodic random variables, time averages and ensemble averages are interchangeable.

In Chapter 7 we introduced the expectation operator $\mathcal{E}\{x(t)\}$ which denotes the ensemble average of the random variable x computed at time t. Under the ergodicity property, $\mathcal{E}\{x(t)\}$ also denotes the value obtained when $x(t)$ is averaged over time,

$$\mathcal{E}\{x(t)\} = \int_{-\infty}^{\infty} x(t) \, dt \tag{20}$$

Equation (128) of Chapter 10 defines the autocorrelation function as a time average. For an ergodic random variable, the autocorrelation function is the same for all member functions, and thus characterizes the ensemble. Therefore, when we say $n(t)$ is an ergodic random variable, we mean it is an unknown function that has a known autocorrelation function. This represents the state of our partial knowledge of $n(t)$. Since the autocorrelation function of $n(t)$

$$R_n(\tau) = \int_{-\infty}^{\infty} n(t)n(t + \tau) \, dt \tag{21}$$

is known, its power spectrum

$$P_n(s) = \mathcal{F}\{R_n(\tau)\} \tag{22}$$

is also known. This means we know the amplitude spectrum of $n(t)$ but do not know its phase spectrum. Indeed, the ensemble is composed of infinitely many functions that differ only in their phase spectra. Any real, even, nonnegative function can be the power spectrum of a random variable, and any real, even function that has a nonnegative spectrum can be the autocorrelation function of a random variable.

Fortunately, ergodic random variables model commonly encountered random signals quite well. For example, repeated observations of "white noise" sources show that the measured power spectrum is constant with frequency to a good approximation.

The Wiener Estimator

Suppose we have an observed signal $x(t)$, which is composed of a desired signal $s(t)$ contaminated by an additive noise function $n(t)$. We would like to design a linear filter to reduce the contaminating noise as much as possible and restore the signal as closely as possible to its original form. The configuration is shown in Figure 11-12. The

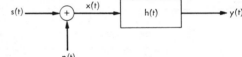

Figure 11-12   Model for the Wiener estimator

impulse response is $h(t)$, and the output of the filter is $y(t)$. We want to select the impulse response so that the output signal $y(t)$ will be as close as possible to $s(t)$. Ideally, we would like $y(t)$ to be equal to $s(t)$, but in general a linear filter is not powerful enough to recover a contaminated signal exactly. What we shall do instead is select the impulse response $h(t)$ so that $y(t)$ will be as close as possible to $s(t)$.

Before we begin, we must decide what knowledge we have about $s(t)$ and $n(t)$. If we know nothing at all about the signal or the noise, we cannot even start on the problem. At the other extreme, if we know one or both of the signals exactly, the solution is trivial. For the purposes of the following analysis, we shall assume that both $s(t)$ and $n(t)$ are ergodic random variables and thus have known power spectra. This means that, although we do not know $n(t)$ exactly, we do know that it comes from an ensemble of functions all having the same autocorrelation function and, hence, the same power spectrum. The same restriction applies to $s(t)$. Furthermore, we assume that either we know the power spectra *a priori* or we can capture samples of $s(t)$ and $n(t)$ and determine their power spectra, which are, in turn, representative of their respective ensembles.

Optimality Criterion. Before we begin the development of the optimal filter, we must establish an objective criterion of optimality. Since asking for $y(t) = s(t)$ is in general asking too much of a linear filter, we shall ask instead for the best job possible under the circumstances. As a criterion of optimality, we shall use the mean-square error.

No matter what $h(t)$ is, optimal or not, we will obtain an output $y(t)$ in response to an input $s(t)$. We define the error signal at the output of the filter as

$$e(t) = s(t) - y(t) \tag{23}$$

that is, the amount by which the actual output differs from the desired output as a function of time. If the impulse response $h(t)$ is well chosen, the error signal will be, on the average, less than it would be with a poor choice of $h(t)$. As a measure of the average error, we use the mean-square error given by

$$\text{MSE} = \mathcal{E}\{e^2(t)\} = \int_{-\infty}^{\infty} e^2(t)\, dt \tag{24}$$

The latter equality holds because the error signal, a linear combination of ergodic random variables, is itself an ergodic random variable.

Notice that $e^2(t)$ is positive for both positive and negative errors. Also, squaring the error causes large errors to be penalized more severely than small errors. For these reasons, minimizing the mean-square error is an intuitively satisfactory choice of an optimality criterion. While other criteria, such as absolute error, could be used, they considerably complicate the analysis and provide, for our purposes, little or no advantage.

The Mean-Square Error. We now approach the problem as follows: Given the power spectra of $s(t)$ and $n(t)$, we must determine the impulse response $h(t)$ that minimizes the mean-square error. Notice that the mean-square error is a functional of $h(t)$, the impulse response, since a function $h(t)$ maps into a real number MSE. The branch of mathematics concerned with functional minimization is the calculus of variations, which we shall use. We shall obtain a functional expression for MSE in terms of $h(t)$, then find an expression for the optimal (minimizing) $h(t)$ in terms of known power spectra, and finally develop an expression for the MSE that results when the optimal $h(t)$ is used. This latter step will show how well the optimal filter works.

We begin by expanding the mean-square error in Eq. (24).

$$\text{MSE} = \mathcal{E}\{e^2(t)\} = \mathcal{E}\{[s(t) - y(t)]^2\} = \mathcal{E}\{s^2(t) - 2s(t)y(t) + y^2(t)\} \tag{25}$$

Since the expectation is an integral operator [Eq. (20)], we can write

$$\text{MSE} = \mathcal{E}\{s^2(t)\} - 2\mathcal{E}\{s(t)y(t)\} + \mathcal{E}\{y^2(t)\} = T_1 + T_2 + T_3 \tag{26}$$

where T_1, T_2, and T_3 are introduced so that we may consider the three terms separately. Writing T_1 as an integral,

$$T_1 = \mathcal{E}\{s^2(t)\} = \int_{-\infty}^{\infty} s^2(t)\, dt = R_s(0) \tag{27}$$

we recognize it as the $\tau = 0$ point on the (known) autocorrelation function of $s(t)$.

Writing $y(t)$ as the convolution of $x(t)$ and $h(t)$ allows us to expand the second term as

$$T_2 = -2\mathcal{E}\{s(t) \int_{-\infty}^{\infty} h(\tau)x(t - \tau)\, d\tau\} \tag{28}$$

Since the expectation operator is actually an integral over time, we can rearrange Eq.

(28) to produce

$$T_2 = -2 \int_{-\infty}^{\infty} h(\tau) \mathcal{E}\{s(t)x(t-\tau)\} \, d\tau \tag{29}$$

Now we recognize the expectation inside the integral as the cross-correlation function of $s(t)$ and $x(t)$ and write

$$T_2 = -2 \int_{-\infty}^{\infty} h(\tau) R_{xs}(\tau) \, d\tau \tag{30}$$

We can expand T_3 as the expectation of the product of two convolutions

$$T_3 = \mathcal{E}\left\{\int_{-\infty}^{\infty} h(\tau)x(t-\tau) \, d\tau \int_{-\infty}^{\infty} h(u)x(t-u) \, du\right\} \tag{31}$$

which may be rearranged as before to yield

$$T_3 = \int_{-\infty}^{\infty} \int_{-\infty}^{\infty} h(\tau)h(u)\mathcal{E}\{x(t-\tau)x(t-u)\} \, d\tau \, du \tag{32}$$

If, inside the expectation operator, we make the variable substitution $v = t - u$, that factor becomes

$$\mathcal{E}\{x(t-\tau)x(t-u)\} = \mathcal{E}\{x(v+u-\tau)x(v)\} \tag{33}$$

which is simply the autocorrelation function of $x(t)$ evaluated at the point $u - \tau$. Now the third term can be written as

$$T_3 = \int_{-\infty}^{\infty} \int_{-\infty}^{\infty} h(\tau)h(u)R_x(u-\tau) \, d\tau \, du \tag{34}$$

The mean-square error of Eq. (26) can now be written as

$$\text{MSE} = R_s(0) - 2 \int_{-\infty}^{\infty} h(\tau)R_{xs}(\tau) \, d\tau + \int_{-\infty}^{\infty} \int_{-\infty}^{\infty} h(\tau)h(u)R_x(u-\tau) \, d\tau \, du \tag{35}$$

This is the mean-square error in terms of the filter's impulse response and known autocorrelation and cross-correlation functions of the two input signal components. As expected, MSE is a functional of $h(t)$. We now wish to select the particular function $h(t)$ that causes MSE to take on its minimum value.

Minimizing MSE. Let us denote by $h_o(t)$ the particular function that minimizes MSE. In general, an arbitrary $h(t)$ will differ from the optimal $h_o(t)$, and we can define a function $g(t)$ to account for this variation from the optimal; that is,

$$h(t) = h_o(t) + g(t) \tag{36}$$

where $h(t)$ is an arbitrarily chosen (suboptimal) impulse response function and $g(t)$ is chosen to make the equality hold. The reason for this seemingly unnecessary complication is not obvious now, but it will allow us to establish a necessary condition upon $h_o(t)$.

If we substitute the definition for $g(t)$ in Eq. (36) into the MSE equation [Eq. (35)], we obtain

$$\begin{aligned} \text{MSE} = R_s(0) - 2 \int_{-\infty}^{\infty} [h_o(\tau) + g(\tau)]R_{xs}(\tau) \, d\tau \\ + \int_{-\infty}^{\infty} \int_{-\infty}^{\infty} [h_o(\tau) + g(\tau)] \, [h_o(u) + g(u)]R_x(u-\tau) \, d\tau \, du \end{aligned} \tag{37}$$

This expression can be expanded, producing seven terms

$$\text{MSE} = R_s(0) - 2 \int_{-\infty}^{\infty} h_o(\tau)R_{xs}(\tau)\,d\tau + \int_{-\infty}^{\infty} \int_{-\infty}^{\infty} h_o(\tau)h_o(u)R_x(u-\tau)\,d\tau\,du$$

$$+ \int_{-\infty}^{\infty} \int_{-\infty}^{\infty} h_o(\tau)g(u)R_x(u-\tau)\,d\tau\,du + \int_{-\infty}^{\infty} \int_{-\infty}^{\infty} h_o(u)g(\tau)R_x(u-\tau)\,d\tau\,du$$

$$- 2 \int_{-\infty}^{\infty} g(\tau)R_{xs}(\tau)\,d\tau + \int_{-\infty}^{\infty} \int_{-\infty}^{\infty} g(\tau)g(u)R_x(u-\tau)\,d\tau\,du \qquad (38)$$

Comparing the first three terms with Eq. (35) we see that their sum represents the mean-square error that results when the optimal impulse response $h_o(t)$ is used. We denote this value by MSE_o. Since the autocorrelation function $R_x(u-\tau)$ is an even function, the fourth and fifth terms of Eq. (38) are equal. We can combine them with the sixth term and write Eq. (38) as

$$\text{MSE} = \text{MSE}_o + 2 \int_{-\infty}^{\infty} g(u) \left[\int_{-\infty}^{\infty} h_o(\tau)R_x(u-\tau)\,d\tau - R_{xs}(u) \right] du$$

$$+ \int_{-\infty}^{\infty} \int_{-\infty}^{\infty} g(u)g(\tau)R_x(u-\tau)\,du\,d\tau = \text{MSE}_o + T_4 + T_5 \qquad (39)$$

where T_4 and T_5 are introduced for compactness of notation.

We shall now show that the term T_5 is nonnegative. Writing the autocorrelation function $R_x(u-\tau)$ as an integral produces

$$T_5 = \int_{-\infty}^{\infty} \int_{-\infty}^{\infty} g(u)g(\tau) \int_{-\infty}^{\infty} x(t-\tau)x(t-u)\,dt\,du\,d\tau \qquad (40)$$

which may be rearranged to yield

$$T_5 = \int_{-\infty}^{\infty} \int_{-\infty}^{\infty} g(u)x(t-u)\,du \int_{-\infty}^{\infty} g(\tau)x(t-\tau)\,d\tau\,dt \qquad (41)$$

If we define $z(t)$ as the function that results from convolving $g(t)$ with $x(t)$, we can recognize Eq. (41) as

$$T_5 = \int_{-\infty}^{\infty} z^2(t)\,dt \geq 0 \qquad (42)$$

which can never be negative.

Returning to the MSE, we can write Eq. (39) as

$$\text{MSE} = \text{MSE}_o + 2 \int_{-\infty}^{\infty} g(u) \left[\int_{-\infty}^{\infty} h_o(\tau)R_x(u-\tau)\,d\tau - R_{xs}(u) \right] du + T_5 \qquad (43)$$

where MSE_o is the mean-square error under optimal conditions and T_5 cannot be negative. We wish to establish a condition on $h_o(\tau)$ that will ensure that MSE_o is the smallest value that MSE can possibly have. One way to do this is to make the quantity in brackets be zero for all values of u. This makes T_4 drop out of Eq. (43) and guarantees that $\text{MSE}_o \leq \text{MSE}$. However, we still must make sure that condition is both necessary and sufficient to optimize the filter.

Suppose that the term in brackets in Eq. (43) were nonzero for some values of u. Since $g(u)$ is an arbitrary function, it could take on large negative values where the bracketed term was positive and vice versa. The integral in T_4 would then take on a large negative value and MSE would become smaller than MSE_o. Since this violates our definition, we conclude that it is a necessary condition that the bracketed term in Eq. (43) must be identically zero. This means that

$$R_{xs}(\tau) = \int_{-\infty}^{\infty} h_o(u)R_x(u - \tau)\,du \tag{44}$$

is a necessary condition in order for the mean-square error to be minimized. Now the complication introduced in Eq. (36) has paid off by giving us a necessary condition for the optimal filter.

It is easy to see that the condition of Eq. (44) is also sufficient to optimize the filter, that is, that no additional conditions are required. Since the necessary condition causes T_4 to drop out of Eq. (43), it becomes

$$\text{MSE} = \text{MSE}_o + T_5 \qquad T_5 \geq 0 \tag{45}$$

from which it is clear that

$$\text{MSE} \geq \text{MSE}_o \tag{46}$$

Thus Eq. (44) defines the impulse response of the linear estimator that is optimal in the mean-square sense.

Notice that the right-hand side of Eq. (44) is a convolution integral, which can be written as

$$R_{xs}(\tau) = h_o(u) * R_x(u) \tag{47}$$

It relates the optimal impulse response to the autocorrelation of the input signal and the cross-correlation of the input and the desired signal.

It is easy to show that, for any linear system, the cross-correlation between input and output is given by

$$R_{xy}(\tau) = h(u) * R_x(u) \tag{48}$$

where $R_x(u)$ is the autocorrelation function of the input signal [see Chapter 13, Eq. (57)]. Comparing this with Eq. (47) illustrates that the Wiener filter makes the input/output cross-correlation function equal to the signal/signal-plus-noise cross-correlation function.

If we take the Fourier transform of both sides of Eq. (47), we are left with

$$P_{xs}(s) = H_o(s)P_x(s) \tag{49}$$

which implies that

$$H_0(s) = \frac{P_{xs}(s)}{P_x(s)} \tag{50}$$

is the frequency domain specification of the Wiener estimator.

Wiener Filter Design. Equation (50) implies that we can design a Wiener estimator in the following way: (1) First, digitize a sample of the input signal $s(t)$. (2) Autocorrelate the input sample to produce an estimate of $R_x(\tau)$. (3) Compute the Fourier transform of $R_x(\tau)$ to produce $P_x(s)$. (4) Obtain and digitize a sample of the signal in the absence of noise. (5) Cross-correlate the signal sample with the input sample to estimate $R_{xs}(\tau)$. (6) Compute the Fourier transform of $R_{xs}(\tau)$ to produce $P_{xs}(s)$. (7)

Compute the transfer function of the optimal filter by Eq. (50). (8) If the filter is to be implemented by convolution, compute the inverse Fourier transform of $H_o(s)$ to produce the impulse response $h_o(t)$ of the optimum linear estimator.

If it is impossible or impractical to obtain samples of the noise-free signal and the input signal, one could assume a functional form for the correlation functions or the power spectra required in Eq. (50). For example, white noise has a constant power spectrum, and some functional form might be assumed for the power spectrum of the desired signal.

Uncorrelated Signal and Noise. The autocorrelation functions in Eq. (45) and the power spectra in Eq. (50) are somewhat difficult to visualize and interpret. The situation is improved considerably, however, if we assume that the desired signal and the noise are uncorrelated. By definition, this means that

$$\mathscr{E}\{s(t)n(t)\} = \mathscr{E}\{s(t)\}\mathscr{E}\{n(t)\} \tag{51}$$

We can transform the numerator of $H_o(s)$ [Eq. (50)] and write

$$R_{xs}(\tau) = \mathscr{E}\{x(t)s(t+\tau)\} = \mathscr{E}\{[s(t) + n(t)]s(t+\tau)\} \tag{52}$$

or

$$R_{xs}(\tau) = \mathscr{E}\{s(t)s(t+\tau)\} + \mathscr{E}\{n(t)s(t+\tau)\} \tag{53}$$

In view of Eq. (51) we can write

$$R_{xs}(\tau) = R_s(\tau) + \mathscr{E}\{n(t)\}\mathscr{E}\{s(t+\tau)\} = R_s(\tau) + \int_{-\infty}^{\infty} n(t)\,dt \int_{-\infty}^{\infty} s(t+\tau)\,dt \tag{54}$$

or

$$R_{xs}(\tau) = R_s(\tau) + N(0)S(0) \tag{55}$$

A similar exercise in the denominator of Eq. (50) produces

$$R_x(\tau) = R_s(\tau) + R_n(\tau) + 2S(0)N(0) \tag{56}$$

Then Eq. (50) becomes

$$H_o(s) = \frac{P_s(s) + N(0)S(0)\,\delta(s)}{P_s(s) + P_n(s) + 2N(0)S(0)\,\delta(s)} \tag{57}$$

or, ignoring zero frequency,

$$H_o(s) = \frac{P_s(s)}{P_s(s) + P_n(s)} \qquad s \neq 0 \tag{58}$$

Notice that if either the signal or noise has zero mean value, then Eq. (58) is valid for all frequencies, including zero. If both signal and noise have nonzero mean values, then

$$H_o(0) = \tfrac{1}{2} \tag{59}$$

Recall that Eq. (35) gives the mean-square error at the filter output. If we install the optimality condition of Eq. (44) in the third term of Eq. (35), it combines with the second term, leaving

$$\text{MSE}_o = R_s(0) - \int_{-\infty}^{\infty} h_o(\tau)R_{xs}(\tau)\,d\tau \tag{60}$$

a simple expression for the mean-square error of the optimal filter. With uncorrelated zero mean noise, Eq. (55) suggests we can replace $R_{xs}(\tau)$ with $R_s(\tau)$, which is the inverse Fourier transform of $P_s(s)$. Then Eq. (60) becomes

$$\text{MSE}_o = R_s(0) - \int_{-\infty}^{\infty} h_o(\tau) \mathcal{F}^{-1}\{P_s(s)\} \, d\tau \tag{61}$$

Writing out the inverse transformation and rearranging integrals produces

$$\text{MSE}_o = R_s(0) - \int_{-\infty}^{\infty} P_s(s) \int_{-\infty}^{\infty} h_o(\tau) e^{j2\pi s\tau} \, d\tau \, ds \tag{62}$$

Recognizing the first term and the second integral as Fourier transforms allows us to write

$$\text{MSE}_o = \int_{-\infty}^{\infty} P_s(s) \, ds - \int_{-\infty}^{\infty} P_s(s) H_o(-s) \, ds \tag{63}$$

Since the transfer function $H_o(s)$ is even, the minus sign on its argument can be ignored. We now substitute Eq. (58) and obtain

$$\text{MSE}_o = \int_{-\infty}^{\infty} P_s(s) \, ds - \int_{-\infty}^{\infty} P_s(s) \frac{P_s(s)}{P_s(s) + P_n(s)} \, ds \tag{64}$$

which may be rearranged to yield

$$\text{MSE}_o = \int_{-\infty}^{\infty} \frac{P_s(s) P_n(s)}{P_s(s) + P_n(s)} \, ds \tag{65}$$

the frequency domain expression for mean-square error in the uncorrelated case.

Figure 11-13 illustrates the frequency domain behavior of the Wiener filter in the uncorrelated case. Notice that the magnitude of the transfer function $H_o(s)$ is bounded by 0 and 1. Also, the transfer function decreases with a decrease in signal power spectrum or an increase in noise power spectrum. When the signal-to-noise ratio is high, the transfer function approaches unity, passing all the energy in the signal. When the signal-to-noise ratio is low, however, the transfer function declines toward zero.

When we assumed that our knowledge of $s(t)$ and $n(t)$ was limited to power spectra, we admitted we had no phase information. Notice that the transfer function $H_o(s)$ is real and even and thus has no phase shift.

The actual mean-square error at the output, which indicates how successfully the filter is able to recover the signal from the contaminating noise, is given by Eq. (65). The integrand is plotted in Figure 11-13. Notice that the contributions to MSE occur in frequency bands where both the signal and noise power spectra are nonzero.

Figure 11-14 illustrates the case where signal and noise are separable in the frequency domain. In this case, the Wiener estimator passes the signal in its entirety and discriminates completely against the noise.

The case of a band-limited signal imbedded in white noise is illustrated in Figure 11-15. If the signal power spectrum is constant, the mean-square error is proportional to the signal bandwidth, $s_2 - s_1$.

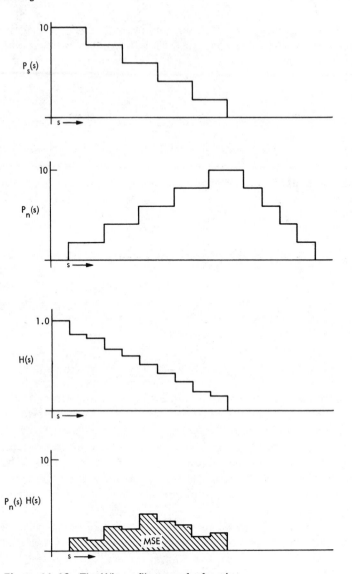

Figure 11-13 The Wiener filter transfer function

If the signal-to-noise ratio is low, Eq. (65) reduces to approximately

$$\text{MSE}_o \approx \int_{-\infty}^{\infty} P_s(s) \, ds = \int_{-\infty}^{\infty} |S(s)|^2 \, ds \tag{66}$$

which is, by Rayleigh's theorem,

$$\text{MSE}_o \approx \int_{-\infty}^{\infty} s^2(t) \, dt = R_s(0) = \text{energy} \tag{67}$$

Thus, in this case, the mean-square error is proportional to the energy in the signal.

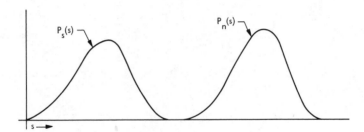

Figure 11-14   Separable signal and noise

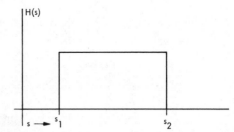

Figure 11-15   Band-limited signal

Wiener Deconvolution. Ordinary deconvolution, as previously discussed, does not account for noise. Thus deconvolution transfer functions, which often take on extremely large magnitudes at high frequencies, are not practical when noise is present. Figure 11-16 illustrates the situation when deconvolution is followed by a Wiener filter. The

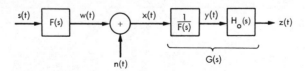

Figure 11-16   Wiener deconvolution

desired signal $s(t)$ is first degraded by a linear system with impulse response $f(t)$. The output of the filter is corrupted by a noise source $n(t)$ to form the observed signal $x(t)$. It is desired to design a linear filter $g(t)$ that will simultaneously deconvolve the undesired impulse response $f(t)$ and discriminate against the noise. In Figure 11-16, $g(t)$ is illustrated as a concatenation of a deconvolution filter and a Wiener filter with impulse response $h_o(t)$.

    Since the deconvolution filter is known, it only remains to determine the impulse response $h_o(t)$ before combining the two linear filters to produce $g(t)$.

    The configuration in Figure 11-16 implies that the spectrum of the observed signal is given by

$$X(s) = F(s)S(s) + N(s) \tag{68}$$

Furthermore, assuming $F(s)$ has no zeros, the input spectrum to the Wiener filter is

$$Y(s) = S(s) + \frac{N(s)}{F(s)} = S(s) + K(s) \tag{69}$$

Equation (58) implies, for uncorrelated signal and noise sources, that the Wiener filter transfer function is given by

$$H_o(s) = \frac{P_s(s)}{P_s(s) + P_k(s)} = \frac{|S(s)|^2}{|S(s)|^2 + \left|\frac{N(s)}{F(s)}\right|^2} \tag{70}$$

Thus the transfer function $G(s)$ of the optimal deconvolution filter in the mean-square sense is given by

$$G(s) = \frac{H_o(s)}{F(s)} = \frac{1}{F(s)}\left[\frac{P_s(s)}{P_s(s) + P_k(s)}\right] = \frac{F^*(s)P_s(s)}{|F(s)|^2 P_s(s) + P_n(s)} \tag{71}$$

The Matched Detector

We now consider a filter that is optimal for a different purpose. Whereas the Wiener filter is optimal for recovering an unknown signal from noise, the matched detector is optimal for locating a known signal in a noisy background (Refs. 4, 5, 6). The matched filter is designed to "detect" the occurrence of a signal of prescribed form in a noisy background. This contrasts with the Wiener filter, which is designed to "estimate" what the signal was before it was contaminated with noise.

The model for the development of the matched detector is shown in Figure 11-17. A signal $m(t)$ is contaminated by additive noise $n(t)$ to form the observed signal

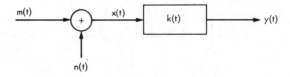

Figure 11-17   Model for the matched detector

$x(t)$, which is input to the linear filter having impulse response $k(t)$, producing the output $y(t)$. We wish to use the filter to detect the presence or absence of $m(t)$. That is to say, we shall monitor $y(t)$ to detect the occurrence of $m(t)$, a specified signal of known form. We wish to select the impulse response $k(t)$ to make this job easy.

For the system in Figure 11-17,

$$y(t) = [m(t) + n(t)] * k(t) = m(t) * k(t) + n(t) * k(t) \tag{72}$$

which means that the system in Figure 11-18 is equivalent. It makes no difference whether $m(t)$ and $n(t)$ are summed before or after passing through the filter. We define the component outputs as

$$u(t) = m(t) * k(t) \quad \text{and} \quad v(t) = n(t) * k(t) \tag{73}$$

Now $u(t)$ is the filtered signal and $v(t)$ is the filtered noise.

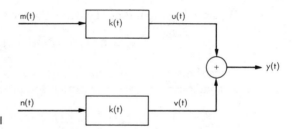

Figure 11-18   An equivalent model

As with the Wiener filter, we must first stipulate what knowledge we have about the signal and noise and establish a criterion of optimality. Suppose we know the functional form of $m(t)$, but we do not know at what point in time it occurs. The classical application of the matched detector has been the detection of reflected radar pulses. In this case, the reflected pulse, identical to the transmitted pulse, is known in form but not in time of arrival. In digital image processing, the matched detector is useful for locating known features in a noisy image.

As with the Wiener filter, we shall assume that the noise is an ergodic random variable of known power spectrum. We wish to design $k(t)$ so that, by observing the output, we may best be able to detect the signal when it does occur.

Optimality Criterion. As a measure of performance for the filter, we shall use the average signal-to-noise power ratio at the output, evaluated at time t_o,

$$\rho = \frac{\mathcal{E}\{u^2(t_0)\}}{\mathcal{E}\{v^2(t_0)\}} \tag{74}$$

assuming that the signal $m(t)$ occurs at $t = 0$.

Since the system is shift invariant, if the signal should choose to occur at some time t_1, then the signal to noise power ratio at the output will be maximized at $t_0 + t_1$. Thus t_0 allows us to introduce offset, if desired, between the occurrences of the signal and the output pulse. Ordinarily, for filter design purposes, the signal $m(t)$ is some relatively narrow function located at the origin, and t_0 is set to zero. Then, when $m(t - t_1)$ arrives at time t_1, the amplitude of the filter output becomes large. Before and after, in the absence of signal, the output amplitude is relatively small.

Clearly, if ρ is large, the amplitude of the output $y(t)$ will be highly dependent on the presence or absence of $m(t)$ and relatively insensitive to fluctuations in the noise $n(t)$. Thus, as a criterion for optimality of $k(t)$, we shall choose the maximization of ρ.

It is important to note that this criterion of optimality makes no guarantee that the output $y(t)$ will resemble $m(t)$. However, since we know the functional form of $m(t)$, we are not interested in fidelity of reproduction, as we were in the case of the Wiener filter. Instead, we want the output to be large when $m(t)$ is present and small when it is not.

Since $u(t)$ is deterministic, we can drop the expectation operator in the numerator and write Eq. (74) as

$$\rho = \frac{u^2(t_0)}{\mathcal{E}\{v^2(t)\}} = \frac{[m(t) * k(t)]^2}{\mathcal{E}\{[n(t) * k(t)]^2\}} = \frac{[\mathcal{F}^{-1}\{M(s)K(s)\}]^2}{\mathcal{E}\{[n(t) * k(t)]^2\}} = \frac{\rho_n}{\rho_d} \tag{75}$$

We begin by expanding the denominator as a product of two convolution integrals:

$$\rho_d = \mathcal{E}\left\{\int_{-\infty}^{\infty} k(q)n(t-q)\,dq \int_{-\infty}^{\infty} k(\tau)n(t-\tau)\,d\tau\right\} \tag{76}$$

Since the expectation is an integral over time and the impulse response $k(t)$ is not a random signal, we can rearrange the integrals in Eq. (76) to produce

$$\rho_d = \int_{-\infty}^{\infty}\int_{-\infty}^{\infty} k(q)k(\tau)\mathcal{E}\{n(t-q)n(t-\tau)\}\,dq\,d\tau \tag{77}$$

We recognize the expectation factor within the integral as the autocorrelation function $R_n(\tau - q)$ of the noise, which is, in turn, the inverse Fourier transform of $P_n(s)$, the noise power spectrum. Thus

$$\mathcal{E}\{n(t-q)n(t-\tau)\} = R_n(\tau - q) = \int_{-\infty}^{\infty} P_n(s)e^{j2\pi s(\tau - q)}\,ds \tag{78}$$

which makes the denominator of ρ

$$\rho_d = \int_{-\infty}^{\infty}\int_{-\infty}^{\infty} k(q)k(\tau) \int_{-\infty}^{\infty} P_n(s)e^{j2\pi s(\tau - q)}\,ds\,dq\,d\tau \tag{79}$$

Now we can factor the exponential and rearrange the integrals to produce

$$\rho_d = \int_{-\infty}^{\infty} P_n(s)\left[\int_{-\infty}^{\infty} k(q)e^{-j2\pi sq}\,dq \int_{-\infty}^{\infty} k(\tau)e^{j2\pi s\tau}\,d\tau\right]ds \tag{80}$$

The term in brackets is the product of two inverse Fourier transforms, namely $K(s)K(-s)$. Furthermore, since the impulse response $k(t)$ is a real function, the transfer function $K(s)$ is Hermite and $K(-s) = K^*(s)$. Thus the term in brackets reduces to

$$K(s)K(-s) = K(s)K^*(s) = |K(s)|^2 \tag{81}$$

Substituting this into Eq. (75) and writing out the Fourier transform in the numerator allows us to write the signal-to-noise power ratio as

$$\rho = \frac{\left[\int_{-\infty}^{\infty} K(s)M(s)e^{j2\pi st_0}\,ds\right]^2}{\int_{-\infty}^{\infty} |K(s)|^2 P_n(s)\,ds} \tag{82}$$

It is this expression we wish to maximize. As with the Wiener filter, we must select a function to optimize a quantity.

Schwartz's Inequality. In this case, we shall make use of Schwartz's inequality. This is the mathematical result which states that

$$\int f^2(t)\,dt \int g^2(t)\,dt \geq \left[\int f(t)g(t)\,dt\right]^2 \tag{83}$$

where $f(t)$ and $g(t)$ are arbitrary real functions and the integration is performed between arbitrary limits. Our approach is to define the functions $f(t)$ and $g(t)$ in terms of factors appearing in Eq. (82) and obtain an inequality involving ρ. We shall then assume a form for the transfer function and show that it maximizes ρ. First, however, we shall prove the Schwartz inequality.

We first define a function of the variable λ by writing

$$Q(\lambda) = \int [\lambda f(t) + g(t)]^2 \, dt \geq 0 \tag{84}$$

Expanding the integrand and collecting terms produces

$$\int [\lambda f(t) + g(t)]^2 \, dt = \lambda^2 \int f^2(t) \, dt + 2\lambda \int f(t)g(t) \, dt + \int g^2(t) \, dt \geq 0 \tag{85}$$

Equation (85) is a quadratic equation in the variable λ. Therefore,

$$\left[2 \int f(t)g(t) \, dt \right]^2 - 4 \int f^2(t) \, dt \int g^2(t) \, dt \leq 0 \tag{86}$$

or

$$\left[\int f(t)g(t) \, dt \right]^2 \leq \int f^2(t) \, dt \int g^2(t) \, dt \tag{87}$$

thus proving Eq. (83).

A Necessary Condition. We shall now use Schwartz's inequality to obtain a condition upon the signal-to-noise ratio ρ. First we define two functions

$$f(s) = e^{j2\pi s t_0} K(s) \sqrt{P_n(s)} \tag{88}$$

and

$$g(s) = \frac{M(s)}{\sqrt{P_n(s)}} \tag{89}$$

Their product is

$$f(s)g(s) = e^{j2\pi s t_0} K(s)M(s) \tag{90}$$

and their squared magnitudes are

$$|f(s)|^2 = |K(s)|^2 P_n(s) \tag{91}$$

and

$$|g(s)|^2 = \frac{|M(s)|^2}{P_n(s)} \tag{92}$$

If we substitute the functions defined in Eqs. (88) and (89) into Schwartz's inequality, using s as the variable of integration, we obtain

$$\left| \int_{-\infty}^{\infty} e^{j2\pi s t_0} K(s)M(s) \, ds \right|^2 \leq \left[\int_{-\infty}^{\infty} |K(s)|^2 P_n(s) \, ds \right]\left[\int_{-\infty}^{\infty} \frac{|M(s)|^2}{P_n(s)} \, ds \right] \tag{93}$$

If we divide both sides by

$$\int_{\infty}^{\infty} |K(s)|^2 P_n(s) \, ds \tag{94}$$

we are left with

$$\frac{\left| \int_{-\infty}^{\infty} e^{j2\pi s t_0} K(s)M(s) \, ds \right|^2}{\int_{-\infty}^{\infty} |K(s)|^2 P_n(s) \, ds} \leq \frac{\int_{-\infty}^{\infty} |K(s)|^2 P_n(s) \, ds \int_{-\infty}^{\infty} \frac{|M(s)|^2}{P_n(s)} \, ds}{\int_{-\infty}^{\infty} |K(s)|^2 P_n(s) \, ds} \tag{95}$$

Recalling Eq. (82), we recognize the left side of the inequality as ρ. Furthermore, the denominator on the right-hand side cancels the first term of the numerator, leaving us with

$$\rho \leq \int_{-\infty}^{\infty} \frac{|M(s)|^2}{P_n(s)} \, ds \tag{96}$$

a relatively simple upper bound on ρ.

Schwartz's inequality has led us to Eq. (96), which states that ρ is less than or equal to an expression involving the power spectrum of the signal and that of the noise. Clearly, ρ will be maximized when equality holds in Eq. (96). Since we want ρ to be as large as possible, we take

$$\rho_{\max} = \int_{-\infty}^{\infty} \frac{|M(s)|^2}{P_n(s)} \, ds \tag{97}$$

as a necessary condition to maximize ρ.

The Transfer Function. We shall now assume a particular form for $K(s)$ and show that it does indeed maximize ρ. We assume that the optimal transfer function is given by

$$K_o(s) = Ce^{-j2\pi st_0} \frac{M^*(s)}{P_n(s)} \tag{98}$$

Substituting that assumed form into the general expression for ρ [Eq. (82)] produces

$$\rho = \frac{\left| \int_{-\infty}^{\infty} Ce^{-j2\pi st_0} e^{j2\pi st_0} \frac{M^*(s)}{P_n(s)} M(s) \, ds \right|^2}{\int_{-\infty}^{\infty} C^2 \frac{M^*(s)M(s)}{P_n^*(s)P_n(s)} P_n(s) \, ds} \tag{99}$$

Canceling the constants, the exponentials, and the $P_n(s)$ in the denominator reduces the expression to

$$\rho = \frac{\left| \int_{-\infty}^{\infty} \frac{M^*(s)}{P_n(s)} M(s) \, ds \right|^2}{\int_{-\infty}^{\infty} \frac{M^*(s)M(s)}{P_n^*(s)} \, ds} \tag{100}$$

Since $P_n(s)$ is real and even, $P_n^*(s) = P_n(s)$, and the numerator is the square of the denominator. Now ρ reduces to

$$\rho = \int_{-\infty}^{\infty} \frac{|M(s)|^2}{P_n(s)} \, ds = \rho_{\max} \tag{101}$$

which satisfies the necessary condition for optimality of Eq. (97). This means that the transfer function assumed in Eq. (98) does indeed maximize the signal-to-noise power ratio at the output of the filter at time t_0 when the signal occurs at $t = 0$.

Notice that the magnitude of the transfer function

$$|K_o(s)| = |C| \frac{|M(s)|}{P_n(s)} \tag{102}$$

is proportional to the signal amplitude to noise power ratio as a function of frequency.

The arbitrary constant C is not surprising, since we originally endeavored to maximize a ratio at the output.

Examples of the Matched Detector

In order to develop insight into the operation of the matched detector, we consider some illustrative examples under particular conditions.

White Noise. In the first case, let us assume that the noise $n(t)$ is spectrally white; that is,

$$P_n(s) = N_0^2 \tag{103}$$

Since C in Eq. (98) is an arbitrary constant, we may set it equal to N_0^2, in which case the matched detector becomes

$$K_o(s) = M^*(s)e^{-j2\pi s t_0} \tag{104}$$

In the time domain, the impulse response is

$$k_o(t) = \mathcal{F}^{-1}\{K(s)\} = \int_{-\infty}^{\infty} M^*(s)e^{-j2\pi s t_0}e^{j2\pi s t}\,ds \tag{105}$$

Since $m(t)$ is real, $M(s)$ is Hermite and

$$k_o(t) = \int_{-\infty}^{\infty} M(-s)e^{j2\pi(-s)(t_0-t)}\,ds = m(t_0 - t) \tag{106}$$

Thus the impulse response for the white noise case is merely a reflected and shifted version of the signal itself. This filter is said to be "matched" to the signal (Ref. 5), and this name has become attached to the more general detector of Eq. (98).

The signal component of the output is given by

$$u(t) = m(t) * k_o(t) = \int_{-\infty}^{\infty} m(\tau)m(t_0 - t + \tau)\,d\tau = R_m(t_0 - t) \tag{107}$$

and the noise component by

$$v(t) = n(t) * k_o(t) = \int_{-\infty}^{\infty} n(\tau)m(t_0 - t + \tau)\,d\tau = R_{mn}(t_0 - t) \tag{108}$$

Since $k_o(t)$ in Eq. (106) is just the reflected signal we are trying to detect, the matched filter $k_o(t)$ is merely a cross-correlator. It cross-correlates the incoming signal plus noise with the known form of the desired signal. The output is

$$y(t) = u(t) + v(t) = R_m(t_0 - t) + R_{mn}(t_0 - t) \tag{109}$$

which has an autocorrelation component only when the signal is present and always a cross-correlation component. If the correlation between the signal and noise is small, then $R_{mn}(\tau)$ is small for all values of τ and the noise component at the output is small. Furthermore, the autocorrelation function $R_m(\tau)$ has a peak at $\tau = 0$ so, clearly,

$$\rho = \frac{u^2(t_0)}{\mathcal{E}\{v^2(t)\}} \tag{110}$$

is large at $t = t_0$ as desired.

The Rectangular Pulse Detector. As a particular example, suppose $m(t) = \Pi(t)$; that is, the matched filter is designed to detect a rectangular pulse in white noise. Suppose also that the input is $x(t) = s(t) + n(t)$, where $s(t) = \Pi(t - T)$ and $n(t)$ is white noise. Recall that the autocorrelation function of the rectangular pulse is given by

$$R_\pi(\tau) = \Pi(t) * \Pi(t) = \Lambda(\tau) \tag{111}$$

Now the output of the filter is

$$y(t) = R_{xm}(t) = R_{sm}(t) + R_{mn}(t) = \Lambda(t - T) + R_{mn}(t) \tag{112}$$

So, for the system shown in Figure 11-19, the components of the input and output are presented in Figure 11-20.

From Figure 11-20, we see how the matched filter discriminates against the noise while responding to the signal. The output has a peak at $t = T$, the time at which the input pulse occurs, but takes on relatively small amplitude otherwise. Thus a simple examination of the output signal indicates when the input pulse occurs.

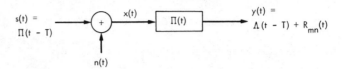

Figure 11-19 Rectangular pulse detector

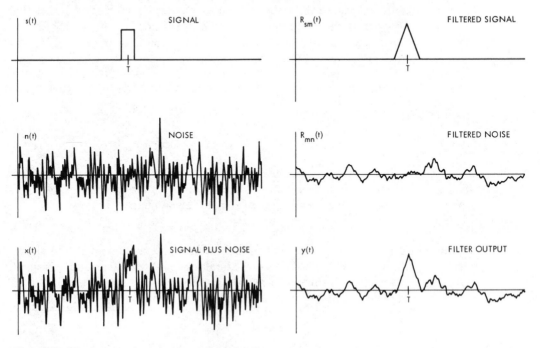

Figure 11-20 Input and output component signals

Notice that the form or shape of the signal is not preserved by the matched detector as it was with the Wiener estimator. This is because we designed the filter to detect the existence of a particular known input signal rather than to estimate its noise-free shape.

Comparison of the Wiener Estimator and the Matched Detector

The Wiener estimator and the matched detector are both optimal filters designed to do specific jobs. It is instructive to compare the two filters. Recall from Eq. (58) that for uncorrelated signal and noise the Wiener estimator transfer function is given by

$$H_o(s) = \frac{P_s(s)}{P_s(s) + P_n(s)} \tag{113}$$

and the mean-square error one can expect when using this filter is, from Eq. (65),

$$MSE_o = \int_{-\infty}^{\infty} \frac{P_s(s)P_n(s)}{P_s(s) + P_n(s)} \, ds \tag{114}$$

If we let $c = 1$, and $t_0 = 0$, the matched detector transfer function is

$$K_o(s) = \frac{S^*(s)}{P_n(s)} \tag{115}$$

and the signal-to-noise power ratio at its output is

$$\rho_{\max} = \int_{-\infty}^{\infty} \frac{P_s(s)}{P_n(s)} \, ds \tag{116}$$

First, notice that while $H_o(s)$ is real and even, hence containing no phase information, $K_o(s)$ is Hermite and does contain phase information. Notice also that $H_o(s)$ is bounded between 0 and $+1$. This means it can never amplify spectral components of the input signal. However $K_o(s)$ has neither positive nor negative bound. Therefore its frequency domain behavior is much less constrained.

Let us define the signal-to-noise power ratio as a function of frequency by

$$R(s) = \frac{|S(s)|^2}{|N(s)|^2} = \frac{P_s(s)}{P_n(s)} \tag{117}$$

In terms of this function, the magnitude of the matched detector transfer function is given by

$$|K_o(s)| = \frac{R(s)}{|S(s)|} = \frac{\sqrt{R(s)}}{|N(s)|} \tag{118}$$

and the signal-to-noise ratio by

$$\rho_{\max} = \int_{-\infty}^{\infty} R(s) \, ds \tag{119}$$

The Wiener filter transfer function is

$$|H_o(s)| = H_o(s) = \frac{R(s)}{1 + R(s)} \tag{120}$$

and the mean-square error is given by

$$\text{MSE}_o = \int_{-\infty}^{\infty} \frac{R(s)P_n(s)}{1 + R(s)} \tag{121}$$

Comparing Eqs. (120) and (121), we see that

$$\text{MSE}_o = \int_{-\infty}^{\infty} P_n(s)H_o(s)\,ds \tag{122}$$

which indicates that the mean-square error is just the noise power that passes through the filter accumulated over all frequencies.

In a sense, estimation is a more difficult task than detection. There are two reasons for this. First, we ask an estimator to recover the signal at all points in time, whereas we ask the detector only to determine when the signal occurs. Second, we have more *a priori* information in a detection problem in that we know the form of the signal exactly instead of having only its power spectrum. Since we are asking a detector to do less with more information, we can expect better performance under the same conditions.

Whether one uses a detector or an estimator is dictated by the problem. Since they are designed for different jobs, they usually do not compete for consideration. Nevertheless, it is instructive to compare the behavior of the two filters under similar conditions. Figure 11-21 presents a computer simulation that illustrates both the Wiener estimator and the matched detector when the signal is a Gaussian pulse embedded in white random noise. In this case, the signal-to-noise ratio is on the order of unity.

Both the estimator [Eq. (113)] and the detector [Eq. (115)] are lowpass filters in

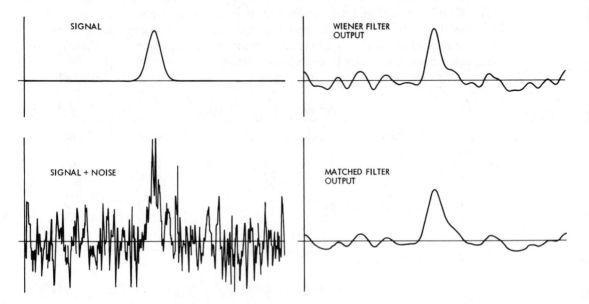

Figure 11-21 Comparison of the Wiener and matched filters

this situation, but they differ somewhat in form. The detector output clearly shows a peak at the point where the input pulse occurs. The estimator recovers the pulse from the noise, but not without residual error. The low-frequency components of the noise penetrate the Wiener filter and prevent exact recovery. One would expect better performance from both filters with improved signal-to-noise ratio, and conversely.

A Practical Example

We conclude this chapter with an example that illustrates how optimal filter theory can guide the design of practical filters. Figure 11-22 shows a digitized X ray of a tube filled with X-ray absorbing dye. This models angiography, the diagnostic technique in

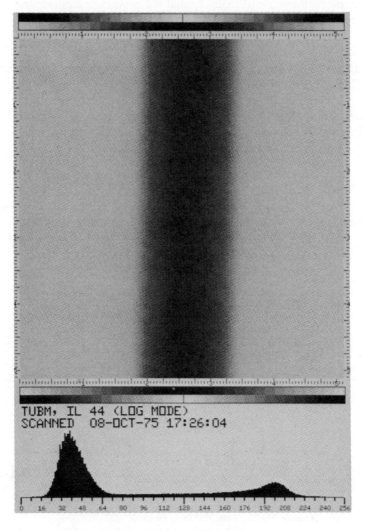

Figure 11-22   Digitized angiogram of a smooth tube

which dye is injected into blood vessels during X-ray exposure. Here the smooth tube substitutes for the vessel. The goal in this example is a processing technique that will find the tube edges in the noisy image of Figure 11-22 and reliably measure tube diameter all along its length. Such a technique is useful for quantifying the blood vessel narrowing which accompanies atherosclerosis and produces heart attacks (Ref. 7).

Since this an edge detection problem, the matched detector seems a natural choice. In this example, however, we shall pose the problem somewhat differently. We shall assume the vessel edges occur, on each image line, at the two points of steepest slope, and locate those by differentiation. Before differentiating, however, we shall use a Wiener filter to estimate the noise-free image. Furthermore, we shall process each line individually so that the procedure can respond to rapid changes in width, should they occur.

Figure 11-23 shows a gray level plot of one line $f_i(x)$ from Figure 11-22. The evident noise is common in radiography, due primarily to film grain and photon statistics in the illuminating beam. Clearly, differentiating this curve would not produce reliable peaks at the inflection points.

Assuming uncorrelated signal $s(x)$ and noise $n(x)$, the specification of the Wiener filter [Eq. (58)] requires the power spectrum of the signal and that of the noise. We can estimate the signal power spectrum by line averaging since, with a smooth tube, all lines $f_i(x)$ should be identical in the absence of noise. Thus

$$P_s(s) = |\mathcal{F}\{s(x)\}|^2 \approx \left|\mathcal{F}\left\{\frac{1}{N}\sum_{1=i}^{N} f_i(x)\right\}\right|^2 \tag{123}$$

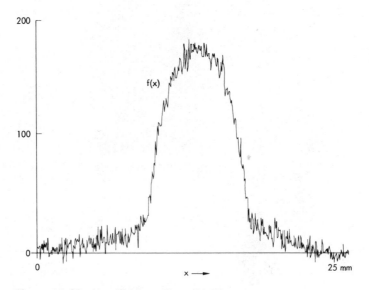

Figure 11-23   Line 100 from Figure 11-22

will reduce the noise by the factor $1/\sqrt{N}$. Figure 11-24 shows the result of averaging 60 lines in Figure 11-22, and the resulting signal amplitude spectrum.

Once the signal has been estimated, the noise power spectrum can be estimated from Figure 11-22 using line by line power spectrum averaging after signal subtraction, that is

$$P_n(s) \approx \frac{1}{N} \sum_{i=1}^{N} |\mathfrak{F}\{f_i(x) - s(x)\}|^2 \tag{124}$$

In this study Eq. (124) showed the noise power spectrum to be essentially constant with frequency (white noise).

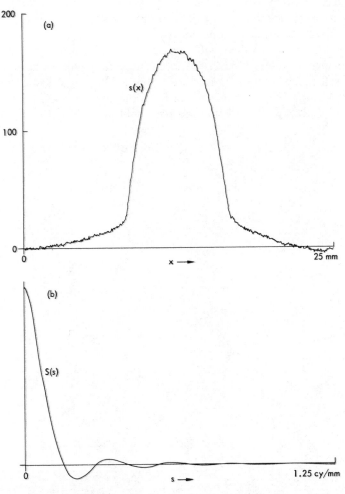

Figure 11-24   (a) Noise-free signal estimate obtained by line averaging in Figure 11-22; (b) Fourier amplitude spectrum of (a)

Figure 11-25(a) shows the Wiener filter transfer function $H_o(s)$ computed by Eq. (58). The transfer function takes on values near unity at the signal dominated low frequencies, and tends to zero at high frequencies. We could inverse transform the transfer function in Figure 11-25(a) to obtain the impulse response for pre-differentiation smoothing. There are, however, some practical considerations worthy of note.

The notches in the transfer function of Figure 11-25(a) are produced by the zero crossings in the signal spectrum [Figure 11-24(b)]. By the similarity theorem, the position of these notches will shift with changes in vessel width. This points out that our signal is not actually an ergodic random process as the Wiener filter development assumes. The member functions in the signal ensemble correspond to vessels of different width and thus do not all have identical power spectra. As it happens, we are forced to violate one of the assumptions on which the Wiener filter is based. We shall nevertheless proceed, acting in the belief that a "near-optimal" technique will prove an adequate substitute for optimality.

If we include the troublesome notches, our filter will be quite sensitive to slight changes in vessel width. We choose instead to ignore the notches by fitting a smooth envelope to the transfer function. Figure 11-25(b) shows a smooth approximation $\tilde{H}(s)$ to the Wiener filter transfer function. This function was chosen because of two desirable properties. First, it is a reasonable approximation to the envelope of Figure

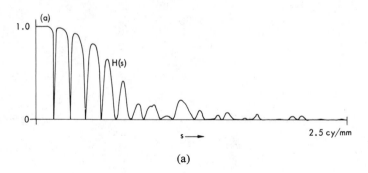

(a)

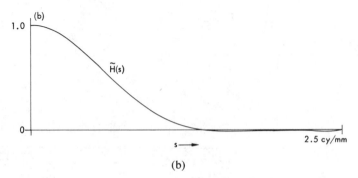

(b)

Figure 11-25 (a) Wiener filter transfer function; (b) Smooth approximation to (a)

11-25(a). Second, its impulse response renders digital convolution quite an efficient computation.

Figure 11-26 shows the corresponding impulse response $\tilde{h}(x)$, which is piecewise parabolic, and its piecewise linear first derivative $\tilde{h}'(x)$. Since differentiation commutes with convolution, using the latter function combines smoothing and differentiation into one step. Furthermore, digital convolution using a piecewise linear impulse response can be programmed to execute very efficiently (Ref. 8).

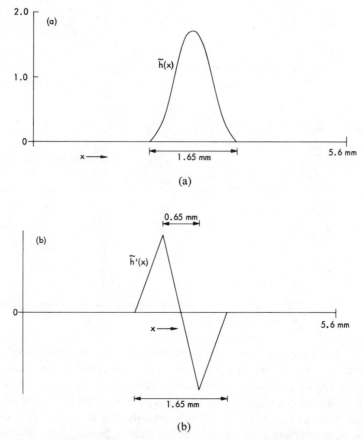

Figure 11-26   (a) Impulse response of Figure 11-25(b); (b) Derivative of (a)

Figure 11-27 shows the results of using the two impulse responses in Figure 11-26 on the image line in Figure 11-23. The first produces smoothing for noise reduction only while the second combines smoothing with differentiation. In this case the degree of noise reduction is gratifying. Notice also that the inflection points in the upper curve give rise to distinct peaks in the lower curve, suggesting that vessel edge detection is now a simple task.

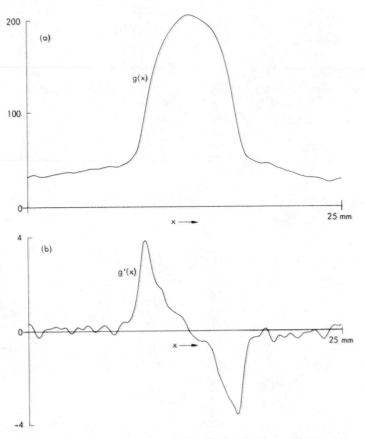

Figure 11-27 (a) Result of smoothing the line in Figure 11-23 with $\tilde{h}(x)$; (b) with $\tilde{h}'(x)$

The piecewise linear impulse response $\tilde{h}'(x)$ is a computationally efficient approximation to the differentiating Wiener filter for this application. Even though our signal is non-ergodic, the notch-free transfer function $\tilde{H}(s)$ should be rather well-behaved under suboptimal conditions since it has no abrupt behavior in the frequency domain. Furthermore, Figure 11-27 strongly suggests that we have a comfortable solution to this edge detection problem. The differentiating Wiener filter designed on the smooth tube has proved useful on routine angiograms (Ref. 8).

SUMMARY OF IMPORTANT POINTS

1. A high-frequency enhancement filter impulse response can be designed as the sum of a narrow positive pulse and a broad negative pulse.

2. The transfer function of such a filter approaches a maximum value that is equal to the area under the narrow positive pulse.

3. The transfer function of such a filter has a zero-frequency response equal to the difference of the areas under the two component pulses.

4. The zero-frequency response of a filter determines how the contrast of large features is affected.

5. Filters designed for ease of computation rather than for optimal performance are likely to introduce artifacts into an image.

6. An ergodic random process is a signal whose known power spectrum and auto-correlation function represent all the available knowledge.

7. The Wiener estimator is optimal, in the mean-square error sense, for recovering a signal of known power spectrum from additive noise.

8. The Wiener filter transfer function takes on values near unity in frequency bands of high signal-to-noise ratio and near zero in bands dominated by noise.

9. The matched detector is optimal for detecting the occurrence of a known signal in a background of additive noise.

10. In the case of white noise, the matched filter correlates the input with the known signal.

11. The Wiener filter transfer function is real and even and bounded by zero and unity.

12. The matched filter transfer function is complex and Hermite and in general unbounded.

REFERENCES

1. N. WIENER, *Extrapolation, Interpolation, and Smoothing of Stationary Time Series*, John Wiley & Sons, New York, 1949.

2. W. B. DAVENPORT and W. L. ROOT, *An Introduction to the Theory of Random Signals and Noise*, McGraw-Hill Book Company, New York, 1958.

3. Y. W. LEE, *Statistical Theory of Communication*, John Wiley & Sons, New York, 1960.

4. L. A. WAINSTEIN and V. D. ZUBAKOV, *Extraction of Signals From Noise*, Prentice-Hall, Inc., Englewood Cliffs, New Jersey, 1962.

5. G. L. TURIN, "An Introduction to Matched Filters," *IRE Transactions on Information Theory*, 311–329, June 1960.

6. D. MIDDLETON, "On New Classes of Matched Filters and Generalizations of the Matched Filter Concept," *IRE Transactions on Information Theory*, 349–360, June 1960.

7. E. S. BECKENBACH, R. H. SELZER, D. W. CRAWFORD, S. H. BROOKS, and D. H. BLANKEN-HORN, "Computer Tracking and Measurement of Blood Vessel Shadows from Arteriograms," *Medical Instrumentation*, **8**, No. 5, September–October, 1974.

8. K. R. CASTLEMAN, R. H. SELZER, and D. H. BLANKENHORN, "Vessel Edge Detection in Angiograms: An Application of the Wiener Filter," in J. K. AGGARWAL, ed., *Digital Signal Processing*, Point Lobos Press, 13000 Raymer St., No. Hollywood, California 91605, 1979.

PROCESSING SAMPLED DATA

▊▊

INTRODUCTION

In previous chapters, we have discussed digital image processing without particular attention to the effects of sampling, something that is inherent in digital processing. We have instead been working under the assumption that sampling, done properly, will not invalidate the results obtained from continuous function analysis. If we had approached the question of sampling earlier, the discussion would necessarily have involved considerable laying of groundwork before sampling effects could be described. In preceding chapters, we have developed powerful tools that now allow us to approach sampling in a concise and effective manner.

In this chapter, we investigate the ramifications of sampling continuous images and of processing sampled data. In particular, we shall address the following questions: (1) To what extent does sampling cause loss of information and what information is lost? (2) Once a continuous function has been sampled, can it be recovered completely? (3) How finely must we sample a signal in order to preserve it? (4) What effect does sampling have upon the spectrum of a function? (5) If we treat a sampled signal as if it were continuous, what assumptions, approximations, and errors are involved?

SAMPLING

Before we can describe quantitatively the effects of sampling, we must establish a mathematical procedure for modeling the sampling process. To do this, we shall use another special function.

The Shah Function

A valuable tool for modeling the sampling process is the infinite impulse train, $III(x)$, pronounced "Shah of x" and defined by

$$III(x) = \sum_{n=-\infty}^{\infty} \delta(x - n) \tag{1}$$

It is a series of unit amplitude impulses that occur at unit spacing along the x-axis. Much to our good fortune, the Shah function is its own Fourier transform (Refs. 1, 2).

$$\mathcal{F}\{III(x)\} = III(s) \tag{2}$$

We shall use this function to model the process of sampling a continuous signal.

Behavior of the Shah Function under Similarity. If we substitute the similarity theorem

$$\mathcal{F}\{f(ax)\} = \frac{1}{|a|} F\left(\frac{s}{a}\right) \tag{3}$$

into Eq. (2), we obtain

$$\mathcal{F}\left\{III\left(\frac{x}{\tau}\right)\right\} = \tau III(\tau s) \tag{4}$$

where the spectrum is a train of impulses spaced every $1/\tau$ along the s-axis (Figure 12-1).

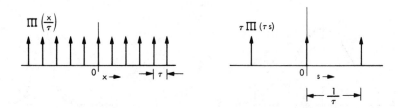

Figure 12-1 The Shah function and its spectrum

Recall that the impulse has, under similarity, the curious property

$$\delta(ax) = \frac{1}{|a|} \delta(x) \tag{5}$$

Since $III(x)$ is an infinite train of equally spaced impulses [Eq. (1)], it also exhibits curious behavior under stretching and compression. In particular,

$$III(ax) = \sum_{n=-\infty}^{\infty} \delta(ax - n) = \sum_{n=-\infty}^{\infty} \delta\left[a\left(x - \frac{n}{a}\right)\right] = \sum_{n=-\infty}^{\infty} \frac{1}{|a|} \delta\left(x - \frac{n}{a}\right) \tag{6}$$

227

which means that

$$\text{III}(ax) = \frac{1}{|a|} \sum_{n=-\infty}^{\infty} \delta\left(x - \frac{n}{a}\right) \qquad (7)$$

If we let $a = 1/\tau$, we have

$$\text{III}\left(\frac{x}{\tau}\right) = \tau \sum_{n=-\infty}^{\infty} \delta(x - n\tau) \qquad (8)$$

or impulses spaced every τ. Notice that spacing the impulses every τ rather than at unit intervals multiplies the strength of the impulses by the factor τ. Transforming Eq. (8) yields

$$\mathscr{F}\left\{\text{III}\left(\frac{x}{\tau}\right)\right\} = \tau\text{III}(\tau s) = \sum_{n=-\infty}^{\infty} \delta\left(s - \frac{n}{\tau}\right) \qquad (9)$$

These last two equations indicate that a train of impulses of strength τ spaced every τ in the time domain produces a train of unit impulses spaced every $1/\tau$ in the frequency domain. We could, of course, divide Eq. (8) by τ to have unit strength impulses in the time domain and, correspondingly, impulses of strength $1/\tau$ in the frequency domain.

Sampling with the Shah Function

Suppose a function $f(x)$ is band-limited at a frequency s_0; that is

$$F(s) = 0 \qquad |s| \geq s_0 \qquad (10)$$

This is shown in Figure 12-2. If we sample $f(x)$ at equal intervals τ, we destroy $f(x)$ everywhere except at $x = n\tau$. We can model the sampling process as simply multiplying the function $f(x)$ by $\text{III}(x/\tau)$ to form $g(x)$, the sampled function. This destroys the

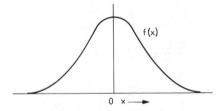
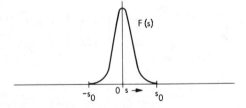

Figure 12-2 A band-limited function

function between sample points by driving it to zero and yet preserves the value of the function at the sample points in the strength of the resulting impulses. This process is illustrated in Figure 12-3. This model for sampling is perhaps not as straightforward as other approaches, but mathematical convenience makes it the method of choice.

We now examine what sampling does to the spectrum of $f(x)$. The convolution theorem dictates that when we multiply $f(x)$ by $\text{III}(x/\tau)$, we convolve $F(s)$ with $\tau\text{III}(\tau s)$. Recall that $\tau\text{III}(\tau s)$ is a series of unit strength impulses spaced every $1/\tau$ along the

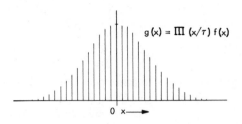

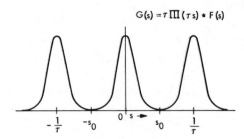

Figure 12-3 A sampled function

s-axis. Thus the convolution in the frequency domain replicates the spectrum $F(s)$ every $1/\tau$ along the *s*-axis. As indicated in Figure 12-3, $G(s)$ consists of infinitely many copies of the spectrum $F(s)$ equally spaced along the *s*-axis from minus infinity to infinity. Notice that the spectrum $G(s)$ of the sampled function is periodic with frequency τ. Thus any function sampled at equal intervals τ has a spectrum that is periodic with frequency τ.

Now that the function $f(x)$ has been sampled, is the information between sample points lost? Can we recover the original function intact from the sample points? Clearly, we can reclaim $f(x)$ from $g(x)$ if we can reclaim $F(s)$ from $G(s)$. We can do this by merely eliminating all the replicas of $F(s)$ except the one centered upon the origin. One way to do this is to multiply $G(s)$ by $\Pi(s/2s_1)$, where

$$s_0 \leq s_1 \leq \frac{1}{\tau} - s_0 \tag{11}$$

Then

$$G(s)\Pi\left(\frac{s}{2s_1}\right) = F(s) \tag{12}$$

and we have recovered the spectrum of $f(x)$ from the spectrum of the sampled signal $g(x)$. The original function is given by

$$f(x) = \mathfrak{F}^{-1}\{F(s)\} = \mathfrak{F}^{-1}\left\{G(s)\Pi\left(\frac{s}{2s_1}\right)\right\} \tag{13}$$

Applying the convolution theorem to the right-hand side of Eq. (13) yields

$$f(x) = g(x) * 2s_1 \frac{\sin(2\pi s_1 x)}{2\pi s_1 x} \tag{14}$$

which tells us how to reconstruct $f(x)$ from $g(x)$. We merely convolve the sampled function with an interpolating function of the form $\sin(x)/x$.

Equation (14) shows us that we can indeed recover $f(x)$ from $g(x)$, and it tells us how to do it. This development, however, is subject to two restrictions. First, $f(x)$ must be band-limited at s_0 [recall Eq. (10)] and, second, the relationship between the sampling interval τ and the band-limit s_0 must satisfy Eq. (11). What we have done is prove the well-known sampling theorem which states that a function sampled at uniform spacing τ can be completely recovered from the sample values, provided that

$$\tau \leq \frac{1}{2s_0} \tag{15}$$

where the function is band-limited at s_0.

Convolving $g(x)$ with the interpolating function, as suggested by Eq. (14), in effect replicates a narrow $\sin(x)/x$ at each sample point, as shown in Figure 12-4. Equation (14) guarantees that the summation of the overlapping $\sin(x)/x$ functions will add up to reproduce the original function exactly.

Figure 12-4 illustrates the case where $s_1 = 1/2\tau$, but Eq. (11) allows considerable arbitration in the frequency of the $\sin(x)/x$ function if the reciprocal of the sampling interval is considerably larger than the band-limit s_0. Equation (11) allows us to place s_1 anywhere between s_0 and $1/\tau - s_0$. For convenience, we may place s_1 at the midway point

$$s_1 = \frac{1}{2\tau} \tag{16}$$

and the interpolating function becomes

$$\frac{1}{\tau} \frac{\sin\left(\pi\dfrac{x}{\tau}\right)}{\pi\dfrac{x}{\tau}} \tag{17}$$

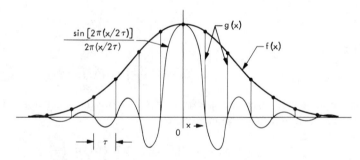

Figure 12-4 Interpolation with $\sin(x)/x$

Undersampling and Aliasing

Equation (15) specifies how finely one must sample a function if it is to be totally recoverable from its sample values. We now examine what happens if that condition is not satisfied.

Suppose $1/\tau < 2s_0$. Then when $F(s)$ is replicated to form $G(s)$, the individual replicas will overlap and sum (Figure 12-5). If we then interpolate, using the function in Eq. (17), we will not recover $f(s)$ exactly because

$$G(s)\Pi\left(\frac{s}{2s_1}\right) \neq F(s) \tag{18}$$

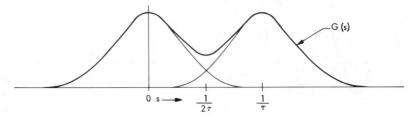

Figure 12-5 Overlap of replicated spectra

The effect of overlap of the spectral replicas can be viewed as follows. Energy above the frequency s_1 is folded back below s_1 and added to the spectrum. This folding back of energy is called *aliasing*, and the difference between $f(x)$ and the interpolated function is due to *aliasing error*. As a general rule, the more energy that falls above s_1, the more energy will be folded down into the spectrum and the worse will be the aliasing error. Notice that if $f(x)$ is even, then $F(s)$ is also even and aliasing effectively increases the energy in the spectrum. If $f(x)$ is odd, the opposite occurs and the energy in the spectrum decreases. If $f(x)$ is neither even nor odd, then aliasing tends to make it more even than it was before.

The following examples illustrate aliasing in the frequency domain and its effect in the time domain. Suppose we have the function

$$f(t) = 2\cos(2\pi f_0 t) \tag{19}$$

which has the spectrum

$$F(s) = \delta(s + f_0) + \delta(s - f_0) \tag{20}$$

as shown in Figure 12-6, and suppose we sample $f(t)$ at equal intervals Δt. The period of $f(t)$ is $1/f_0$. For case 1, suppose that

$$\Delta t = \frac{1}{4}\left(\frac{1}{f_0}\right) \tag{21}$$

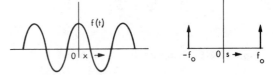

Figure 12-6 The cosine and its spectrum

which means that the folding or Nyquist frequency is

$$f_N = \frac{1}{2\Delta t} = 2f_0 \tag{22}$$

and we are taking four sample points per cycle of $f(t)$.

Figure 12-7 shows the sampled function and its spectrum. It also shows the interpolating function and its spectrum. Since $F(s)$ contains no energy above f_N, $f(t)$ can be completely recovered from its sample points.

$$g(t) = f(t) \frac{1}{\Delta t} III\left(\frac{t}{\Delta t}\right)$$

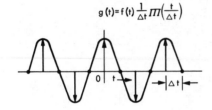

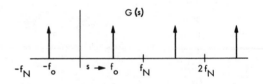

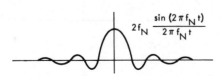

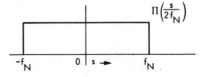

Figure 12-7 Sampling the cosine, case 1

In case 2, assume that

$$\Delta t = \frac{1}{2}\left(\frac{1}{f_0}\right) \tag{23}$$

which means that

$$f_N = f_0 \tag{24}$$

and we have two sample points per cycle. This case is illustrated in Figure 12-8. Here we are sampling the cosine at its positive and negative peaks, and the function still

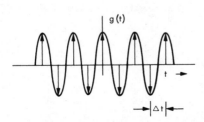

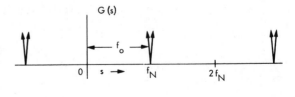

Figure 12-8 Sampling the cosine, case 2

can be completely recovered by interpolation as in case 1. In the frequency domain, the impulses from adjacent replicas combine at $s = f_0$, but the spectrum of the interpolating function takes on the value $\frac{1}{2}$ at that point, so the function is recovered intact.

For case 3, we let

$$\Delta t = \frac{2}{3}\left(\frac{1}{f_0}\right) \tag{25}$$

which means that

$$f_N = \tfrac{3}{4}f_0 \tag{26}$$

This case is illustrated in Figure 12-9. Here the left-hand impulse from the spectral replicate centered upon $s = 2f_N$ falls between zero and f_N at the point $s = f_0/2$. Upon interpolation, the energy at $s = f_0$ is aliased down to the frequency $f_0/2$. Figure 12-9 illustrates how interpolation fits a cosine of frequency $f_0/2$ through the sample points. This illustrates graphically how high-frequency information is aliased to appear as low-frequency information.

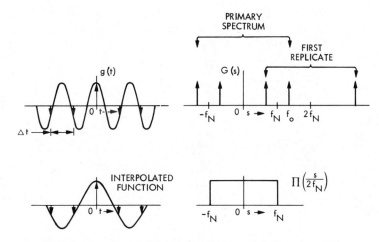

Figure 12-9 Sampling the cosine, case 3

In case 4, we let

$$\Delta t = \frac{1}{f_0} \tag{27}$$

so that

$$f_N = \tfrac{1}{2} f_0 \tag{28}$$

This case is illustrated in Figure 12-10. The energy at f_0 is aliased all the way down to

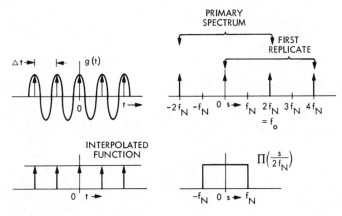

Figure 12-10 Sampling the cosine, case 4

zero frequency. The cosine is sampled only at its positive peaks, and when these sample points are interpolated, the resulting function is constant with unit amplitude.

Case 5 is the same as case 2, except the function is

$$f(t) = 2 \sin (2\pi f_0 t) \tag{29}$$

as shown in Figure 12-11. In this case, the odd impulse pairs from adjacent spectral replicas overlap at $s = f_N$, where they cancel. Figure 12-11 illustrates why the interpolated function is zero. This case corresponds to sampling the sine at its zero crossings.

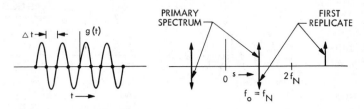

Figure 12-11   Sampling the sine, case 5

COMPUTING SPECTRA

One important application of digital processing is merely to compute the spectrum of a signal or an image. In this section, we describe how to compute the spectrum of a signal and how the computed spectrum compares with the actual spectrum of the signal.

Suppose a signal $f(t)$ is represented by N sample points separated by constant spacing Δt, as shown in Figure 12-12. The total interval over which the signal is sampled is

$$T = N\Delta t \tag{30}$$

where T is the width of the "truncation window." Since a signal can be sampled with only a finite number of points, the sampling process truncates the signal by ignoring

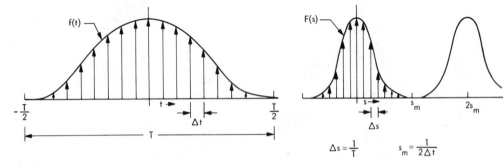

Figure 12-12   Computing spectra

it outside the truncation window. This amounts to assuming the signal is zero outside the truncation window.

We want to use the sample values of $f(t)$ to compute points on its spectrum $F(s)$. We may do this by programming the Fourier transform as a numerical integration. First we must decide how many points we shall compute on the spectrum, the spacing between those sample points, and the frequency range over which we shall compute the spectrum.

Since the sampled signal consists of N independent measurements, it is reasonable to compute a total of N points on the spectrum. To compute more points would introduce redundancy, while computing fewer points would not take advantage of all the information we have on $f(t)$. Thus a general-purpose computer program for calculating the Fourier transform should take N (complex) sample points into N (complex) points on the spectrum. For convenience, the computed points are usually spaced equally along the s-axis.

Since $f(t)$ is a sampled function, with sample spacing Δt, its spectrum $F(s)$ is periodic with period $1/\Delta t$. Clearly, we should confine our computation to cover only one cycle of $F(s)$. It is common practice to spread the N sample points evenly across the cycle of $F(s)$ which is centered upon the origin. This means we compute points on $F(s)$ only over the range

$$-\frac{1}{2\Delta t} \le s \le \frac{1}{2\Delta t} \tag{31}$$

If we spread N equally spaced sample points over one cycle of $F(s)$, this means that

$$N\Delta s = \frac{1}{\Delta t} \tag{32}$$

where Δs is the sample spacing in the frequency domain. It is given by

$$\Delta s = \frac{1}{N\Delta t} = \frac{1}{T} \tag{33}$$

Thus, for our purposes, the best choice for computing the spectrum of $f(t)$ is to compute points with equal spacing, given by Eq. (33), over a frequency range from $-s_{max}$ to s_{max}, where

$$s_{max} = \frac{1}{2\Delta t} \tag{34}$$

Notice that the maximum frequency we can compute is inversely related to the time domain sample spacing [Eq. (34)]. The frequency domain sample spacing, which determines how finely we can compute the spectrum, is inversely related to the width of the time domain truncation window [Eq. (33)].

In summary, the sample spacing in one domain dictates (or is dictated by) the truncation width in the other domain. If we desire to compute high-frequency components of the spectrum, then we must sample finely in the time domain. Furthermore, if we insist upon high resolution in the spectrum (small Δs), we must use a large truncation window in the time domain. The relationships between the time and frequency domain sampling and truncation parameters are summarized in Table 12-1.

Table 12-1.   Summary of sampling and truncation parameters

Parameter	Domain	Relations
Number of sample points	Both	$N = \dfrac{T}{\Delta t} = \dfrac{2s_m}{\Delta s}$
Sample spacing	Time	$\Delta t = \dfrac{T}{N} = \dfrac{1}{2s_m}$
Sample spacing	Frequency	$\Delta s = \dfrac{2s_m}{N} = \dfrac{1}{T}$
Truncation window width	Time	$T = N \Delta t$
Maximum computed frequency (also Nyquist or folding frequency)	Frequency	$s_m = \dfrac{1}{2\Delta t} = \dfrac{1}{2} N \Delta s$

If $f(t)$ is complex and we compute its spectrum, then N real and N imaginary numbers are transformed to produce N real and N imaginary numbers of the spectrum. If $f(t)$ is real, then N real numbers and N zeros (the imaginary part) give rise to $N/2$ real and $N/2$ imaginary numbers in the right-hand half of the spectrum. Since $F(s)$ is Hermite, the left half of the spectrum is a mirror image of the right. Thus the $N/2$ real and the $N/2$ imaginary numbers in the left half of the spectrum are, from an information point of view, redundant. Notice that, in both cases, the number of unconstrained sample points in the two domains is the same.

The Unavoidability of Aliasing

The sampling theorem indicates that judicious choice of sample spacing can avoid aliasing when sampling a band-limited function. Clearly, if we are forced to work with inherently non-band-limited functions, we are condemned to certain unavoidable aliasing. One might expect that if judicious selection or good fortune allowed us to work with band-limited functions, then aliasing might be avoided. This opportunity, however, is foiled by the process of truncation.

Suppose a band-limited function is truncated to a finite duration T. This process may be modeled as multiplying the function by a rectangular pulse of width T. Recall that this has the effect of convolving the spectrum with a $\sin(x)/x$ function that has infinite duration in the frequency domain. Since the convolution of two functions can be no narrower than either, we conclude that the spectrum of the truncated function is of infinite extent in the frequency domain. Thus truncation destroys band-limitedness and condemns digital processing to produce aliasing in all cases. While aliasing cannot be avoided, the resulting error can be bounded and reduced to the point of an acceptable approximation.

Bounding Aliasing Error

The following example illustrates how one can place a bound on aliasing error and select digitizing parameters to produce a desired accuracy in spite of aliasing. Suppose we wish to identify the linear system shown in Figure 12-13 by computing the spectrum

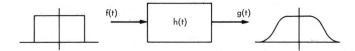

Figure 12-13   Linear system identification

of its response to a rectangular pulse. If $f(t)$ is the input pulse and $g(t)$ is the system's output, then the transfer function is given by

$$H(s) = \frac{G(s)}{F(s)} \tag{35}$$

Assume, for this case, that we know that the system is a lowpass filter and thus its output is a rectangular pulse with slightly rounded corners.

If we are to evaluate Eq. (35) by digital computation, we must digitize $f(t)$ and $g(t)$ and compute their spectra. We must select the sample spacing Δt and the sampling period T so as to yield good spectral resolution with reasonably small aliasing errors. To do this, we must define a measure of spectral resolution and a measure of aliasing error and relate them to the sampling parameters. Then we can make an intelligent choice of N, T, and Δt.

The input signal and its spectrum are shown in Figure 12-14. Since $F(s)$ extends from minus to plus infinity, no choice of Δt will completely avoid aliasing. $F(s)$ is

Figure 12-14   The input signal and its spectrum

enclosed in an envelope of the form $1/s$, however, and this assures that the peak amplitude of the function dies out with increasing frequency. If we ignore the sinusoidal variations and consider only the envelope, we note that the largest spectral amplitude possible to be aliased occurs at the frequency s_{max}. We can take this to be worst case for aliasing and define, as a measure of aliasing error, the ratio of $F(s_{max})$ to $F(0)$. Since $F(0)$ is unity and the envelope is $1/2\pi as$, we can write an upper bound on aliasing as

$$A \leq \frac{1}{2\pi a s_0} = \frac{2\Delta t}{2\pi a} = \frac{\Delta t}{\pi a} \tag{36}$$

Notice that this bound on aliasing error, as we have defined it, is proportional to Δt but independent of T. Thus we can make the aliasing error as small as desired by making Δt small with respect to the pulse width $2a$.

$F(s)$ has sinusoidal variations of frequency a. Let us denote by M the number of sample points per cycle of $F(s)$ on the computed spectrum and use it as a measure of spectral resolution. The parameter M indicates how finely we are computing the sampled spectrum $F(s)$. The period of the sinusoidal variations of $F(s)$ is $1/a$ and

$$M\Delta S = \frac{1}{a} \tag{37}$$

or

$$M = \frac{1}{a\Delta s} = \frac{T}{a} \tag{38}$$

This means that we may have as many sample points per cycle of $F(s)$ as desired if we make the sampling period T large compared to the half width of the pulse. Notice that if we insist upon both small aliasing error and high spectral resolution, Δt is small, T is large, and the required number of sample points is very large. As it frequently happens, one must purchase accuracy with computer time.

TRUNCATION

Like sampling, truncation can also cause a computed spectrum to differ from the actual spectrum of a function. Like the sample spacing, the truncation window must be selected wisely to produce suitably accurate results. The following example illustrates the effect of truncation.

Suppose we wish to calculate the spectrum of the step function. In this example, we shall use the function sign (x) shown in Figure 12-15. In order to calculate the

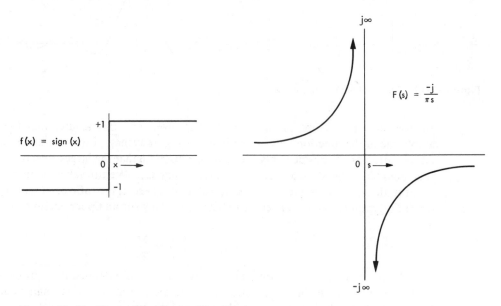

Figure 12-15 The step function and its spectrum

spectrum, we must first truncate $f(x)$ to a finite duration T. Since the sign (x) function goes to infinity with constant amplitude, we recognize that this example is sensitive to truncation. If we truncate the function with a truncation window of width T, the resulting function is given by

$$g(x) = f(x)\Pi\left(\frac{x}{T}\right) = \Pi\left(\frac{x}{T/2} - \frac{1}{2}\right) - \Pi\left(\frac{x}{T/2} + \frac{1}{2}\right) \tag{39}$$

as shown in Figure 12-16. Since the truncated function is an odd pair of rectangular pulses, it can be written as

$$g(x) = \Pi\left(\frac{x}{T/2}\right) * \left[\delta\left(x - \frac{T}{4}\right) - \delta\left(x + \frac{T}{4}\right)\right] \tag{40}$$

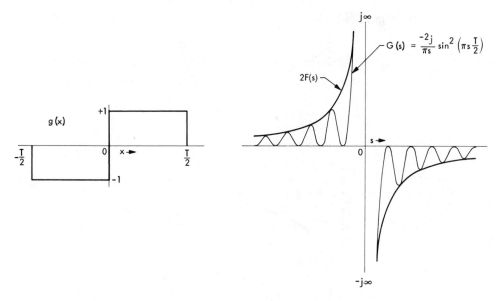

Figure 12-16 The truncated step function and its spectrum

Transforming Eq. (40) produces the spectrum of the truncated edge function

$$G(s) = -2j \sin\left(\pi s \frac{T}{2}\right)\frac{\sin(\pi s T/2)}{\pi s} \tag{41}$$

This may be rearranged to produce

$$G(s) = \frac{-2j}{\pi s}\sin^2\left(\frac{\pi s T}{2}\right) = 2F(s)\left[\frac{1}{2} - \frac{1}{2}\cos(\pi s T)\right] \tag{42}$$

This is shown in Figure 12-16. The spectrum of the truncated signal is a sinusoid enclosed under an envelope that is twice the desired spectrum $F(s)$. This considerable change in the nature of the spectrum is a result of truncation—in this case a relatively radical modification of the original function.

Since we actually compute points on $G(s)$, we can ask where those points fall with respect to the sinusoidal variations in $G(s)$. The sample points on $G(s)$ will be computed at discrete frequencies s_i, where

$$s_i = i\Delta s = \frac{i}{T} \qquad i = 0, 1, 2, \ldots, \frac{N}{2} \tag{43}$$

and the computed points will be

$$G(s_i) = 2F(s_i)[\tfrac{1}{2} - \tfrac{1}{2}\cos(i\pi)] \tag{44}$$

The cosine term takes on the value $+1$ for even i and -1 for odd i, so

$$G(s_i) = \begin{cases} 2F(s_i) & i \text{ odd} \\ 0 & i \text{ even} \end{cases} \tag{45}$$

This is shown in Figure 12-17.

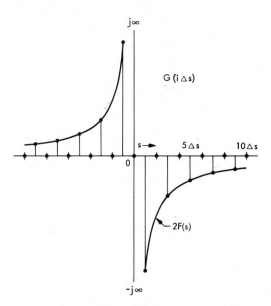

Figure 12-17 The computed spectrum of the step function

In the foregoing analysis, it was assumed that the edge was centered within the truncation window. The reader is invited to determine the effect upon the sample points of $G(s)$ if the edge is slightly off center in the truncation window.

THE EFFECTS OF DIGITAL PROCESSING

We are now in a position to examine the cumulative effects of digital processing upon a continuous signal or image. We shall examine the effects of sampling, truncation, interpolation, digitally implemented convolution, and Fourier transformation.

Sampling and Interpolating a Function

Suppose we begin with a continuous function $f(t)$, as shown in Figure 12-18. This example has a triangular amplitude spectrum but random phase. For the present, assume that we desire only to digitize the function and then reconstruct it without

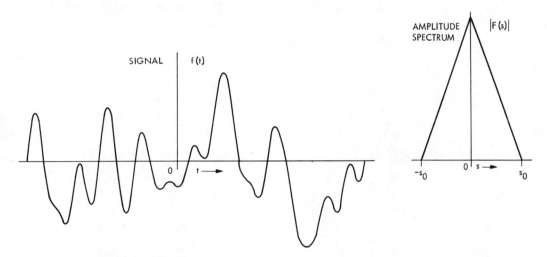

Figure 12-18 A signal and its spectrum

processing. When we digitize the signal, we must necessarily truncate it to a finite duration T. The truncation window $\Pi(t/T)$ and its spectrum are shown in Figure 12-19. Also shown are the truncated function and its spectrum. Truncating $f(t)$ convolves its spectrum with a narrow $\sin(x)/x$ function.

The digitizer will have a finite-width sampling aperture over which the signal is averaged at each sample point. As discussed in Chapter 9, this local averaging can be modeled by convolution with a suitable aperture function. For an image digitizer, the sampling aperture function models the sensitivity of the scanning spot. Electrical signals are usually sampled with a circuit that integrates over a fixed period.

In Figure 12-20, we model the sampling aperture with a small rectangular pulse of width τ. As shown in the figure, convolving the truncated signal with the sampling aperture function multiplies the spectrum by a broad $\sin(x)/x$ function. If the sampling aperture were, for instance, a Gaussian, the spectrum of the truncated signal would be multiplied by a broad Gaussian. In either case, the effect of the sampling aperture is to reduce the high-frequency energy in the signal. Notice in Figure 12-20 that at frequencies beyond $s = 1/\tau$, the polarity of the energy will be reversed.

The sampling process is illustrated in Figure 12-21. The truncated signal, smoothed by the sampling aperture, is multiplied by $III(t/\Delta t)$ to effect sampling. As illustrated, sampling the signal makes its spectrum periodic by replicating the original spectrum at intervals $1/\Delta t$.

TRUNCATION WINDOW $\Pi\left(\frac{t}{T}\right)$

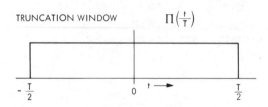

SPECTRUM OF
TRUNCATION WINDOW $T\frac{\sin(\pi Ts)}{\pi Ts}$

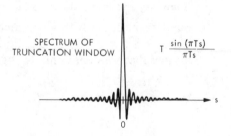

TRUNCATED SIGNAL $f(t)\Pi\left(\frac{t}{T}\right)$

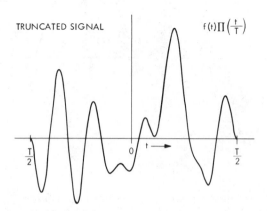

TRUNCATED SIGNAL
SPECTRUM $\left|F(s) * T\frac{\sin(\pi sT)}{\pi sT}\right|$

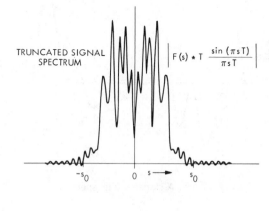

Figure 12-19 Truncating the signal

Suppose we wish merely to interpolate the sampled function to regain, as well as possible, $f(t)$. Figure 12-22 illustrates interpolation by convolution of the sampled function with a triangular pulse. In the figure, the width of the triangular pulse is $2t_0$. Convolving the sampled function with the interpolating function multiplies its spectrum by a function of the form $\sin^2(x)/x^2$. Since this function generally decreases with increasing frequency, it tends to drive all replicas, except the primary replica at $s = 0$, to zero. Recall that the ideal interpolating function is $\sin(x)/x$, which multiplies the spectrum by a rectangular pulse centered on $s = 0$. However, the triangular pulse of Figure 12-22 produces approximately the same effect.

If we denote by $h(t)$ the function obtained by interpolating the truncated sampled function, it is given by

$$h(t) = \left(\left\{\left[f(t)\Pi\frac{t}{T}\right] * \frac{1}{\tau}\Pi\left(\frac{t}{\tau}\right)\right\}\mathrm{III}\left(\frac{t}{\Delta t}\right)\right) * \frac{1}{t_0}\Lambda\left(\frac{t}{t_0}\right) \qquad (46)$$

and its spectrum by

$$H(s) = \left(\left\{\left[F(s) * T\frac{\sin(\pi sT)}{\pi sT}\right]\frac{\sin(\pi s\tau)}{\pi s\tau}\right\} * \Delta t\,\mathrm{III}(s\Delta t)\right)\left[\frac{\sin(\pi st_0)}{\pi st_0}\right]^2 \qquad (47)$$

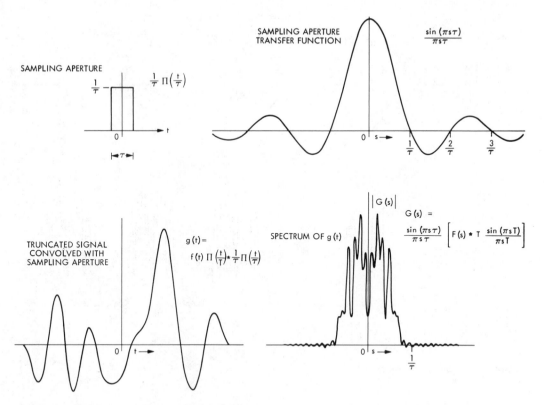

Figure 12-20 Convolving the sampling aperture

Clearly, the question is not whether digital processing has an effect on the signal but rather how much effect it has. In the preceding example, the sampling aperture and interpolating function were chosen rather wide to exaggerate their effect. Specifically, $\tau = t_0 = 2\Delta t$. These parameters, while arbitrary, should be chosen in proper relationship to each other. For example, the sampling aperture should have width τ roughly equal to the sample spacing Δt. Also, for linear interpolation, $t_0 = \Delta t$.

Truncation convolves the spectrum with a narrow $\sin (x)/x$. If the truncation window is wide, its spectrum becomes narrow, approximating an impulse, and this reduces its effect. Also, if the function is already zero outside the truncation window, there is no effect.

The sampling aperture, as illustrated in Figure 12-20, tends to reduce the high-frequency energy in the spectrum. In so doing, it can reduce subsequent aliasing. The sampling aperture can also reverse the polarity of the high-frequency energy.

Sampling, of course, makes the spectrum periodic. This produces aliasing of energy above the folding frequency, $1/2\Delta t$.

Interpolation restores the spectrum to a single replica centered upon the origin. This is done accurately, however, only if $\sin (x)/x$ is used as the interpolating function.

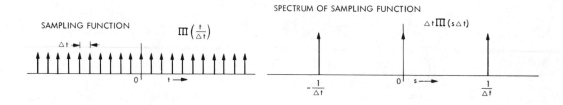

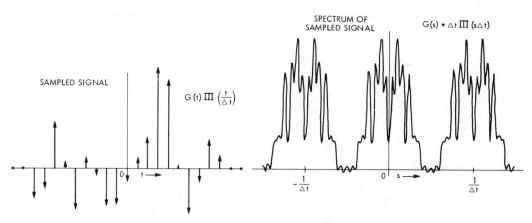

Figure 12-21 Sampling the signal

Other interpolating functions incompletely remove spectral replicas and reduce the high-frequency energy content of the primary replica.

The digitizing parameters usually result from limitations of the digitizing equipment. The truncation window, for example, represents the maximum field of view of the image digitizer. The sampling aperture is merely the sensitivity function of the scanning spot. The sample spacing is usually adjustable and should be set in relation to the spot diameter. The interpolating function, for image display, is the display spot itself.

Figure 12-23 illustrates how one can use a rectangular sampling aperture to reduce aliasing. The width of the aperture is twice the sample spacing. This places the first zero crossing of its transfer function at $f_N = 1/2\Delta t$. The triangular sampling aperture used in Figure 12-24 is four sample points wide and also has its first zero crossing at f_N. Since its spectrum dies out with frequency more rapidly than that of the rectangular pulse, it is more effective against aliasing. Like the rectangular pulse, however, it reduces the energy in $F(s)$ below f_N.

Equations (46) and (47) might appear to suggest that a continuous function cannot be processed digitally without severe distortion and that our previous development has been in vain. There is, however, a way out—and that is by oversampling. If we make the sample spacing suitably small, we can place f_N far beyond the frequencies of interest in the spectrum. Then when aliasing contaminates the upper part

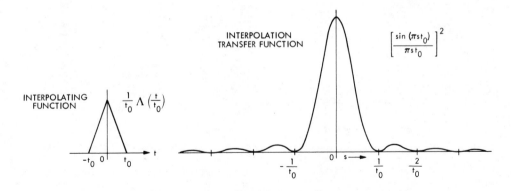

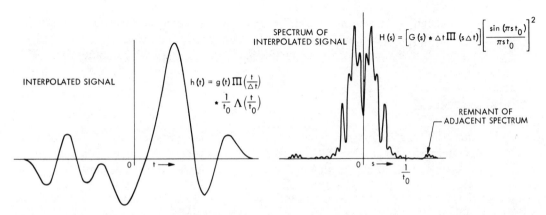

Figure 12-22 Interpolating the sampled signal

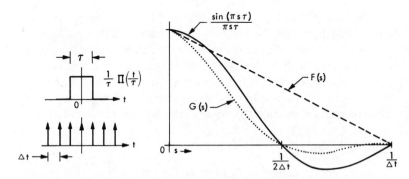

Figure 12-23 Aliasing reduction with a rectangular aperture

of the spectrum, it will have little or no effect upon the data of interest. As a rule of thumb, oversampling by a factor of 2 is adequate for most applications, although an analysis should be performed in each case. The truncation window should also be

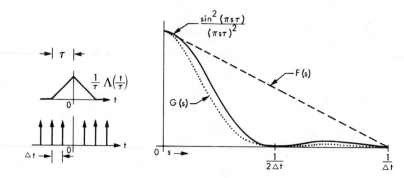

Figure 12-24 Aliasing reduction with a triangular aperture

large enough to produce minimum contamination of the signal spectrum. By suitably oversampling, one can reduce aliasing and truncation effects to any desired order of magnitude. The piper, of course, must be paid—in this case with computer time.

DIGITAL FILTERING

Linear filtering can be implemented digitally in two different ways. The filtering operation implied in Figure 12-25 could be implemented by digital convolution of the sampled function $f(t)$ with $h(t)$ to produce $g(t)$. Alternatively, one could transform

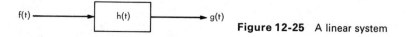

Figure 12-25 A linear system

$f(t)$ and $h(t)$ to the frequency domain with a Fourier transform algorithm employing numerical integration. Then the output spectrum $G(s)$ could be formed by multiplication, and the output signal generated by an inverse transformation.

   If one or both of the convolution input signals are of short duration, then the method of digital convolution is computationally more efficient. Otherwise, efficient Fourier transform algorithms make the second method more practical. In this section, we shall compare the two approaches in terms of aliasing and truncation error.

Convolution Filtering

As noted before, sampling $f(t)$ and $h(t)$ makes their spectra periodic. If both signals are sampled at the same interval Δt, their spectra will be periodic with the same frequency, $1/\Delta t$. Convolution of the sampled signals multiplies the two spectra in the frequency domain to form $G(s)$, which is also periodic with frequency Δt. When $g(t)$ is interpolated, its spectrum is reduced to a single replica at the origin as in the previous discussion.

If either $f(t)$ or $h(t)$ is band-limited below $s = 1/2\Delta t$, then $g(t)$ will be similarly band-limited, and interpolation will reconstruct it exactly. Truncation, however, destroys band-limitedness, and some aliasing is unavoidable. This aliasing will express itself in $g(t)$ in a straightforward manner. Thus the effects introduced by digital convolution are the same ones produced by sampling, truncation, and interpolation.

Frequency Domain Filtering

Figure 12-26 illustrates what happens when we compute a Fourier transform. The input signal $f(t)$ is sampled to form $x(t)$, which has a continuous periodic spectrum. When we compute its Fourier transform, we actually calculate equally spaced points on the primary cycle of its periodic spectrum, as illustrated in Figure 12-26. We com-

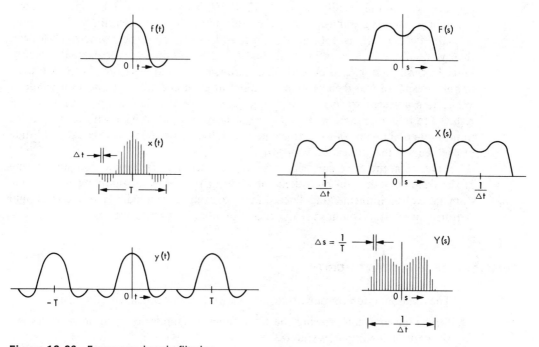

Figure 12-26 Frequency domain filtering

pute N points equally spaced every Δs over the frequency range from $-1/2\Delta t$ to $1/2\Delta t$. We denote the computed spectrum by $Y(s)$ because it is, in fact, not $X(s)$, the spectrum of $x(t)$.

Since $Y(s)$ is sampled, its inverse transform $y(t)$ is a continuous (unsampled) periodic function of infinite duration. Thus the computed spectrum $Y(s)$ is not the spectrum of $x(t)$ or even that of $f(t)$, the underlying unsampled function, but it is rather the spectrum of a continuous periodic function having period T. All the sample points of $x(t)$ fall exactly upon the primary cycle of $y(t)$ and, barring aliasing, the primary cycle of $y(t)$ is exactly $f(t)$, the function that was sampled to form $x(t)$.

By computing the spectrum of $x(t)$ digitally, we have necessarily sampled that spectrum to produce $Y(s)$. This in turn is the spectrum of a continuous periodic function $y(t)$. We now have in the frequency domain the equivalent of spectral replication due to sampling in the time domain. If we implement digitally the inverse transform, we can, of course, reclaim $x(t)$ from $Y(s)$. If we then interpolate $x(t)$, we can recover f. The fact that $Y(s)$ corresponds to a periodic function produces no ill effect. If we implement digital filtering in the frequency domain, however, the situation is not so simple.

Suppose we implement frequency domain filtering by multiplying $Y(s)$ by some transfer function $H(s)$. This convolves $y(t)$ with the impulse response $h(t)$. Since $y(t)$ is periodic, the convolution will tend to smear adjacent cycles of $y(t)$ down into the primary cycle in the vicinity of $t = \pm T/2$. If $h(t)$ is narrow and $y(t)$ is approximately constant in the area about $t = T/2$, then this adjacent cycle smearing will have only a small effect. If $x(t)$ is not equal at each end of the truncation window, however, then $y(t)$ will have a discontinuity at $t = T/2$. This produces an artifactual discontinuity in the function at each end of the truncation window. Convolution with the impulse response $h(t)$ then produces artifact at each end of the truncation window. While this adjacent cycle smearing effect cannot be avoided completely, it can be reduced to tolerable levels by making the truncation window wide with respect to the important components of the signal and by arranging for $x(t)$ to have equal amplitude at each end of the truncation window.

The adjacent cycle smearing effect encountered in frequency domain filtering is the frequency domain equivalent of aliasing produced by sampling in the time domain. When implementing linear filtering using computed spectra, one should perform an analysis to quantify the effects of adjacent cycle smearing.

SUMMARY OF IMPORTANT POINTS

1. The Shah function (impulse train) is its own Fourier transform [Eq. (2)].

2. Stretching and compressing the Shah function (similarity operations) alters the strength of the impulses [Eq. (8)].

3. Sampling a continuous function can be modeled by multiplication with the Shah function.

4. A function band-limited at frequency s_0 can be completely recovered from its sample values if they are taken no farther than $1/2s_0$ apart.

5. Undersampling causes aliasing, wherein energy above the folding frequency ($s = 1/2\Delta t$) appears an equal distance below the folding frequency.

6. Truncation destroys band-limitedness and makes aliasing unavoidable in digital processing.

7. The effects of aliasing can be reduced to tolerable levels by oversampling.

8. Frequency domain filtering can produce an adjacent cycle smearing effect near the ends of the truncation window.

REFERENCES

1. R. BRACEWELL, *The Fourier Transform and Its Applications*, McGraw-Hill Book Company, New York, 1965.

2. E. O. BRIGHAM, *The Fast Fourier Transform*, Prentice-Hall, Inc., Englewood Cliffs, New Jersey, 1974.

OPTICS AND SYSTEM ANALYSIS

IIIIIIIIIIIIIIIIIIIIIIIIIIIIIIIIIIIIIIIIIIIIIIIIIIIIIIIIIIIIIIIIIIIIIIIIIIIIIIIIIIIIIIIIIIIIIIIIIIIIIIIIIIIIIIIIIIIIIIII

INTRODUCTION

So far in Part II, we have developed a set of analytical tools that will allow us to analyze completely most digital image processing systems. The detailed analysis of all aspects of an image processing system can become extremely complex and is therefore outside our scope. We are prepared, however, to perform a first approximation analysis, which is adequate for most cases of practical concern.

Ordinarily, the user has control over only one link in the image processing chain—the computer program that performs the digital processing operation. The performance of the other links in the chain, from the digitizer to the display, is usually fixed, although proper maintenance is required for best performance. We must be able to specify the effect that the hardware portion of the system has upon an image so that we can correct for this effect in software. In this way, the processing program can be defined to accomplish a given goal.

In this chapter, we shall use linear system theory to model the operation of an image processing system. We have the tools to describe the effects of sampling, interpolation, and filtering. Now we must add one other element to describe the effects of the optical systems usually situated at each end of the system. In the first part of this chapter, we develop linear system techniques for analyzing optical systems, and in the remainder, we apply linear system techniques to the analysis of complete digital image processing systems.

OPTICS AND IMAGING SYSTEMS

Imaging systems play an important role in digital image processing because they almost always appear at the front end (and frequently at both ends) of an image processing system. If photography is involved, then another lens system must be included in the analysis.

Optical systems produce two effects upon an image: projection, as discussed in Chapter 2, and a degradation due to the effects of diffraction and lens aberrations. The projection accounts for inversion of the image in its coordinate system and for magnification. The field of physical optics, particularly diffraction theory, provides the tools to describe image degradation due to (1) the wave nature of light with finite wavelength and (2) the aberrations of imperfect optical systems. We now present a brief development of important points from physical optics. For a more detailed treatment of optical system analysis the reader should consult an optics text (Refs. 1–3).

Linearity of Optical Systems

Figure 13-1 shows an optical system consisting of a simple lens. A point source at the origin of the object plane produces a small spot image at the origin of the image plane, provided that the system is in focus; that is,

$$\frac{1}{d_o} + \frac{1}{d_i} = \frac{1}{f} \tag{1}$$

where f is the focal length of the lens. It is intuitively clear that increasing the intensity of the point source causes a proportional increase in the intensity of the spot image. This means that the lens is a two-dimensional linear system. It follows that two point sources produce an image in which the two spots combine by addition.

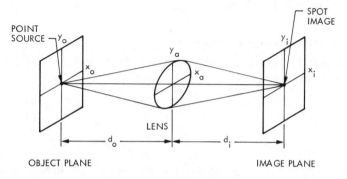

Figure 13.1 A simple lens

If the point source moves off the z-axis to a position (x_o, y_o), then the spot image moves to a new position given by

$$x_i = -\frac{d_i}{d_o} x_o = -M x_o \qquad y_i = -M y_o \qquad (2)$$

For reasonable off-axis distances in well-designed optical systems, the shape of the spot image undergoes essentially no change. Thus the system can be assumed to be shift invariant or, in optics terminology, "isoplanatic." This is an assumption rather than a property of optical systems because the spot shape does indeed undergo change, particularly in the peripheral field of low-quality optical systems. To an approximation that we shall use routinely, however, an optical system is a two-dimensional, shift invariant linear system. The spot image produced by a point source is the impulse response, or "point spread function" (psf) in optical terms.

An opaque object illuminated from the front (epi-illuminated) and a light-absorbing object illuminated from behind (trans-illuminated) can be thought of as a two-dimensional distribution of point sources of light. The image of such an object is a summation of spatially distributed psf spots. This means the image can be described as a convolution of the object with the point spread function of the optical system. Furthermore, an isoplanatic optical system can be completely specified by either its two-dimensional psf or its two-dimensional "optical transfer function" (OTF). Equation (2) accounts for the projection of the optical system, while convolution with the point spread function accounts for the inherent loss of detail.

Physical lens systems are not truly shift invariant, but the shift variance is a gradual phenomenon. Typically, focal sharpness degrades gradually as one moves off-axis. For a quality lens the psf, although not an impulse, is at least nonzero only over a small region. Since the shift variance is a gradual phenomenon, we can assume that each point is surrounded by a neighborhood of shift invariance. In the field of optics, these neighborhoods are called *isoplanatic patches*. Thus, if not globally shift invariant, the optical system can at least be assumed locally shift invariant over the small extent of the psf, and convolution is still a valid model locally. If necessary, we can model the system with a psf having a spatially variant parameter. While this technique can account for most typically encountered anisoplanatism, it is usually unnecessary in the analysis of high-quality lens systems.

Coherent and Incoherent Illumination

In Figure 13-1 the point source emits a spherical wave. The E-field amplitude as a function of time and space can be written as

$$u(x, y, z, t) = \frac{a}{r} \cos\left[2\pi \frac{r}{\lambda} + 2\pi\left(\frac{c}{\lambda} t + \frac{\phi(t)}{\lambda}\right)\right] \qquad (3)$$

where

$$r = \sqrt{x^2 + y^2 + z^2} \qquad (4)$$

λ is the mean wavelength of the light, c is the speed of light, and $\phi(t)$ accounts for the phase fluctuation with time. Usually this is random. It also accounts for the bandwidth of quasi-monochromatic light. For convenience, we define the "wave number" as

$$k = 2\frac{\pi}{\lambda} \tag{5}$$

and move to complex exponentials as before. Now Eq. (3) becomes

$$u(x, y, z, t) = \Re e \left\{ \frac{A}{r} e^{jkr} e^{jk[ct + \phi(t)]} \right\} \tag{6}$$

In this section, we are concerned with the spatial distribution of light intensity in the spot image. We shall, for the time being, drop the $\Re e\{\ \}$ and the time-varying components as being understood.

Under monochromatic illumination, the object is a spatial distribution of point sources at the same temporal frequency c/λ. If all the point sources have a fixed phase relationship, the illumination is called *coherent*. They may still fluctuate randomly but remain in synchrony to preserve fixed relative phase. If, on the other hand, the point sources vary in phase independently of each other, the illumination is called *incoherent*. In this case, the phase of each point source varies independently of its neighbors.

In most cases, the human eye or some other time-averaging sensor makes ultimate use of the image. Under time-averaging, the random fluctuations of $\phi(t)$ are averaged out. In coherent light, since the point sources fluctuate in unison, the fixed phase relationship permits stable patterns of constructive and destructive interference to exist in the point images. These stable patterns of interference are apparent to a time-averaging sensor. Thus, for coherent illumination, the convolution operation must be performed on the complex amplitude of the electromagnetic waves.

Under incoherent illumination, the random relative phase relationships cause interference phenomena to average out to no net effect. Thus the point images add statistically. This behavior is modeled accurately if the convolution is performed on an intensity (amplitude squared or power) basis. Hence, under coherent illumination, an optical system is linear in complex amplitude, while in incoherent light the system is linear in intensity.

Image Quality Factors

The two factors that limit the image quality of an optical system are lens aberrations and diffraction effects. Careful lens design can minimize, although never completely eliminate, aberrations. Diffraction effects result from the wave nature of light and are beyond our control. Since image processing equipment usually employs high-quality optics with relatively low aberration levels, it is often diffraction that places an upper bound on image quality. In the following section, we derive the point spread function of an aberration-free (diffraction-limited) optical system and indicate how to account for aberrations. A complete study of lens design to minimize aberrations is beyond our scope. We shall be content to specify an optical system by its diffraction-limited psf, by manufacturer-supplied psf data, or by an experimentally determined psf.

Since we have argued that, to a reasonable approximation, an optical system is a shift invariant linear system, we need only to find an expression for either the point spread function or the transfer function. In Figure 13-1, the point source emits an expanding spherical wave, part of which enters the lens. The high refractive index of the lens slows the wave. Since the lens is thicker near the axis than near the edges, axial rays are slowed more than peripheral rays. In the ideal case, the thickness variation is just right to convert the expanding spherical wave into another spherical wave converging toward the image point. Any deviation of the exit wave from spherical form is, by definition, due to aberration. Thus a diffraction-limited optical system produces a converging spherical exit wave in response to the diverging spherical wave of a point source.

Lens Shape

For a thin, double-convex lens having a diameter that is small compared to the focal length f, the surfaces of the lens must be spherical to produce a spherical exit wave. Furthermore, the focal length f of the lens is given by the equation

$$\frac{1}{f} = (n - 1)\left(\frac{1}{R_1} + \frac{1}{R_2}\right) \tag{7}$$

where n is the refractive index of the glass and R_1 and R_2 are the radii of the front and rear spherical surfaces of the lens.

For lens diameters that are not small in comparison to f, spherical lens surfaces are not adequate to produce a spherical exit wave. Such a lens does not converge peripheral rays to the same point on the z-axis as it does near-axial rays. This phenomenon is called *spherical aberration* since it is an aberration resulting from the (inappropriate) spherical shape of the lens surfaces. High-quality optical systems employ aspheric surfaces and multiple lens elements to reduce spherical aberration.

Apertures and the Pupil Function

In Figure 13-1, the spot image formed by the converging spherical wave is exactly the point spread function of the system. That spot is formed by a truncated converging spherical wave. Figure 13-2 shows a different but equivalent way to create the same image. Here a converging spherical wave is truncated by an opaque screen containing an aperture. That aperture represents the extent of the lens in Figure 13-1. More complicated optical systems may contain several lenses and apertures or "stops"; however, all apertures can be projected through to the exit pupil to establish an effective exit aperture of the system. In Figure 13-2, the aperture represents the effective exit aperture of any aberration-free lens system.

The spatial distribution of transmittance in the screen containing the aperture is the pupil function. Thus, for a circular aperture of diameter a centered on a coor-

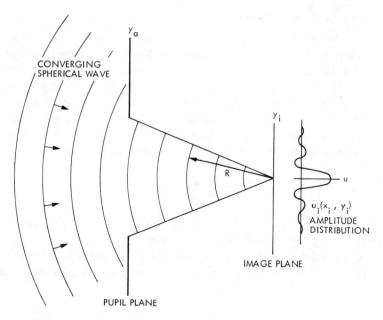

Figure 13-2 The truncated spherical exit wave

dinate system (x_a, y_a), the pupil function is

$$p(x_a, y_a) = \Pi\left(\frac{\sqrt{x_a^2 + y_a^2}}{a}\right) \tag{8}$$

For ordinary apertures, the pupil function assumes only the values 0 and 1. It is possible, however, to implement variable transmittance pupils using photographic or metal film deposition techniques.

For aberration-free systems, the pupil function is real-valued; otherwise it would disturb the spherical shape of the exit wave. Complex-valued pupil functions are used to model optical systems with aberrations. While the following analysis permits the use of arbitrary pupil functions, the case of most practical importance is that of the circular aperture.

The E-field of the unit-amplitude converging spherical wave in Figure 13-2 can be written as

$$u(x_i, y_i, z_i) = \frac{1}{R}e^{-jkR} \tag{9}$$

using the conventions described in connection with Eq. (6). R is the distance of the point (x_i, y_i, z_i) from the origin of the image plane. In order to determine the light distribution on the image plane, we shall make use of an important principle of wave motion.

The Huygens-Fresnel Principle

One of the most interesting and useful properties of optical wave propagation is the Huygens-Fresnel principle. It states that the field produced by a propagating wave front is the same as that which would be produced by an infinity of "secondary" point sources distributed all along that wave front. In the case of a wave propagating through an aperture, the field at any point behind the aperture is the same as that which would be produced by filling the aperture with secondary point sources of the proper amplitude and phase. Mathematically the Huygens-Fresnel principle says that the field at the point (x_i, y_i) in the image plane is given by

$$u_i(x_i, y_i) = \frac{1}{j\lambda} \iint_A u_a(x_a, y_a) \frac{1}{r} e^{jkr} \cos(\theta)\, dx_a\, dy_a \tag{10}$$

with reference to Figure 13-3. The term $u_a(x_a, y_a)$ is the field in the aperture, and the integration is performed over the aperture. The distance from the point of interest at x_i, y_i to the point x_a, y_a in the aperture is r, while θ is the angle between the line connecting those two points and the normal to the plane of the aperture.

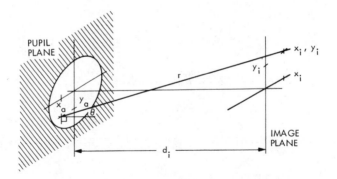

Figure 13-3 Imaging geometry

For our purposes, the angle θ is small enough that $\cos(\theta)$ can be assumed unity. We can extend the integration limits of Eq. (10) to infinity if we multiply the converging wave by the pupil function. This effects truncation by driving the field to zero everywhere in the pupil plane except inside the aperture. Under these conditions, Eq. (10) becomes

$$u_i(x_i, y_i) = \frac{1}{j\lambda} \int_{-\infty}^{\infty} \int_{-\infty}^{\infty} p(x_a, y_a) \frac{1}{R} e^{-jkR} \frac{1}{r} e^{jkr}\, dx_a\, dy_a \tag{11}$$

This distance from the convergence point at the origin of the image plane to the point (x_a, y_a) in the aperture is

$$R = \sqrt{x_a^2 + y_a^2 + d_i^2} \tag{12}$$

and the distance from (x_a, y_a) to (x_i, y_i) is

$$r = \sqrt{(x_i - x_a)^2 + (y_i - y_a)^2 + d_i^2} \tag{13}$$

In Eq. (11), the terms $1/R$ and $1/r$ are well approximated by $1/d_i$. In the exponentials, however, the terms R and r have the large coefficient k, and we must use a better approximation.

The Fresnel Approximation

We can factor d_i out of Eqs. (12) and (13) and write them as

$$R = d_i\sqrt{1 + \left(\frac{x_a}{d_i}\right)^2 + \left(\frac{y_a}{d_i}\right)^2} \tag{14}$$

and

$$r = d_i\sqrt{1 + \left(\frac{x_i - x_a}{d_i}\right)^2 + \left(\frac{y_i - y_a}{d_i}\right)^2} \tag{15}$$

The binomial series expansion of the square root is

$$\sqrt{1 + q} = 1 + \frac{q}{2} - \frac{q^2}{8} + \cdots \qquad |q| < 1 \tag{16}$$

If we use only the first two terms of the binomial series expansion, we produce the Fresnel approximation to the distances in Eqs. (14) and (15), which are

$$R \approx d_i\left[1 + \frac{1}{2}\left(\frac{x_a}{d_i}\right)^2 + \frac{1}{2}\left(\frac{y_a}{d_i}\right)^2\right] \tag{17}$$

and

$$r \approx d_i\left[1 + \frac{1}{2}\left(\frac{x_i - x_a}{d_i}\right)^2 + \frac{1}{2}\left(\frac{y_i - y_a}{d_i}\right)\right]^2 \tag{18}$$

The Coherent Point Spread Function

Substituting the foregoing approximations into Eq. (11) produces

$$u_i(x_i, y_i) = \frac{1}{j\lambda d_i^2} \int_{-\infty}^{\infty}\int_{-\infty}^{\infty} p(x_a, y_a)e^{-jkd_i}\left[1 + \frac{1}{2}\left(\frac{x_a}{d_i}\right)^2 + \left(\frac{y_a}{d_i}\right)\right]^2$$
$$\times e^{jkd_i}\left[1 + \frac{1}{2}\left(\frac{x_i - x_a}{d_i}\right)^2 + \left(\frac{y_i - y_a}{d_i}\right)\right]^2 dx_a\, dy_a \tag{19}$$

After expanding the exponents and collecting terms, Eq. (19) can be written as

$$u_i(x_i, y_i) = \frac{e^{(jk/2d_i)(x_i^2 + y_i^2)}}{j\lambda d_i^2} \int_{-\infty}^{\infty}\int_{-\infty}^{\infty} p(x_a, y_a)e^{(-j2\pi/\lambda d_i)(x_ix_a + y_iy_a)}\, dx_a\, dy_a \tag{20}$$

If we make the variable substitutions

$$x_a' = \frac{x_a}{\lambda d_i} \qquad y_a' = \frac{y_a}{\lambda d_i} \tag{21}$$

then Eq. (20) becomes

$$u_i(x_i, y_i) = \frac{\lambda}{j} e^{(jk/2d_i)(x_i^2 + y_i^2)} \int_{-\infty}^{\infty} \int_{-\infty}^{\infty} p(\lambda d_i x_a', \lambda d_i y_a') e^{-j2\pi(x_i x_a' + y_i y_a')} dx_a' dy_a' \quad (22)$$

We now have the extremely important result that the coherent point spread function is, aside from a complex coefficient, merely the two-dimensional Fourier transform of the pupil function.

The complex exponential coefficient in Eq. (22) affects only the phase in the image plane, and this is ignored by commonly used image sensors. Thus, for our purposes, the term in front of the integral is merely a complex constant.

In Figure 13-2, the point source is on the z-axis. The preceding development can be done with the source located off-axis, and it produces the same result, although shifted as dictated by Eq. (2). This means that, under our assumptions, the system is indeed shift invariant. As the image point moves off-axis, however, the assumptions begin to break down. Thus the point spread function of an imaging system does change (for the worse) in the periphery of the field. It is customary, however, to specify an imaging system by its on-axis point spread function.

Equation (22) gives the amplitude distribution in the image plane produced in response to a point source at the origin of the object plane. The complex terms in front of the integral relate the brightness of the image to that of the point source and describe the phase variations in the image plane. Since typical image sensors ignore phase information, it is of little interest to us here. Furthermore, the overall brightness of the image is most easily determined by a separate analysis, taking into account that portion of the source radiation which is intercepted by the lens. Thus the only parameters of interest to us are those that affect image quality, namely the shape of the point spread function. We can considerably simplify the notation if we give up absolute amplitude calibration and ignore the terms in front of the integral. Then we can write the convolution relation between the object and the image as

$$u_i(x_i, y_i) = \int_{-\infty}^{\infty} \int_{-\infty}^{\infty} h(x_i - x_o, y_i - y_o) u_o(M x_o, M y_o) dx_o dy_o \quad (23)$$

where the impulse response is given by

$$h(x, y) = \mathcal{F}\{p(\lambda d_i x_a, \lambda d_i y_a)\} \quad (24)$$

In Eq. (23), the term $u_o(x_o, y_o)$ is the amplitude distribution of the object and $u_o(M x_o, M y_o)$ is the object after projection without degradation into the image plane. Thus we can consider imaging as a two-step process: projection followed by convolution in the image plane with the point spread function. The magnification factor M is negative unless the coordinate axes in the image plane and object plane are rotated 180° with respect to each other.

Frequently it is most convenient to perform our analysis in the object plane. We can assume that convolution with the psf occurs in the object plane and merely substitute d_o for d_i in Eq. (24). We then convolve the resulting psf with the unprojected object $u_o(x_o, y_o)$.

The Coherent Transfer Function

The transfer function of an optical system is merely the Fourier transform of the impulse response in Eq. (24); however, this is itself a Fourier transform—that of the pupil function. Twice transforming a function merely reflects it about the origin, so the coherent transfer function is given by

$$H(u, v) = p(-\lambda d_i u, -\lambda d_i v) \tag{25}$$

In the common case of symmetrical apertures, the 180° rotation has no effect whatsoever. Thus the pupil function, properly scaled, is the coherent transfer function.

The Incoherent Point Spread Function

A distribution of point sources described by Eq. (6) is adequate to model three kinds of illumination: monochromatic, narrow band spatially coherent, and narrow band incoherent. For monochromatic illumination, $\phi(t)$ is constant. If the light is spatially coherent, $\phi(t)$ is random but bears a fixed relationship to all other points in the image. In the case of incoherent light, $\phi(t)$ is random at each point and independent of its neighbors. In the incoherent case, the observed intensity at a point (x_i, y_i) is given by

$$I_i(x_i, y_i) = \mathcal{E}\{u_i(x_i, y_i)u_i^*(x_i, y_i)\} \tag{26}$$

where the expectation operator $\mathcal{E}\{\ \}$ represents the time average over a period that is long compared with the vibration period of the light source. Since the $u_i(x_i, y_i)$ that results from a point source at the origin of the object plane is given by Eq. (23), we can substitute into Eq. (26) to obtain

$$I_i(x_i, y_i) = \mathcal{E}\left\{\int_{-\infty}^{\infty}\int_{-\infty}^{\infty} h(x_i - x_1, y_i - y_1)u_o(Mx_1, My_1)\, dx_1\, dy_1 \right.$$
$$\left. \times \int_{-\infty}^{\infty}\int_{-\infty}^{\infty} h^*(x_i - x_2, y_i - y_2)u_0^*(Mx_2, My_2)\, dx_2\, dy_2\right\} \tag{27}$$

Since $h(x, y)$ is independent of time, we can rearrange Eq. (27) to yield

$$I_i(x_i, y_i) = \int_{-\infty}^{\infty}\int_{-\infty}^{\infty}\int_{-\infty}^{\infty}\int_{-\infty}^{\infty} h(x_i - x_1, y_i - y_1)h^*(x_i - x_2, y_i - y_2)$$
$$\times \mathcal{E}\{u_o(Mx_1, My_1)u_o^*(Mx_2, My_2)\}\, dx_1\, dy_1\, dx_2\, dy_2 \tag{28}$$

The expectation term is merely the temporal cross-correlation function of u_o at the points (x_1, y_1) and (x_2, y_2). Since, in the incoherent case, the cross-correlation of distinct image point sources is zero, this is a spatial impulse. Furthermore, if $x_1 = x_2$ and $y_1 = y_2$, the value of the expectation term is merely the intensity of the image at that point. This means that

$$\mathcal{E}\{u_o(Mx_1, My_1)u_o^*(Mx_2, My_2)\} = I_o(Mx_1, My_1)\delta(x_1 - x_2, y_1 - y_2) \tag{29}$$

Substituting this into Eq. (28) and carrying out the integration to eliminate the variables x_2 and y_2 produces

$$I_i(x_i, y_i) = \iint |h(x_i - x_o, y_i - y_o)|^2 I_o(Mx_o, My_o)\, dx_o,\, dy_o \tag{30}$$

where the variables x_o and y_o have been substituted for x_1 and y_1. Equation (30) indicates that, under incoherent illumination, the system is linear in intensity, and the point spread function is the squared modulus of the coherent psf. Since $h(x, y)$ is the Fourier transform of the pupil function, the incoherent psf is the power spectrum of the pupil function.

The Optical Transfer Function

The normalized Fourier transform of the incoherent psf is called the *optical transfer function* (OTF). The modulus of the OTF is the modulation transfer function (MTF) discussed in Chapter 2. Since the incoherent psf is the power spectrum of the pupil function, the autocorrelation theorem implies that the incoherent OTF is the normalized autocorrelation function of the pupil function.

$$\text{OTF}(u, v) = \frac{R_p(u, v)}{R_p(0, 0)} = \frac{\int_{-\infty}^{\infty} \int_{-\infty}^{\infty} p(\lambda d_i x, \lambda d_i y) p(\lambda d_i x - u, \lambda d_i y - v) \, dx \, dy}{\int_{-\infty}^{\infty} \int_{-\infty}^{\infty} p^2(\lambda d_i x, \lambda d_i y) \, dx \, dy} \quad (31)$$

Figure 13-4 illustrates, for circular and rectangular apertures, the relationship between the pupil function and the coherent and incoherent point spread and transfer functions.

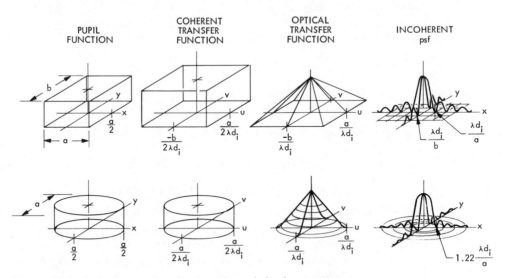

Figure 13-4 Optical properties of rectangular and circular apertures

OTF Design

If the exit pupil of an optical system is an aperture, the pupil function $p(x, y)$ takes on only the values 0 and 1. We can exert some control over the OTF by careful selection of the aperture shape. Since photographic or metal film deposition techniques can be used to implement pupil functions that take on intermediate values, one can exert considerable control over the OTF. For example, Frieden (Ref. 6) has computed the circular pupil functions that maximize the OTF at particular frequencies. Several of these pupil functions and the corresponding OTFs are shown in Figure 13-5. Notice that the circular aperture is very nearly optimal for maximizing the OTF at midrange frequencies. To maximize the OTF at lower frequencies, the transmittance of the pupil falls off with increasing radius. This is called *apodisation*. A central stop of appropriate diameter is near optimal for maximizing the OTF at frequencies above the midrange.

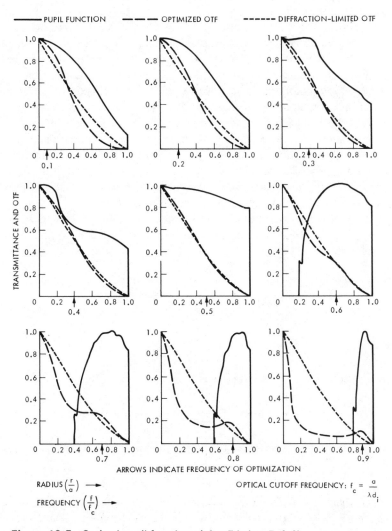

Figure 13-5 Optimal pupil functions (after Frieden, Ref. 6)

In earlier discussions, it was mentioned that an aberration-free optical system produces a spherical exit wave. Aberrations in the optical system cause the exit wave to depart from its ideal spherical shape. This can be modeled as before using Figure 13-2 if we generalize the pupil function by defining it as

$$p(x, y) = T(x, y)e^{jkW(x,y)} \tag{32}$$

where $T(x, y)$ is the transmittance of the pupil as before and $W(x, y)$ accounts for the aberrations. $W(x, y)$ is the path length difference in wavelengths between the actual and the ideal (spherical wave) propagation paths from the point x, y in the aperture to the origin of the image plane.

Proper choice of the aberration function $W(x, y)$ allows one to model the effects of spherical aberration, defocus, astigmatism, coma, field curvature, and image distortion (Ref. 3). Field curvature refers to the situation where the surface of proper focus is a curved surface rather than the image plane. Astigmatism refers to the condition wherein rays coming through the exit pupil on the x_a-axis are not focused to the same point as those coming through on the y_a-axis. Distortion causes straight lines in the object plane to be imaged as curved lines in the image plane. Coma refers to the situation where rays from a single point in the object plane, but passing through opposite sides of the aperture, are converged to different points in the image plane.

A study of optical aberrations is beyond our scope, but two results from that field are of interest. First, there exists no transmittance function $T(x, y)$ that can drive the OTF negative. Second, no aberration function $W(x, y)$ can increase the OTF at any frequency, but aberrations can indeed drive the OTF negative (Ref. 3).

Figure 13-6 illustrates the effect of spherical aberration on the OTF. In this

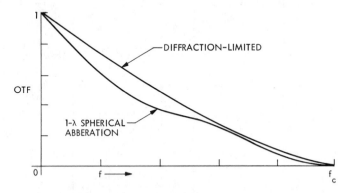

Figure 13-6 Effect of spherical aberration on the OTF (after O'Neil, Ref. 2)

case, there is a path length difference of λ between axial and marginal rays. The image plane is located midway between the marginal and axial focal distances.

Figure 13-7 illustrates the effect of various amounts of defocus (Ref. 7, 8). Here

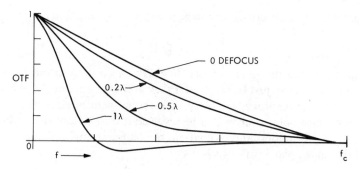

Figure 13-7 Effect of defocus on the OTF (after Stokseth, Ref. 7)

defocus is measured in units of wavelengths of path difference between axial and marginal rays, not by the out-of-focus distance.

Figure 13-8 illustrates, in more detail, the point spread function of diffraction-limited optical systems with circular and rectangular exit pupils. For the circular

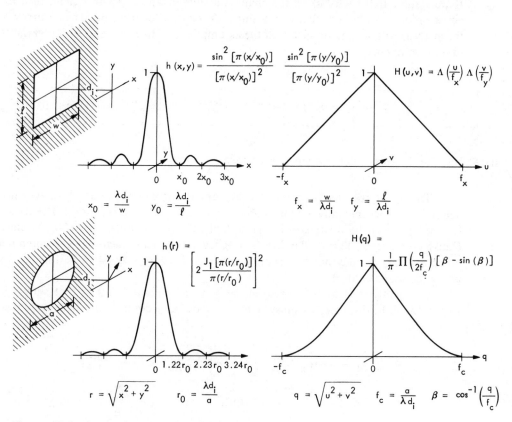

Figure 13-8 Summary of aperture properties

aperture, the first zero of the image plane point spread function occurs at a radius

$$r_o = 1.22\frac{\lambda d_i}{a} \tag{33}$$

which is called the radius of the *Airy disk*. According to the *Rayleigh criterion*, two point sources can just be resolved if they are separated, in the image, by that distance. In optics terminology, the Airy disk radius defines circular *resolution cells* in the image, since point sources can be resolved if they do not fall within the same resolution cell. To a good approximation, the half-amplitude diameter of the central peak of the image plane psf is given by

$$D = \frac{\lambda d_i}{a} \tag{34}$$

THE ANALYSIS OF COMPLETE SYSTEMS

Figure 13-9 shows a linear system model of a typical digital image processing system. If we assume that each link in the chain is a shift invariant linear system, then the entire process can be modeled with a single psf or transfer function. The accuracy of the analysis will depend on how well the assumptions of linearity and shift invariance fit each component.

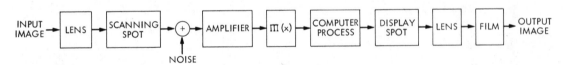

Figure 13-9 The elements of an image processing system

The psf or transfer function of each component can be modeled analytically, determined experimentally, or taken from manufacturer's specifications. The lenses, for example, can be assumed diffraction-limited, the display spot can be assumed Gaussian, and the MTF of the film is supplied by the manufacturer. The computer operation may or may not be linear, but this is the only subsystem in Figure 13-9 under the user's control.

Frequently it is useful to reduce the system in Figure 13-9 to that shown in Figure 13-10. In this case, all subsystems not under user control have been combined

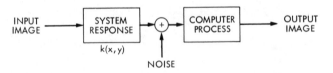

Figure 13-10 The equivalent system

into $k(x, y)$, the overall system psf. Here the computer processes not the original input image but that image degraded by the system characteristics and contaminated with noise. In actual fact, noise is introduced at every step in the process. Image sensor noise is usually the prime offender, excluding noise that may already be present in the input image. We can assume that the noise is introduced at any position in Figure 13-9 provided we account for the modification of its power spectrum as it is filtered by the various linear subsystems.

In the remainder of this chapter, we consider several examples of image processing systems and determine their overall psf's. We shall, for this exercise, ignore the effects of sampling and truncation.

A Film-to-Film System

Figure 13-11 diagrams an image processing system that uses photographic film for both input and output. We shall reduce the system in Figure 13-11 to the system in

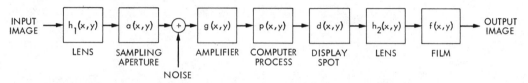

Figure 13-11   A film-to-film system

Figure 13-10 by combining all the transfer functions, except the computer process, into a single equivalent transfer function. We shall assume that the input and output lens systems are diffraction-limited, the sampling aperture is square, the amplifier has a single pole lowpass characteristic, the display spot is Gaussian, and the MTF of the film is supplied by the manufacturer. The overall transfer function is merely the product of the individual transfer functions, and the overall psf $k(x, y)$ is its inverse Fourier transform. Figure 13-12 shows the component psfs and transfer functions and the overall transfer function and psf. The equations for the component point spread functions and transfer functions are listed in Table 13-1.

The point spread functions, and thus the transfer functions, of the two lenses and the display spot are circularly symmetric. We assume that the MTF of the film can be approximated by a product of hyperbolic secant functions. The sampling aperture and the amplifier are characterized by impulse responses separable in the x- and y-directions. Since the image is scanned in the x-direction, $g(x, y)$ is a lowpass filter in the x-direction and an impulse in the y-direction.

Before the transfer functions and psfs of the various components can be compared, they must be projected into a consistent frame of reference. Figure 13-13 illustrates how the various intermediate image planes can be projected back into the input plane. Magnification factors, based on overall image size, allow projection of the psfs into the image plane. Since the amplifier processes an electrical signal, its

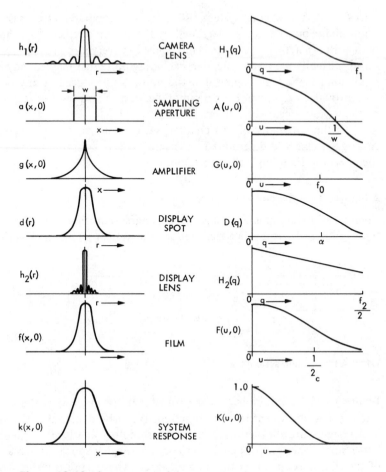

Figure 13-12 Components of the system response

Table 13-1. Point spread functions and transfer functions

$$h_1(r) = \left[2\frac{J_1(\pi r/r_1)}{\pi r/r_1} \right]^2 \quad r_1 = \frac{\lambda d_1}{a_1} \qquad H_1(q) = \frac{1}{\pi} \Pi\left(\frac{q}{2f_1}\right)[\beta_1 - \sin(\beta_1)] \qquad \beta_1 = \cos^{-1}\left(\frac{q}{f_1}\right) \qquad f_1 = \frac{a_1}{\lambda d_1}$$

$$a(x, y) = \frac{1}{w^2}\Pi\left(\frac{x}{w}\right)\Pi\left(\frac{y}{w}\right) \qquad A(u, v) = \frac{\sin(\pi wu)}{\pi wu}\frac{\sin(\pi wv)}{\pi wv}$$

$$g(x, y) = \frac{f_o}{2}e^{-|f_o x|} \qquad G(u, v) = \frac{\delta(v)}{1 + (2\pi u/f_o)^2}$$

$$d(r) = [2\pi\sigma^2]^{-1/2}e^{-r^2/2\sigma^2} \qquad D(q) = e^{-q^2/2\alpha^2} \qquad \alpha = \frac{1}{2\pi\sigma}$$

$$h_2(r) = \left[2\frac{J_1(\pi r/r_2)}{\pi r/r_2} \right]^2 \quad r_2 = \frac{\lambda d_2}{a_2} \qquad H_2(q) = \frac{1}{\pi} \Pi\left(\frac{q}{2f_2}\right)[\beta_2 - \sin(\beta_2)] \qquad \beta_2 = \cos^{-1}\left(\frac{q}{f_2}\right) \qquad f_2 = \frac{a_2}{\lambda d_2}$$

$$f(x, y) = \frac{1}{c^2}\operatorname{sech}\left(\pi\frac{x}{c}\right)\operatorname{sech}\left(\pi\frac{y}{c}\right) \qquad F(u, v) = \operatorname{sech}(\pi cu)\operatorname{sech}(\pi cv)$$

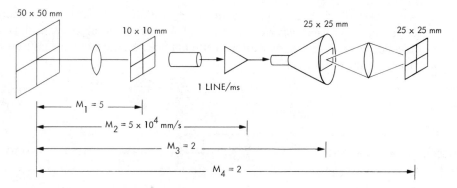

Figure 13-13 Magnification factors

magnification factor reflects a change from time to space. If the scanning mechanism operates at 1 line per millisecond, then every second at the amplifier corresponds to 50 meters at the input plane.

Using the magnification factors in Figure 13-13 to project the assumed point spread functions and transfer functions into the image plane produces the functions summarized in Figure 13-12. As Figure 13-12 shows, the transfer function of the complete system is narrower, and the psf broader, than that of any system component.

In Figure 13-12, the OTF of the entire system is the product of the component transfer functions. It is obvious, for example, that while the camera lens plays a significant role in limiting the overall frequency response, the display lens does not. Also, the amplifier, which has little effect in the x-direction, need not appear at all in an analysis of the y-direction.

A Microscope Digitizing System

As a second example, consider the system shown in Figure 13-14. It consists of a digitizing vidicon television camera mounted on a microscope and can be modeled

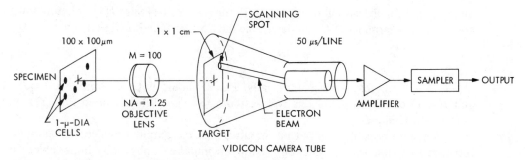

Figure 13-14 A microscope/television digitizing system

as in Figure 13-15. The specimen is imaged by a 100X, 1.25 numerical aperture microscope objective. The *numerical aperture* of a microscope objective is related to its object distance d_o and equivalent aperture diameter a, and to its magnification power M and image distance d_i by

$$\text{N.A.} = \frac{a}{2d_o} = \frac{aM}{2d_i} \tag{35}$$

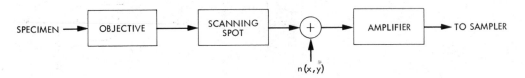

Figure 13-15 Linear components of the microscope digitizer

The Rayleigh resolution criterion (Airy disk radius at the object plane) is conveniently expressed by

$$\delta = \frac{0.61\lambda}{\text{N.A.}} \tag{36}$$

where λ is the mean wavelength of the narrow band incoherent illumination.

Figure 13-16 shows a two-dimensional analysis producing the overall transfer function and point spread function of the system. If noise is introduced at the sensor, its power spectrum will be modified by the transfer function of the amplifier.

Figure 13-17 shows the u-axis components of the various transfer functions in the microscope digitizing system. If the specimens of interest are circular spots that can be modeled by a 1-micron-diameter Gaussian spot, their spectrum is shown in Figure 13-17 as $S(u)$. Since the transfer function of the system stays above 0.5 out to f_s, the frequency limit of the specimen, we would conclude that this system is probably adequate to digitize these specimens.

System Identification

Suppose that for the system in Figure 13-18 the impulse response $h(x, y)$ is unknown and must be determined. We can determine the transfer function directly by

$$H(u, v) = \frac{G(u, v)}{F(u, v)} \tag{37}$$

if $f(x, y)$ is a suitable test signal. Ideally, $F(u, v)$ should not have zeros. If it does, and $H(u, v)$ can be assumed relatively smooth, we can still determine Eq. (37) by numerical techniques.

If it were possible to input an impulse, then the output would be the impulse response (or point spread function). While an impulse is physically impossible, we could get by with a pulse that is narrow compared to the point spread function itself. Even this is difficult to obtain in many cases. Direct measurement of the psf of optical

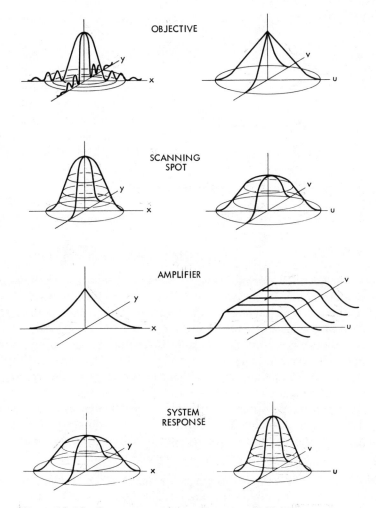

OBJECTIVE

SCANNING
SPOT

AMPLIFIER

SYSTEM
RESPONSE

Figure 13-16 Components of the microscope digitizer response

systems, particularly microscopes, is difficult to perform accurately, and some other means must be used.

The Line Spread Function. Suppose the input to the system is a line of infinitesimal width lying along the y-axis. We can express this as

$$f(x, y) = \delta(x) \tag{38}$$

which may be thought of as the product of a delta function in the x-direction and a constant (unity) in the y-direction. Then the output is given by the convolution

$$g(x, y) = \int_{-\infty}^{\infty} \int_{-\infty}^{\infty} h(p, q)\delta(x - p) \, dp \, dq = \int_{-\infty}^{\infty} h(x, y) \, dy \tag{39}$$

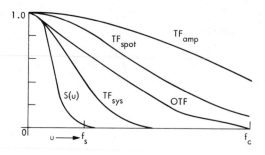

Figure 13-17 A one-dimensional analysis of the microscope digitizer

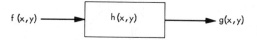

Figure 13-18 A linear system

and the output spectrum is

$$G(u, v) = H(u, v)\delta(v) = H(u, 0)\delta(v) \tag{40}$$

Thus the line input function has the effect of integrating out the y-component of the impulse response. By the projection property mentioned in Chapter 10, the spectrum of the output is merely the transfer function evaluated along the u-axis.

If $h(x, y)$ has circular symmetry, then the transfer function $H(u, v)$ can be completely determined from the line spread function produced by an input line at any orientation. If $h(x, y)$ is separable into a product of a function of x times a function of y, the vertical and horizontal line spread functions of the system are adequate to determine the transfer function. If $h(x, y)$ is asymmetrical, the rotation property of the two-dimensional Fourier transform implies that we can take line spread functions at every angle of orientation, transform them to obtain profiles of $H(u, v)$ at every angle, and thus reconstruct the transfer function. This technique forms the basis of computerized axial tomography, discussed in Chapter 17.

The Edge Spread Function. Suppose the input contains, along the y-axis, an abrupt transition from low to high amplitude. This input can be expressed as a step function in the x-direction times a constant in the y-direction and can be written as

$$f(x, y) = u(x) \tag{41}$$

where $u(x)$ is the unit step function introduced in Chapter 9. Since the edge function is the integral of the line input, and since convolution commutes with differentiation and integration, the edge spread function is the integral of the line spread function. Thus one can differentiate the edge spread function and proceed as before. Alternatively, we can make use of the property that integration merely introduces a coefficient $1/j2\pi s$ into the Fourier transform. Thus

$$G(u, v) = \frac{H(u, 0)\delta(v)}{j2\pi u} \tag{42}$$

from which the transfer function may be determined for nonzero u.

Sine Wave Targets. Perhaps the most reliable means for determining the transfer function involves the use of sinusoidal input functions. Suppose the input is

$$f(x, y) = \cos(2\pi s_0 x) \tag{43}$$

a vertical bar pattern with cosinusoidal profile. Since this input is also constant in the y-direction, the output is

$$g(x, y) = H(s_0, 0) \cos(2\pi s_0 x) \tag{44}$$

and the output spectrum is given by

$$G(u, v) = H(s_0, 0)[\delta(u - s_0) + \delta(u + s_0)]\delta(v) \tag{45}$$

This is an even impulse pair located on the u-axis at $u = \pm s_0$. By repeating this procedure with many different frequencies at many different orientations, one can determine the transfer function to any extent desired. Again, for circularly symmetric or separable transfer functions, the required amount of work is considerably reduced. In fact, the entire job can be done with one input image containing vertically and horizontally oriented sinusoidal bar patterns at several different frequencies. Such an input image is called a *sine wave target*.

Frequency Sweep Targets. Another input that avoids the necessity of transforming the output to determine the transfer function is the frequency sweep target. For purposes of illustration, consider the one-dimensional linear system in Figure 13-19.

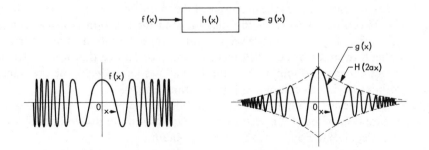

Figure 13-19 System identification with a frequency sweep input

The input is a harmonic signal whose frequency increases linearly with distance from the origin. A harmonic signal with frequency ax is given by

$$f(x) = e^{j2\pi a x^2} \tag{46}$$

The output signal is given by the convolution

$$g(x) = \int_{-\infty}^{\infty} h(\tau)e^{j2\pi a(x-\tau)^2} \, d\tau = e^{j2\pi a x^2} \int_{-\infty}^{\infty} h(\tau)e^{j2\pi a \tau^2} e^{-j4\pi a x \tau} \, d\tau \tag{47}$$

where the second form is obtained by expanding the square in the exponential. If we make the substitution

$$s = 2a\tau \qquad ds = 2a \, d\tau \qquad \tau = \frac{s}{2a} \qquad d\tau = \frac{1}{2a} \, ds \tag{48}$$

and recognize the input signal in front of the integral sign, we obtain

$$g(x) = \frac{1}{2a} f(x) \int_{-\infty}^{\infty} h\left(\frac{s}{2a}\right) e^{j\pi s^2/2a} e^{-j2\pi sx}\, ds \tag{49}$$

We can now recognize the integral in Eq. (49) as the Fourier transform of a product. This can be written

$$g(x) = \frac{1}{2a} f(x) \mathcal{F}\left\{ h\left(\frac{s}{2a}\right) e^{j\pi s^2/2a}\right\} \tag{50}$$

If the impulse response goes to zero outside the interval $-T$ to T, then

$$h\left(\frac{s}{2a}\right) \approx 0 \qquad |s| > 2aT \tag{51}$$

Furthermore, if

$$\frac{(2aT)^2}{2a} = 2aT^2 \ll 1 \tag{52}$$

then

$$e^{j\pi s^2/2a} \approx 1 \qquad |s| \le 2aT \tag{53}$$

and the output reduces to

$$g(x) = \frac{1}{2a} f(x) 2aH(2ax) = f(x)H(2ax) \tag{54}$$

which is merely the input in an envelope that is the transfer function. The assumption in Eq. (52) can be interpreted in two ways. First, it implies that the impulse response is narrow compared to the first cycle of the frequency sweep. By the similarity theorem, this is equivalent to assuming that the transfer function is broad compared to the first cycle of the frequency sweep. If this second condition were not true, it would be difficult to observe the envelope of the output. Notice that using a frequency sweep target under the assumption of Eq. (52) allows us to determine the transfer function without having to compute a Fourier transform.

If we are willing to compute a Fourier transform, we can avoid making the assumption of Eq. (52). Returning to Eq. (50), if we divide both sides by $f(x)$ and take the inverse Fourier transform, we obtain

$$\mathcal{F}^{-1}\left\{ \frac{g(x)}{f(x)}\right\} = \frac{1}{2a} h\left(\frac{s}{2a}\right) e^{j\pi s^2/2a} \tag{55}$$

If we take the magnitude of both sides of Eq. (55), the complex exponential disappears and

$$\left| \mathcal{F}^{-1}\left\{ \frac{g(x)}{f(x)}\right\}\right| = \frac{1}{2a} h\left(\frac{s}{2a}\right) \tag{56}$$

which is easily solved for the impulse response $h(x)$. It is curious that transforming the ratio of the output to the input produces the impulse response rather than the transfer function.

Cross-Correlation. Suppose we cross-correlate the output of a linear system with its input as shown in Figure 13-20. The spectrum of the output of the cross-correlation is

$$Z(s) = G(s)F^*(s) = H(s)F(s)F^*(s) = H(s)P_f(s) \tag{57}$$

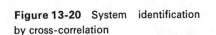 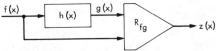

Figure 13-20 System identification by cross-correlation

where $P_f(s)$ is the power spectrum of the input signal. If $f(x)$ is uncorrelated white noise, then $P_f(s)$ is merely a constant and the output of the cross-correlator is the inpulse response of the system.

SUMMARY OF IMPORTANT POINTS

1. Lenses and other optical imaging systems can be treated as two-dimensional shift invariant linear systems.

2. The assumptions involved in the development of linear analysis of optical systems begin to break down as one moves off-axis, particularly for wide-aperture or poorly designed systems.

3. Coherent illumination can be thought of as a distribution of point sources whose amplitudes maintain fixed phase relationships among themselves.

4. Incoherent illumination may be viewed as a distribution of point sources, each having random phase that is uncorrelated with its neighbors.

5. Under coherent illumination, an optical system is linear in complex amplitude.

6. Under incoherent illumination, an optical system is linear in intensity (amplitude squared).

7. The point spread function of an optical system is finitely broad because of two effects: aberrations in the optical system and the wave nature of light.

8. An optical system having no aberrations is called *diffraction-limited* because its resolution is limited only by the wave nature of light (diffraction effects).

9. A diffraction-limited optical system transforms a diverging spherical entrance wave into a converging spherical exit wave.

10. The pupil function gives the transmittance of the plane containing the aperture of the optical system.

11. The coherent point spread function is merely the Fourier transform of the pupil function [Eq. (24)].

12. The coherent transfer function has the same shape as the pupil function [Eq. (25)].

13. The incoherent point spread function is the power spectrum of the pupil function [Eq. (30)].

14. The optical transfer function is the autocorrelation function of the pupil function [Eq. (31)].

15. A diffraction-limited optical system has a real-valued pupil function.

16. Aberrations in an optical system can be modeled by introducing a complex component into the pupil function [Eq. (32)].

17. Careful selection of the pupil function can increase the transfer function at specific spatial frequencies (Figure 13-6).

18. The transfer function of a diffraction-limited optical system can never go negative.

19. Aberrations in an optical system can never increase the modulation transfer function but can drive the optical transfer function negative.

20. Complete image processing systems may be modeled as a cascade of linear subsystems, each having an assumed or experimentally determined point spread function.

21. A linear system can be identified with an input that is an impulse, a line, an edge, a sine wave target, or a frequency sweep target.

22. A linear system can be identified by cross-correlating a white random noise input signal with the system output.

REFERENCES

1. J. W. GOODMAN, *Introduction to Fourier Optics*, McGraw-Hill Book Company, New York, 1968.

2. E. L. O'NEILL, *Introduction to Statistical Optics*, Addison-Wesley, Reading, Massachusetts, 1963.

3. M. BORN and E. WOLF, *Principles of Optics*, Pergamon Press, Oxford, 1964.

4. A. PAPOULIS, *Systems and Transforms with Applications in Optics*, McGraw-Hill Book Company, New York, 1968.

5. R. BRACEWELL, *The Fourier Transform and Its Applications*, McGraw-Hill Book Company, New York, 1965.

6. B. R. FRIEDEN, "Maximum Attainable MTF of Rotationally Symmetric Lens Systems," *Journal of the Optical Society of America*, **59**, 402–406, 1969.

7. P. A. STOKSETH, "Properties of a Defocused Optical System," *Journal of the Optical Society of America*, **59**, 1314–1321, 1969.

8. H. H. HOPKINS, "The Frequency Response of a Defocused Optical System," *Proceedings of the Royal Society (London)*, **A231**, 91–103, 1955.

APPLICATIONS

IMAGE RESTORATION

|||||||||||||||||||||||||||||||||||||||||||||||||||||||||||||||||||||||||||||||||||||||||||||||||||||||||||||||||||||||||||||||||

INTRODUCTION

A very large proportion of digital image processing activity is devoted to image restoration. This includes both research in algorithm development, and routine goal-directed image processing. Many of the important contributions of digital image processing have been made in this area. In this chapter, we summarize a few of the most important techniques.

By image restoration we mean the removal or reduction of degradations that were incurred while the image was being obtained. These include the blurring introduced by optical systems and by image motion, as well as noise due to electronic and photometric sources. While image restoration could be defined to include many of the techniques discussed in Part I, it is generally taken to signify a more restricted class of operations.

In this chapter, we shall use the system shown in Figure 14-1 to model image degradation and restoration. The image $s(x, y)$ is blurred by a linear operation $f(x, y)$, and noise $n(x, y)$ is added to form the degraded image $w(x, y)$. This is convolved with the restoration filter $g(x, y)$ to produce the restored image $z(x, y)$. We shall, in some cases, generalize this model slightly.

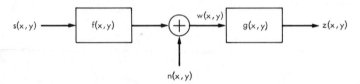

Figure 14-1 Image restoration model

The aim of image restoration is to bring the image toward what it would have been if it had been recorded without degradation. Each element in the imaging chain (lenses, film, digitizer, etc.) contributes to the degradation. Partially restoring the lost image quality can be either a cosmetic frill or a matter of vital importance. The latter case is typified by the lunar and planetary imaging missions in the space program.

Restoring a degraded image can be approached in two different ways. If little is known about the image, one can attempt to model and characterize the sources of degradation (blurring and noise) and subsequently remove or reduce their effects. This is an estimation approach, since one attempts to estimate what the image must have been before it was degraded by relatively well-characterized processes.

If a great deal of prior knowledge of the image is available, it might be more fruitful to model the original image and fit the model to the observed image. As an example of the second case, assume that the image is known to contain only circular objects of fixed size (stars, grains, cells, etc.). Here the task is a detection process, since only a few parameters of the original image are unknown (position, orientation, etc.).

In this chapter, we shall consider several variations of the model in Figure 14-1 and their corresponding approaches to image restoration. We shall also consider the system identification and noise modeling problems. A treatment of this scope is necessarily incomplete. For more thorough coverage, the reader should consult a textbook on the subject (Refs. 1, 2) or a survey article (Refs. 3–9).

APPROACHES AND MODELS

The development of image restoration techniques has been marked by increasing sophistication of the approaches employed. The more recent approaches include attempts to model more accurately the degradation process itself. In this section, we review several of the more important image restoration techniques that have proved useful in actual applications.

Deconvolution

Linear system theory is well developed and has been routinely used in electrical filter design for many years. It has been applied to optics where the Fourier transform property of lenses has been used in crystallography and other fields. The technique of deconvolution has been well-known in electrical filter design and time series analysis. Even the minimum mean-square estimator was worked out by Norbert Wiener in 1942 (Ref. 10). It was not until the mid-1960s, however, that this technique was applied to digital image restoration. This period essentially marks the beginning of the current boom in modern techniques for digital image restoration.

In 1966, Robert Nathan published the digital processing techniques that had been developed for and applied to the early Ranger and Mariner missions (Ref. 11). This brief report outlined techniques for geometric and photometric decalibration, image averaging, periodic noise removal, linear filtering for scan line noise removal, and two-dimensional deconvolution. He used the curves in Figure 14-2 to illustrate the technique.

Since the signal spectrum usually dies out with frequency faster than the noise, the high frequencies, at which deconvolution filters commonly have high gain, are

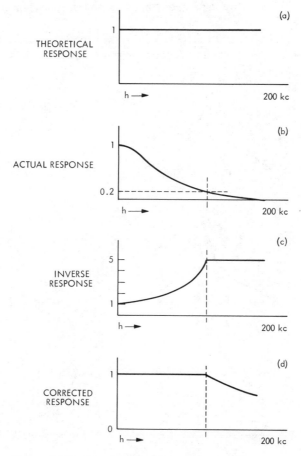

Figure 14-2 Deconvolution (from Nathan, Ref. 11)

usually dominated by noise. Nathan's solution to this problem was to prevent the transfer function of the deconvolution filter from going above some limiting value such as 5.0.

Later that year, Harris (Ref. 4) presented a deconvolution technique based on an analytical model for the point spread function due to atmospheric turbulence in telescope images. The following year, McGlammery (Ref. 12) deconvolved atmospheric turbulence by determining the psf experimentally. Mueller and Reynolds (Ref. 13) implemented a similar technique with coherent optical processing.

Wiener Deconvolution

In most images, adjacent pixels are highly correlated, while the gray level of widely separated pixels is only loosely correlated. From this, we can argue that the auto-correlation function of typical images generally decreases away from the origin (Figure 14-3). Since the power spectrum of an image is the Fourier transform of its autocorrelation function, we can argue that the power spectrum generally decreases with frequency.

Typical noise sources have a flat power spectrum or one that decreases with frequency more slowly than typical image power spectra. Thus the expected situation is

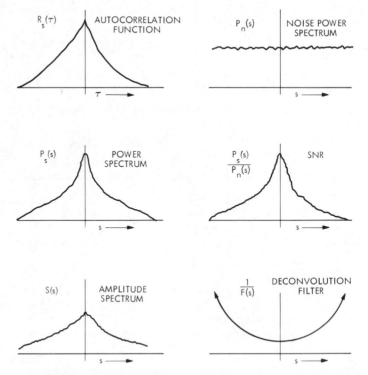

Figure 14-3 The noise problem in deconvolution

for the signal to dominate the spectrum at low frequencies and the noise to dominate at high frequencies. Since the magnitude of the deconvolution filter generally increases with frequency, it enhances high-frequency noise. The early attempts at deconvolution handled the noise problem by *ad hoc* and intuitive methods.

In 1967, Helstrom (Ref. 14) adopted the minimum mean-square error estimation procedure and presented the Wiener deconvolution filter

$$G(u, v) = \frac{F^*(u, v)P_s(u, v)}{|F(u, v)|^2 P_s(u, v) + P_n(u, v)} \tag{1}$$

which may also be written as

$$G(u, v) = \frac{F^*(u, v)}{|F(u, v)|^2 + P_n(u, v)/P_s(u, v)} \tag{2}$$

Recall that this filter was developed in Chapter 11 for one dimension.

Slepian (Ref. 15) extended Wiener deconvolution to account for a stochastic point spread function (atmospheric turbulence). Later, Pratt (Ref. 16) and Habibi (Ref. 17) developed means to increase the computational efficiency of Wiener deconvolution. A review of these techniques has been published by Sondhi (Ref. 6).

Wiener deconvolution affords an optimal method for rolling off the deconvolution transfer function in the presence of noise. However, it is plagued with three problems that limit its effectiveness. First, the mean-square error (MSE) criterion of optimality is not particularly good if the image is being restored for the human eye (Refs. 6, 18). The MSE criterion weights all errors equally regardless of their location in the image. The eye, however, is considerably more tolerant of errors in dark areas and high gradient areas than elsewhere. Secondly, classical Wiener deconvolution cannot handle a spatially variant blurring psf such as that which occurs with coma, astigmatism, curvature of field, and rotary motion blur. Finally, the technique cannot handle the common cases of nonstationary signals and noise. Most images are highly nonstationary, having large flat areas separated by sharp transitions (edges). Furthermore, several important noise sources are highly dependent on local gray level (signal-dependent noise). In the following sections, we examine alternatives to and improvements upon Wiener deconvolution.

Power Spectrum Equalization

Cannon (Ref. 19) showed that the filter which restores the power spectrum of the degraded image is given by

$$G(u, v) = \left[\frac{P_s(u, v)}{|F(u, v)|^2 P_s(u, v) + P_n(u, v)} \right]^{1/2} \tag{3}$$

This restores the power spectrum of the image to what it was before degradation. Like the Wiener filter, the power spectrum equalization (PSE) filter is phaseless (real and even). It is applicable for phaseless blurring functions, or phase may be determined by other methods (Ref. 1).

The similarity between the power spectrum equalization filter [Eq. (3)] and the Wiener deconvolution filter [Eq. (1)] is clear. Both filters reduce to straight deconvolution in the absence of noise, and both cut off completely in the absence of signal.

However, the PSE filter does not cut off at zeroes in the blurring function $F(u, v)$. The image restoration power of the PSE filter is quite good, and, in some cases, it may be preferable to Wiener deconvolution. The PSE filter is sometimes called a homomorphic filter, since it originally resulted from an earlier development (Ref. 20).

Geometrical Mean Filters

Consider the restoration filter given by

$$G(u, v) = \left[\frac{F^*(u, v)}{|F(u, v)|^2} \right]^\alpha \left[\frac{F^*(u, v)}{|F(u, v)|^2 + \gamma P_n(u, v)/P_s(u, v)} \right]^{1-\alpha} \tag{4}$$

where α and γ are positive real constants. This is a generalization of the filters previously discussed. The filter is parameterized in α and γ. Notice that if $\alpha = 1$, Eq. (4) reduces to a deconvolution filter. If $\alpha = 0$ and $\gamma = 1$, it becomes the Wiener deconvolution filter of Eq. (2). If γ is not restricted to unity, the result is called a *parametric Wiener filter*. The constant γ may be selected for any desired amount of Wiener-type smoothing. Also, $\gamma = 0$ reduces the parametric Wiener filter to straight deconvolution. Finally, if $\alpha = \frac{1}{2}$ and $\gamma = 1$, Eq. (4) reduces to the PSE filter of Eq. (3). Notice that $\alpha = \frac{1}{2}$ in Eq. (4) defines a filter that is the geometric mean between ordinary deconvolution and Wiener deconvolution. Thus a third name for the filter in Eq. (3) is the geometric mean filter. It is more common, however, to refer to the more general filter in Eq. (4) as the geometric mean filter.

Equation (4) represents a very general class of restoration filters applicable in cases involving linear, space invariant blurring functions and additive uncorrelated noise. Andrews and Hunt (Ref. 1) have examined the restoration power of the filter in Eq. (4) under conditions of slight blurring and moderate noise. They show that, under these conditions, straight deconvolution is least desirable and Wiener deconvolution produces lowpass filtering more severe than the human eye desires. The parametric Wiener filter with γ less than unity and the geometric mean filter with the same constraint produce more pleasing results.

Spatially Variant Blurring

While optical defocus and linear motion blur are spatially invariant linear operations, astigmatism, coma, curvature of field, and rotary motion are spatially variant. The most direct and, to date, most effective restoration technique is coordinate transformation restoration (CTR). This technique involves a geometric transformation on the degraded image that makes the resultant blurring function spatially invariant. This is followed by an ordinary spatially invariant restoration technique and, finally, by a geometric transformation that inverts the first such operation and puts the image back into its original format. Robbins and Huang (Refs. 21, 22) have applied this technique to coma and Sawchuk has applied it to nonlinear motion blur (Refs. 23, 24) and to astigmatism and curvature of field (Ref. 25). For these spatially variant degradation sources, the required geometric transformations are known, and the restoration results are impressive.

Temporally Variant Blurring

The diffraction-limited resolution of a 200-in. telescope is approximately 0.05″ of arc. Under unfavorable conditions, however, atmospheric turbulence can reduce this resolution to about 2″ of arc. Viewing stars through a turbulent atmosphere is similar to watching a point source of light through a moving textured glass shower door. In short exposure, atmospheric turbulence produces a speckle pattern due to phase distortion in the nonuniform atmosphere above the telescope. In longer exposures, atmospheric turbulence causes the speckle pattern to "dance" as the atmosphere undergoes change. Thus long exposures integrate the dancing speckles to produce a large blur, much larger than the diffraction limited psf of the telescope. Since long exposures are required for photographing faint stars, atmospheric blurring (the so-called "seeing" conditions) place a limit on astronomical resolution.

Time averaging in the spatial domain is equivalent in the frequency domain to averaging complex spectra. The time-averaged transfer function so obtained goes to zero at frequencies well below the diffraction limit of the telescope. Thus, under random phase distortion, time-averaging does more harm than good.

Labeyrie (Refs. 26, 27) has shown experimentally that the time-averaged power spectrum of a point star image goes out to the diffraction limit. This means that the random phase fluctuations in the atmosphere average out in the power spectrum of the image. His restoration technique (speckle interferometry) consists of obtaining time-averaged power spectra of both the astronomical object of interest and a reference point star. He effects deconvolution by dividing the object power spectrum by that of the point star. The result is a diffraction-limited power spectrum of the unknown object. This can be inverse transformed to obtain the autocorrelation function of the object. Since phase information has been lost in the power spectrum, the object cannot be reconstructed exactly, but the autocorrelation function is adequate for identifying double stars and for limb darkening studies.

Knox has extended Labeyrie's technique to recover the phase information and obtain diffraction-limited images even under relatively poor seeing conditions (Refs. 28, 29). Like Labeyrie, he uses an ensemble average of short-exposure power spectra to determine the power spectrum of the object. Phase information is taken from the ensemble autocorrelation of the instantaneous power spectra (Refs. 29, 30). A computer simulation strongly suggests that near-diffraction-limited imaging can be obtained by this technique.

Nonstationary Signals and Noise

The filters discussed earlier in this section all involve the assumption of stationary signal and noise. For an image to be stationary, the locally computed power spectrum would have to be the same (or approximately so) over the entire image. Unfortunately, this is seldom the case. Most images are, in fact, highly nonstationary. Consider, for example, a photograph of the human face. Clearly, the power spectrum of a local area containing the forehead would show much less high-frequency energy than the power spectrum of an area containing the eyes. A very large class of images can be modeled as a collection of regions of relatively constant gray level, separated by boundaries with relatively high gradient. Aerial photographs of farmland are but one example.

Several common noise sources cannot be modeled accurately as stationary random processes. Film grain noise, for example, is almost nonexistent in the low density (least exposed) areas of a photographic negative, but the noise level increases with increasing density. Density digitizers, which follow an intensity detector with a logarithmic amplifier, produce a higher noise level in dark areas, where the small signal gain of the logarithmic amplifier is highest.

It is clear that, while the generalized Wiener filter is superior to straight deconvolution, it does not represent an upper limit on image restoration performance.

Matrix Formulation. In Chapter 9, imposing the constraint of shift invariance allowed us to reduce the superposition integral to a simple convolution. If we do not impose shift invariance, the superposition that models image degradation can be written in matrix notation as

$$\mathbf{W} = \mathbf{FS} + \mathbf{N} \tag{5}$$

where the model of Figure 14-1 has been discretized. For digital images of N by N pixels, the matrices $\mathbf{X}$, $\mathbf{S}$, and $\mathbf{N}$ are N^2 by 1 matrices formed by placing the lines end to end (Ref. 1). The degradation matrix $\mathbf{F}$ is N^2 by N^2. This means that each pixel of $\mathbf{S}$ is degraded by convolution with a different N by N blurring function. A minimum mean-square estimator can be derived for this formulation. For the generalized geometric mean filter, the restored image is given by

$$\mathbf{Z} = [(\mathbf{F}^{*t}\mathbf{F})^{-1}\mathbf{F}^{*t}]^{\alpha}[(\mathbf{F}^{*t}\mathbf{F} + \gamma[\boldsymbol{\phi}_s]^{-1}[\boldsymbol{\phi}_n])^{-1}\mathbf{F}^{*t}]^{1-\alpha}\mathbf{W} \tag{6}$$

where $\boldsymbol{\phi}_s$ and $\boldsymbol{\phi}_n$ are the signal and noise covariance matrices, respectively.

Notice that Eq. (6) is the matrix algebraic equivalent of Eq. (4). Notice also that if $N = 1000$, the matrix F has a trillion (10^{12}) elements. Furthermore, if the degrading function has zeros, F will be singular. Clearly, Eq. (6) represents a formidable computation task. Under certain simplifying assumptions, Eq. (6) can be reduced to manageable computations, and some impressive examples have been generated (Ref. 1). However, the full power of this formulation has yet to be exploited in routine applications.

Local Stationarity. While images are seldom stationary in a global sense, they frequently can be assumed locally stationary. This means that the local power spectrum changes slowly as one moves a small window about within the image. In certain images this assumption might be quite good and in others marginal or questionable, but it represents a significant improvement over the assumption of global stationarity. In most practical image restoration applications, the restoring psf is relatively small compared to the size of the image. If the image is locally stationary in regions covering at least the extent of the psf, then the assumption is reasonably well justified.

One way to implement restoration under a model of local stationarity is to use the Wiener filter or its generalization [Eq. (4)] where the power spectra of the signal and/or the noise are functions of position in the image. Unless these power spectra can be modeled by simple formulas having few spatially variant parameters, the computational expense will be relatively high. Furthermore, it is necessary to determine the local power spectra throughout the image before the filter can be spatially parameterized.

A simple approach is to use the generalized geometric mean filter of Eq. (4) where the parameters γ and α are spatially variant and derived from the image. However, Eq. (4) is written in the frequency domain. If the restoration is implemented by convolution, α and γ do not appear as simple parameters in the convolution kernel.

A simpler method is to specify analytically a convolution kernel having one or more image-derived spatially variant parameters. This represents a computational simplification, since only a new convolution kernel need be computed, from a stored equation, at each pixel position.

Power Spectrum Parameters. Let us model the signal and noise as being space-variant but locally stationary. By this we mean that there are two scales in the image. On a small scale the image is stationary, but on a large scale it is not. To illustrate, suppose we estimate the local power spectrum of the image at the point x_1, y_1 by computing the squared magnitude of the two-dimensional Fourier transform of a relatively small rectangular piece of the image centered on x_1, y_1. Suppose we do the same thing using an identical window centered on a second point x_2, y_2. If the two points are relatively close together the estimated power spectra will be approximately equal, even if the two windows do not overlap. On the other hand, if the two points are widely separated in the image, the estimated power spectra will not necessarily agree. While this concept involves some approximations, it does allow us to extend previously developed techniques to account for common forms of space-variance.

If the signal and noise are uncorrelated, then the local variance of the observed image is the sum of the signal and noise component variances

$$\sigma_w^2(x, y) = \sigma_s^2(x, y) + \sigma_n^2(x, y) \tag{7}$$

where the variances are computed over a relatively small local window centered on x, y. Let us assume the noise is locally white with zero mean and power (mean square amplitude or variance) proportional to local mean gray level. Then the noise power spectrum and variance are related by

$$P_n(u, v, x, y) = P_n(0, 0, x, y) = \sigma_n^2(x, y) = N_0 \mu_w(x, y) \tag{8}$$

where N_0 is a constant and $\mu_w(x, y)$ is the average gray level computed over some local window centered on x, y.

Let us also assume that the signal power spectrum is separable into a prototype power spectrum $P_0(u, v)$ times a spatially variant factor, that is,

$$P_s(u, v, x, y) = f(x, y)P_0(u, v) \tag{9}$$

The resulting signal variance is

$$\sigma_s^2(x, y) = R_s(0, 0, x, y) = \int_{-\infty}^{\infty} \int_{-\infty}^{\infty} f(x, y)P_0(u, v) \, du \, dv = f(x, y)R_0 \tag{10}$$

where R_0 is the volume under the prototype power spectrum. Solving for the space-variant factor produces

$$f(x, y) = \frac{\sigma_s^2(x, y)}{R_0} = \frac{\sigma_w^2(x, y) - N_0 \mu_w(x, y)}{R_0} \tag{11}$$

where Eqs. (7) and (8) have been used. The signal power spectrum can now be written in terms of the local mean and variance of the observed image

$$P_s(u, v, x, y) = \frac{P_0(u, v)}{R_0}\left[\sigma_w^2(x, y) - N_0\mu_w(x, y)\right] \qquad (12)$$

We can now write a signal-dependent spatially variant generalized Wiener filter by substituting Eqs. (8) and (12) into

$$G(u, v, x, y) = \left[\frac{F^*(u, v)}{|F(u, v)|^2}\right]^\alpha \left[\frac{F^*(u, v)P_s}{|F(u, v)|^2 P_s + \gamma P_n}\right]^{1-\alpha} \qquad (13)$$

The spatially variant parameters $\mu_w(x, y)$ and $\sigma_w^2(x, y)$ must be computed from the input image. This means that the restoration must be preceded by a step that computes a mean image and a variance image from the input image. To implement Eq. (13) directly would represent considerable computational expense.

A more practical solution is to produce a two-dimensional histogram of $\mu_w(x, y)$ versus $\sigma_s^2(x, y)$ and look for clusters of pixels in mean versus variance space. The space could then be partitioned into regions containing these clusters. The resulting regions could be mapped back into the image to define areas of relatively constant mean and variance. Then a restoration filter could be designed and implemented on each such area. In this way, spatially variant restoration would be only a few times more expensive than simple stationary restoration.

For example, one could partition the degraded image into disjoint regions having four types of image content. The four regions would correspond to the four possible combinations of high and low mean gray level with high and low signal variance. Four image restoration filters would be used, each in its appropriate region. If the restoration filters all had equal zero frequency response, the region boundaries would have, at most, slope discontinuities. This means that the region boundaries would not be highly visible in the processed image. If more exact restoration were required, one could divide the range of mean and signal variance into smaller intervals. While this technique is several steps removed from full-fledged spatially variant restoration, it can represent a significant improvement over the assumption of global stationarity.

Noise Power Ratio. Recall from Eq. (4) that the generalized Wiener filter responds only to the ratio of noise-to-signal power. The signal and noise power spectra do not appear independently in the filter equation. A simplified restoration procedure results if we assume that the signal and noise power spectra change in amplitude throughout the image but not in functional form. This means that the signal-to-noise ratio function also changes only in amplitude throughout the image.

If the noise is locally white and the signal-dependent amplitude is given by Eq. (8), we can write the ratio of noise-to-signal power spectra as

$$\frac{P_n(u, v, x, y)}{P_s(u, v, x, y)} = \frac{R_0 N_0}{P_0(u, v)}\left[\frac{\mu_w(x, y)}{\sigma_w^2(x, y) - N_0\mu_w(x, y)}\right] = \frac{R_0 N_0}{P_0(u, v)}\text{NPR}(x, y) \qquad (14)$$

which is written as a product of frequency-dependent and position-dependent terms. We shall call the term in brackets the *noise power ratio*, NPR(x, y). It represents the spatial variability of the ratio of power spectra. It is easily computed from the mean and variance images of the degraded image.

The function $NPR(x, y)$ can be viewed as an image itself. Its gray level represents the spatially variant noise-to-signal power ratio. This, in turn, is sufficient to specify the spatial variation of a restoration filter. One could threshold $NPR(x, y)$ at several gray levels to partition the degraded image into regions of roughly similar signal-to-noise ratio. A different restoration filter could then be used in each region.

Linear Combination Filters. There is another way to use the noise power ratio image to guide spatially variant restoration. This technique is relatively inexpensive and implements a smoothly space-variant restoration impulse response. Suppose we generate a "mask" function $m(x, y)$ by normalizing $NPR(x, y)$ to the range $[0, 1]$. Then the value zero corresponds to the minimum, and unity to the maximum, noise-to-signal power ratio in the image. Next we design two restoration filters, $g_1(x, y)$ and $g_2(x, y)$ which correspond to the cases of low and high $NPR(x, y)$, respectively.

We now convolve the image with the two restoration filters. These operations are given by

$$z_1(x, y) = w(x, y) * g_1(x, y) \tag{15}$$

and

$$z_2(x, y) = w(x, y) * g_2(x, y) \tag{16}$$

where $g_1(x, y)$ and $g_2(x, y)$ are the stationary restoration filters resulting from Eq. (4) under high noise and low noise conditions, respectively. The restored image is

$$z(x, y) = m(x, y)z_1(x, y) + [1 - m]z_2(x, y) \tag{17}$$

The final restoration can be written as

$$z(x, y) = w(x, y) * \{m(x, y)g_1(x, y) + [1 - m(x, y)]g_2(x, y)\} \tag{18}$$

If $m(x, y)$ is slowly varying compared to the extent of the restoration filter impulse responses, it may be assumed locally constant. Under this assumption, multiplication approximately commutes with convolution and

$$g(x, y) = m(x, y)[g_1(x, y) - g_2(x, y)] + g_2(x, y) \tag{19}$$

Linear combination restoration consists of the following steps. First the degraded image is processed to obtain a local mean gray level image and a local variance image. Next the mask function $m(x, y)$ is formed by normalizing $NPR(x, y)$ [see Eq. (14)]. Then stationary restoration filters $g_1(x, y)$ and $g_2(x, y)$ are designed for the two cases corresponding to the lowest and highest signal-to-noise ratios existing in the image. Two partially restored images $z_1(x, y)$ and $z_2(x, y)$ are formed by convolving the input image with each of the restoration filters. The final restored output is formed by

$$z(x, y) = m(x, y)[z_1(x, y) - z_2(x, y)] + z_2(x, y) \tag{20}$$

Linear combination restoration implements the smoothly spatially variant impulse response of Eq. (19) and avoids partitioning the image and the risk of visible region boundaries. It is somewhat more complex than globally stationary restoration, since it involves four local operations (mean, variance, and two convolutions) and the algebraic operations of Eqs. (14) and (20). While it is not the optimal filter, it does have the desired behavior, smoothing most in areas of low signal-to-noise ratio.

SUPERRESOLUTION

In Chapter 13, we showed that the incoherent transfer function of an optical system is the autocorrelation function of its pupil function. This implies that the transfer function is necessarily band-limited; that is, it goes to zero for all frequencies above some cutoff frequency. This cutoff frequency corresponds to the diffraction limit of resolution. Clearly, with deconvolution we could hope to restore the spectrum of an object only out to, but not beyond, the diffraction limit. Energy at frequencies beyond the diffraction limit appears hopelessly lost. Resolution beyond the diffraction limit is theoretically possible, however, due to a useful property of the Fourier transform. Restoration procedures that seek to recover information beyond the diffraction limit are referred to as "superresolution" techniques.

If a function $f(x)$ is spatially bounded (zero outside some finite interval), then its spectrum $F(s)$ is an analytic function. Analyticity imposes a severe constraint upon how "wiggly" a function can be. A well-known property of analytic functions is that if such a function is known over a finite interval, then it is known everywhere (Ref. 31). This means that if two analytic functions agree exactly over any given interval, then they must agree everywhere and be identically the same function. Stated in yet a different way, this property means that, given a curve defined over a particular interval, no more than one analytic function can be fitted exactly to the given curve over that interval. Thus, there cannot exist two analytic functions that agree over a given interval and yet disagree outside the interval. The process of reconstructing an analytic function in its entirety, given the values of that function over a specified interval, is called *analytic continuation*.

Since an image is necessarily spatially bounded, its spectrum must be analytic. Ignoring noise for the moment, the spectrum of an image can be determined over the interval from zero to the diffraction limit. Thus it should be possible to reconstruct the analytic spectrum at frequencies beyond the diffraction limit.

It was pointed out in Chapter 12 that a truncated (spatially bounded) function cannot be band-limited. However, diffraction-limited optical systems attempt to enforce band-limitedness on truncated functions. It is the incompatibility between spatial bounding and band-limiting that superresolution techniques attempt to exploit.

Harris (Ref. 32) questioned whether the diffraction limit is a theoretical upper limit on resolution or merely a practical limitation. He showed that no two spatially bounded objects produce identical images unless the objects themselves are identical. From this, it follows that, under noiseless conditions, any recorded image corresponds to one and only one object, and thus it should be possible to reconstruct the object in infinite detail from its diffraction-limited image. He presented a technique, based on the sampling theorem, for such reconstruction and performed a one-dimensional simulation to illustrate the technique. In his example, he resolved two point sources separated by one-fifth the Rayleigh resolution criterion distance.

Of the various ways to reconstruct an imaged object with resolution beyond the diffraction limit, the most popular has involved the use of prolate spheroidal wavefunctions (Refs. 33–38), although one technique involving superimposed sinusoidal masks has been proposed (Ref. 39). Several authors have considered the effect of noise on the reconstruction process (Refs. 36–38). While some authors present impressive one-dimensional simulations of the technique, practical applications are not forthcoming. In fact, Andrews and Hunt (Ref. 1) refer to "the myth of superresolution" and argue that noise constraints preclude any practical extension of resolution beyond the diffraction limit.

In this section, we present the superresolution technique advanced by Harris (Ref. 32) and restated by Goodman (Ref. 40). The technique involves applying the sampling theorem, with domains reversed, to obtain a system of linear equations that can be solved for values of the signal spectrum outside the diffraction-limited passband.

Consider the function and its spectrum shown in Figure 14-4. Since $f(x)$ is spatially bounded, we can apply the sampling theorem as before but with the time

Figure 14-4 A spatially bounded function and its spectrum

and frequency domains reversed. The sampling theorem states that $F(s)$ can be completely reconstructed from a series of equally spaced sample points provided they are taken no more than $1/2T$ apart. This reconstruction can be expressed as

$$F(s) = [\mathrm{III}(2sT)F(s)] * \frac{\sin(2\pi sT)}{2\pi sT} \qquad (21)$$

which accounts for sampling $F(s)$ every $1/2T$ and then interpolating to recover the function. Writing the Shah function as an infinite sum of impulses,

$$F(s) = \left[\sum_{n=-\infty}^{\infty} \delta(s - 2nT)F(s) \right] * \frac{\sin(2\pi sT)}{2\pi sT} \qquad (22)$$

and exploiting the sifting property of the impulse produces

$$F(s) = \sum_{n=-\infty}^{\infty} F(2nT) \frac{\sin(2\pi sT - 2nT)}{2\pi sT - 2nT} \qquad (23)$$

Suppose $f(x)$ is passed through a linear system that passes no energy above some frequency s_m. Deconvolution will recover the signal such that the spectrum is known exactly for frequencies out to s_m. Thus direct measurement followed by deconvolution (if necessary) recovers the spectrum for frequencies less than s_m.

Suppose that $F(s)$ is sampled so that M sample points fall within the passband $-s_m \leq s \leq s_m$ (Figure 14-5). Assume further that we desire to determine $F(s)$ over

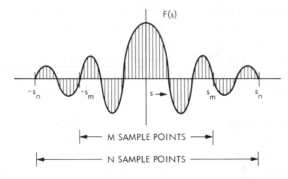

Figure 14-5 Sampled spectrum representation

the range $-s_n$ to s_n, where n implies some larger number N of sample points. Then an estimate of the spectrum, band-limited at s_n (which is larger than s_m) can be computed from

$$F(s) \approx \hat{F}(s) = \sum_{n=-N}^{N} F(2nT) \frac{\sin(2\pi sT - 2nT)}{2\pi sT - 2nT} \qquad (24)$$

If we can compute $\hat{F}(s)$, we have successfully extended the band limit of the function from s_m out to s_n.

Equation (24) may be viewed as a linear equation in $2N + 1$ unknowns. The unknowns are the values of $F(2nT)$ at the sample points. Since the spectrum is known for $|s| \leq s_m$, we can generate a system of $2N + 1$ linear equations in $2N + 1$ unknowns by selecting $2N + 1$ frequencies within the passband and substituting the known values of $F(s)$ into Eq. (24). Classical techniques can be used to solve the system of linear equations for the unknown values of the spectrum. These values can then be substituted into Eq. (24) to generate an estimate of the spectrum that is band-limited at a frequency higher than the diffraction limit of the imaging system.

In practical cases, N may be relatively large and the linear equation solution computationally expensive. Since the spectrum of real functions is Hermite, however, this cuts the number of equations in half. Furthermore, since the spectrum is known below the diffraction limit, only those points falling between s_m and s_n need be computed.

SYSTEM IDENTIFICATION

Before image restoration can be accomplished, the psf of the blurring function must be known. In some cases, this is known in advance, but in others it must be determined experimentally from the degraded image.

Direct psf Measurement

Calibration Targets. In many cases, the transfer function of a system can be measured directly, once and for all, before the system is put into use. This is customary for high-quality systems intended specifically for digital image processing. As described in Chapter 13, this can be done with a sine wave target or a frequency sweep target. This approach has been used successfully for the lunar and planetary exploration missions. The major complication that arises is the change in camera characteristics during the trauma of blastoff. Since sine wave targets are difficult to generate, particularly in the small sizes required for microscope psf measurements, bar targets are sometimes used to measure the square wave response of the system, and this is used to approximate the transfer function.

The psf from the Image. In some cases, it is impractical or impossible to calibrate the imaging system in the same conditions under which a particular degraded image was recorded. This is true for motion blur and stochastic degradation such as atmospheric turbulence, and in cases in which a photograph is presented for restoration and the original camera system is unavailable. In such instances, one must attempt to determine the degrading psf from the image itself.

If one can arrange for the degraded image to include a point source of light or a vanishingly small dark spot on a white background, then the psf is available directly. If the point source or spot is of nonnegligible extent, then it can be modeled with a Gaussian, a flat-topped circular pulse, or some other suitable function that can be deconvolved to yield the point spread function.

This technique is perhaps most valuable in astronomical photographs severely degraded by atmospheric turbulence. Here point sources (stars) are readily available and the degradation is severe enough that deconvolution can be a great help. In high-quality images, such as diffraction-limited camera images, it is difficult to find a point source or speck small enough to show the psf and still large enough to come through the system with sufficient energy to be measured. Since the psf occupies a very small portion of the image, it is particularly vulnerable to corruption by system noise. For this reason, direct measurement of the psf using a point source in the image is of limited use.

If the image contains any feature that can be modeled analytically, then theoretically the psf can be obtained by deconvolution of the model. Particularly valuable in this regard are lines and edges. Under the projection theorem of the two-dimensional Fourier transform, the Fourier transform of the line spread function (lsf) gives a one-dimensional component of the two-dimensional transfer function. The line spread function approach has the advantage that a line source in the image can be averaged along its extent to generate a relatively noise-free estimate. For high-quality systems, however, the line object in the image must be extremely thin. Therefore it must be extremely bright (or dark) relative to its background in order to come through with sufficient amplitude.

Most images of ordinary scenes contain features that can be modeled as ideal edges. Such an edge can be averaged along its extent to produce a relatively noise-free estimate of the system edge spread function (esf). This can be differentiated to produce the lsf, and that can be transformed to produce a component of the transfer function (Refs. 1, 41, 42). If the psf is known to be circularly symmetric, the lsf may be simply rotated to produce the psf. If the psf, and hence the transfer function, are separable into a product of functions, then one vertical and one horizontal edge are sufficient to determine the transfer function. In the general case, edges at many different orientations may be required to determine the transfer function adequately.

Proper use of the esf is not as simple as it might seem, and accurate determination of the transfer function requires considerable care. Since differentiation is a highpass operation, noise in the esf will appear amplified in the resulting lsf. Thus maximum use of averaging on the edge should be employed. If the edge is not perfectly straight and parallel to the sampling raster, however, averaging will blur the edge and make the transfer function appear more of a lowpass filter than it really is. A skewed linear edge may be brought parallel to the sampling raster by a geometric transformation that effects rotation. Unless the scene is considerably oversampled, however, the inherent interpolation will also tend to blur the edge.

Another common mistake in the esf method is to consider the edge spread function over only a narrow region in the vicinity of the edge. Rabedeau (Ref. 43) shows that for a diffraction-limited system, the edge spread function must be considered over a width of almost 10 Airy disk diameters before the transfer function errors due to esf truncation drop below 2%. In general, psf determination from the degraded image must be done very carefully.

OTF FROM THE DEGRADED IMAGE SPECTRUM

Ordinarily, images of complex scenes have reasonably smooth amplitude spectra. If the degrading transfer function has zeros (linear motion blur, for example), these zeros will tend to force the spectrum of the degraded image to zero. If the blurring function is adequately modeled, the locations of the zeros (or near-zeros) in the spatial frequency plane allow determination of the unknown parameters of the blurring OTF. Visualization of the zeros in the spectrum is sometimes aided by preprocessing with a highpass filter (Ref. 44).

By taking the logarithm of the degraded image power spectrum, one can enhance the amplitude of the dips due to the degrading transfer function. If the zeros are equally spaced, they produce a series of periodic spikes in the log power spectrum. The power spectrum of the log power spectrum, sometimes called the *cepstrum* (Ref. 20), is useful for determining the exact spacing of the spikes and consequently the zeros of the degrading transfer function (Refs. 19, 45).

Image Segmentation. A perhaps more powerful technique involves segmenting the degraded image into square regions that are large compared to the extent of the degrading psf (Ref. 46) and averaging the log power spectrum of all such regions (Ref. 47). For complex scenes, the signal components tend to average out in the average log spectrum where the degrading transfer function, which is constant throughout the image, does not. The average log power spectrum converges approximately to the logarithm of the squared magnitude of the degrading transfer function (Ref. 1).

NOISE MODELING

The common noise sources that corrupt images can be divided into three categories. Images originally recorded on photographic film are subject to degradation by film grain noise. Secondly, the conversion of an image from optical to electrical form is a statistical process since, in reality, each picture element receives a finite number of photons. Finally, electronic amplifiers that process the signal introduce thermal noise. Considerable effort has been devoted to modeling noise from these three sources. Those efforts are reviewed in this section.

Electronic Noise

Electronic noise due to the random thermal motion of electrons in resistive circuit elements is the simplest of the three sources to model. This type of noise has been successfully modeled by circuit designers for a long time. It is usually modeled as white Gaussian noise with zero mean value. Thus it has a flat power spectrum and is completely specified by its RMS value (standard deviation). Sometimes electronic circuits exhibit so-called "one over f noise." This is random noise with an intensity that dies out inversely with frequency. However, image processing problems seldom require modeling of this component (Ref. 48).

Photoelectronic Noise

Photoelectronic noise is due to the statistical nature of light and of the photoelectronic conversion process inherent in image sensors. At low light levels, where the effect is relatively severe, photoelectronic noise is often modeled as random with a Poisson density function (Ref. 1). The standard deviation of this distribution is equal to the square root of the mean.

At high light levels, the Poisson distribution approaches the Gaussian, which is simpler to model. Again, the standard deviation (RMS amplitude) is equal to the square root of the mean. This implies that the noise amplitude is signal-dependent.

Film Grain Noise

As described in Chapter 2, the photographic emulsion consists of silver halide crystals suspended in gelatin. Photographic exposure is a binary process, with each grain being either totally exposed or unexposed. During development, exposed grains are reduced to opaque pure silver grains, while unexposed grains are washed off. Thus the variable density of a photographic negative is due to variations in the concentration of silver grains. Under microscopic examination, the smooth tones of a photographic image assume a random "grainy" appearance. Randomness is further introduced by the variable number of photons required to expose a particular grain and by the varying size of the grains themselves. The subjective appearance of these factors is termed *graininess*.

For most practical purposes, film grain noise can be effectively modeled as a Gaussian process (white noise). Like photoelectronic noise, the underlying distribution is Poisson. Since mean grain diameter for specific films is published by the manufacturer, only the standard deviation of film grain noise as a function of grain size and the local image density remain to be determined.

In 1913, Nutting modeled the photographic emulsion as a sandwich of layers approximately one grain diameter thick. He showed that the measured optical density is given by

$$D = 0.43 \frac{na}{A} \tag{25}$$

where a is the cross-sectional area of a single grain, A is the area of the aperture used to measure optical density, and n is the total number of grains that fall within the aperture (Ref. 49). For fixed values of a and A, n is a random variable with a binomial distribution. Taking the expectation of Eq. (25) produces

$$\mathcal{E}\{D\} = 0.43 \frac{\mathcal{E}\{n\}a}{A} \tag{26}$$

and, since Eq. (25) is linear, the variance is given by

$$\sigma_D^2 = 0.43 \frac{\sigma_n^2 a}{A} \tag{27}$$

If a is small compared to A, the binomial distribution of n can be modeled by a Poisson distribution, and hence the variance is equal to the mean (Ref. 50). Making this substitution produces

$$\sigma_D = \sqrt{0.43 \frac{a}{A}} [\mathcal{E}\{D\}]^{1/2} \tag{28}$$

This equation indicates that film grain noise is worse with large-grain (high-speed) emulsions and small scanning apertures and in dense areas of the image. Thus film grain noise is also signal-dependent.

The previous analysis assumes uniform grain size. Haugh (Ref. 51) showed that if the grain size is distributed, the exponent in Eq. (28) should be somewhat less than $\frac{1}{2}$. Using sensitometric data from Higgins and Stultz (Ref. 52), Naderi (Ref. 53) showed, for a reasonably small aperture, that the exponent lies between 0.3 and 0.4. Thus film grain noise can be modeled as a zero-mean white Gaussian process with RMS amplitude (standard deviation) proportional to the cube root of the local average density.

Of the three common noise sources, two are signal-dependent. The signal dependence may be ignored for common restoration work, but for high levels of accuracy it must be taken into account.

SUMMARY OF IMPORTANT POINTS

1. Spatially invariant restoration can be accomplished with deconvolution, Wiener deconvolution, power spectrum equalization (PSE), or geometric mean filters.

2. The geometric mean filter [Eq. (4)] contains the Wiener deconvolution and PSE filters as special cases.

3. Noise usually restricts restoration, particularly at high spatial frequencies.

4. Coordinate transformation restoration (CTR) is useful with known spatially variant blurring functions.

5. Speckle interferometry can reduce the effects of temporally variant blurring functions.

6. While most images are generally nonstationary, many can be assumed locally stationary.

7. One can partition an image into regions based on signal-to-noise ratio and restore each region with a separate filter.

8. The linear combination filter produces a smoothly space-variant impulse response with modest computational complexity.

9. Superresolution techniques exploit the incompatibility between spatial bounding and band-limiting in order to reconstruct the spectrum beyond the diffraction limit.

10. The blurring function can be determined from features in the image or from the degraded image spectrum.

11. Electronic noise is white with a Gaussian histogram.

12. Photoelectronic noise can be modeled as white and Gaussian with the RMS amplitude equal to the square root of the mean.

13. Film grain noise can be modeled as white and Gaussian with RMS amplitude proportional to the cube root of local average density.

REFERENCES

1. H. C. ANDREWS and B. R. HUNT, *Digital Image Restoration*, Prentice-Hall, Inc., Englewood Cliffs, New Jersey, 1977.

2. W. K. PRATT, *Digital Image Processing*, John Wiley & Sons, Inc., New York, 1978.

3. H. C. ANDREWS, "Digital Image Restoration: A Survey," *IEEE Computer*, **7**, No. 5, 36–45, May 1974.

4. J. L. HARRIS, SR., "Image Evaluation and Restoration," *J. Opt. Soc. Amer.*, **56**, 569–574, May 1966.

5. B. R. HUNT, "Digital Signal Processing," *Proc. IEEE*, **63**, No. 4, 693–708, April 1975.

6. M. M. SONDHI, "Image Restoration: The Removal of Spatially Invariant Degradations," *Proc. IEEE*, **60**, No. 7, 842–853, July 1972.

7. H. C. ANDREWS, "Image Processing by Digital Computers," *IEEE Spectrum*, 20–32, July 1972.

8. "Digital Picture Processing," *Computer*, **7**, No. 6, May 1974.

9. "Digital Picture Processing," *Proc. IEEE*, **60**, No. 7, July 1972.

10. N. WIENER, *Extrapolation, Interpolation, and Smoothing of Stationary Time Series*, MIT Press, Cambridge, Massachusetts, 1942.

11. R. NATHAN, *Digital Video Handling*, Technical Report 32-877, Jet Propulsion Laboratory, Pasadena, California, January 5, 1966.

12. B. L. McGLAMERY, "Restoration of Turbulence Degraded Images," *J. Opt. Soc. Amer.*, **57**, No. 3, 293–297, March 1967.

13. P. F. MUELLER, and G. O. REYNOLDS, "Image Restoration by Removal of Random Media Degradations," *J. Opt. Soc. Amer.*, **57**, 1338–1344, November 1967.

14. C. W. HELSTROM, "Image Restoration by the Method of Least Squares," *J. Opt. Soc. Amer.*, **57**, 297–303, March 1967.

15. D. SLEPIAN, "Linear Least-Squares Filtering of Distorted Images," *J. Opt. Soc. Amer.*, **57**, 918–922, July 1967.

16. W. K. PRATT, "Generalized Wiener Filter Computation Techniques," *IEEE Trans. Computers*, 636–641, July 1972.

17. A. HABIBI, *Fast Suboptimal Wiener Filtering of Markov Processes*, University of Southern California, USCIPI Report 530, Los Angeles, California, 75–80, March 1974.

18. T. N. CORNSWEET, *Visual Perception*, Academic Press, New York, 1970.

19. T. M. CANNON, *Digital Image Deblurring by Nonlinear Homomorphic Filtering*, Ph.D. Thesis, Computer Science Department, University of Utah, Salt Lake City, Utah, 1974.

20. A. V. OPPENHEIM, R. W. SCHAFER, and T. G. STOCKHAM, "Nonlinear Filtering of Multiplied and Convolved Signals," *Proc. IEEE*, **56**, 1264–1291, August 1968.

21. G. M. ROBBINS and T. S. HUANG, "Image Restoration for a Class of Linear Spatially-Variant Degradations," *Pattern Recognition*, **2**, No. 2, 91–105, 1970.

22. G. M. ROBBINS and T. S. HUANG, "Inverse Filtering for Linear Shift-Variant Imaging Systems," *Proc. IEEE*, **60**, No. 7, 862–872, July 1972.

23. A. A. SAWCHUK, "Space-Variant Image Motion Degradation and Restoration," *Proc. IEEE*, **60**, No. 7, 854–861, July 1972.

24. A. A. SAWCHUK and M. J. PEYROVIAN, "Space-Variant Image Restoration by Coordinate Transformations," *J. Opt. Soc. Amer.*, **64**, No. 2, 138–144, February 1974.

25. A. A. SAWCHUK and M. J. PEYROVIAN, "Restoration of Astigmatism and Curvature of Field," *J. Opt. Soc. Amer.*, **65**, No. 6, 712–715, June 1975.

26. A. LABEYRIE, "Attainment of Diffraction Limited Resolution in Large Telescopes by Fourier Analysis Speckle Patterns in Star Images," *Astron. & Astrophys.*, **6**, No. 1, 85–87, 1970.

27. D. Y. GEZARI, A. LABEYRIE, and R. V. STACHNIK, "Speckle Interferometry: Diffraction-Limited Measurements of Nine Stars with the 200-Inch Telescope, *Astrophys. J.*, **173**, No. 1, L1–L5, April 1, 1972.

28. K. T. KNOX and B. J. THOMPSON, "Recovery of Images from Atmospherically Degraded Short-Exposure Photographs," *Astrophys. J.*, **193**, L45–L48, October 1, 1974.

29. K. T. KNOX, "Image Retrieval from Astronomical Speckle Patterns," *J. Opt. Soc. Amer.*, **66**, No. 11, 1236–1239, November 1976.

30. K. T. KNOX, *Diffraction Limited Imaging of Astronomical Telescopes*, Ph.D. Thesis, University of Rochester, Rochester, New York, 1975.

31. H. CARTAN, *Elementary Theory of Analytic Functions of One or Several Complex Variables*, Addison Wesley, Reading, Massachusetts, 1963.

32. J. L. HARRIS, "Diffraction and Resolving Power," *J. Opt. Soc. Amer.*, **54**, No. 7, 931–936, July 1964.

33. C. W. BARNES, "Object Restoration in a Diffraction-Limited Imaging System," *J. Opt. Soc. Amer.*, **56**, No. 5, 575–578, May 1966.

34. H. A. BROWN, "Effect of Truncation on Image Enhancement by Prolate Spheroidal Functions," *J. Opt. Soc. Amer.*, **59**, 228–229, 1969.

35. C. PASK, "Simple Optical Theory of Super-Resolution," JOSA Letters, *J. Opt. Soc. Amer.*, **66**, No. 1, 68–70, January 1976.

36. B. R. FRIEDEN, "Band-Unlimited Reconstruction of Optical Objects and Spectra," *J. Opt. Soc. Amer.*, **57**, No. 8, 1013–1019, August 1967.

37. C. K. RUSHFORTH, "Restoration, Resolution, and Noise," *J. Opt. Soc. Amer.*, **58**, No. 4, 539–545, April 1968.

38. C. L. RINO, "Bandlimited Image Restoration by Linear Mean-Square Estimation," *J. Opt. Soc. Amer.*, **59**, No. 5, 547–553, May 1969.

39. S. WADAKA and T. SATO, "Superresolution in Incoherent Imaging System," *J. Opt. Soc. Amer.*, **65**, No. 3, 354–355, March 1975.

40. J. W. GOODMAN, *Introduction to Fourier Optics*, McGraw-Hill Book Company, New York, 1968.

41. B. TATIAN, "Method for Obtaining the Transfer Function from the Edge Response Function," *J. Opt. Soc. Amer.*, **55**, 1014–1019, August 1965.

42. A. G. TESCHER, "Data Compression and Enhancement of Sampled Images," *Applied Optics*, **11**, No. 4, 919–925, April 1972.

43. M. E. RABEDEAU, "Effect of Truncation of Line-Spread and Edge-Response Functions on the Computed Optical Transfer Function," *J. Opt. Soc. Amer.*, **59**, No. 10, 1309–1314, October 1969.

44. D. B. GENNERY, "Determination of Optical Transfer Function by Inspection of Frequency-Domain Plot." *J. Opt. Soc. Amer.*, **63**, No. 12, 1571–1577, December 1973.

45. E. R. COLE, *The Removal of Unknown Image Blurs by Homomorphic Filtering*, Ph.D. Thesis, Department of Electrical Engineering, University of Utah, Salt Lake City, Utah, 1973.

46. T. S. HUANG, W. F. SCHREIBER, and O. J. TRETIAK, "Image Processing," *Proc. IEEE*, **59**, No. 11, 1586–1609, 1971.

47. T. R. STOCKHAM, T. M. CANNON, and R. B. INGEBRETSEN, "Blind Deconvolution by Digital Signal Processing," *Proc. IEEE*, **63**, 679–692, April 1975.

48. W. B. DAVENPORT, and W. L. ROOT, *An Introduction to the Theory of Random Signals and Noise*, McGraw-Hill Book Company, New York, 1958.

49. J. C. DAINTY and R. SHAW, *Image Science*, Academic Press, London, 1974.

50. D. G. FALCONER, "Image Enhancement and Film-Grain Noise," *Optica Acta*, **17**, 693–705, 1970.

51. E. F. HAUGH, "A Structural Theory for the Selwyn Granularity Coefficient," *J. Photo. Soc.*, **11**, p. 65, 1963.

52. G. C. HIGGINS and K. F. STULTZ, "Experimental Study of rms Granularity as a Function of Scanning Spot Size," *J. Opt. Soc. Amer.*, **49**, 925, 1959.

53. F. NADERI, *Estimation and Detection of Images Degraded by Film-Grain Noise*, University of Southern California, USCIPI Report 690, Los Angeles, California, September 1976.

IMAGE SEGMENTATION

||||||||||||||||||||||||||||||||||||||||||||||||||||||||||||||||||||||||||||||||||||||||||||||||||||||||||||||||||||||||||||||||||||||||||||||||

INTRODUCTION

So far in this book we have primarily considered ways to improve images for visual display. In Chapter 14, we were content merely to retrieve an image that resembled the original undegraded version. In this chapter and the next, we address some aspects of analyzing image content. This means we shall attempt to find out what is in the picture. Since we have only room enough to examine one approach, we shall deal with statistical pattern recognition as applied to digital images. Volumes have been written in this field, and we can do little more here than present an introduction to the subject.

Statistical Pattern Recognition

In this chapter and the next, we address a collection of topics from the field of statistical pattern recognition. In particular, we shall consider optical pattern recognition implemented by digital image processing techniques. This involves first finding the objects within an image and then identifying (classifying) those objects using the techniques of statistical decision theory.

Optical Pattern Recognition. The computer vision branch of the artificial intelligence field is concerned with developing algorithms for analyzing image content. While a variety of approaches toward "image understanding" have been proposed, we shall consider only one, optical pattern recognition. This approach assumes that the image may contain one or more objects and that each object belongs to one of several predetermined types or classes. While optical pattern recognition can be implemented in several different ways, we are concerned only with its implementation by digital image processing techniques. The name is unduly restrictive, however, since we can just as easily process digitized nonoptical images.

Given a digitized image containing several objects, the pattern recognition process consists of three major phases (see Figure 15-1). The first phase is *object*

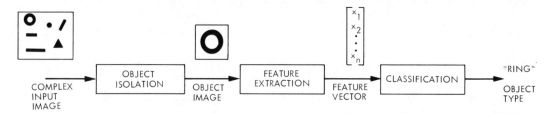

Figure 15-1 The three phases of pattern recognition

isolation, in which each object must be found and its image isolated from the rest of the scene. The second phase is called *feature extraction*. The features are a set of measurable properties. The feature extraction phase measures these properties, producing a set of measurements called the *feature vector*. This drastically reduced amount of information represents all the knowledge upon which the subsequent classification must be based.

The third phase is *object classification*, and its output is merely a decision regarding the class to which the object belongs. The object is thus recognized as being one particular type of object, and the recognition is implemented as a classification process. Each object is assigned to one of several preestablished groups (classes) that represent the possible types of objects expected to be encountered. The classification is based solely on the feature vector. In the next chapter, we shall consider several classification techniques derived from the field of statistical decision theory.

An Example. The basic concepts of statistical pattern recognition can best be illustrated by an example. Suppose we desire to implement a sorting system for fruit coming down a conveyor belt. The actual sorting can be effected with movable partitions that drop down and deflect the items of fruit off the conveyor belt and into the appropriate shipping box, as illustrated in Figure 15-2. What we need is an image processing system that can observe the approaching fruit, classify each item, and drop the appropriate partition.

Let us suppose that the fruit of interest are cherries, apples, lemons, and grapefruit. We can install a digitizing television camera above the conveyor belt and implement the classification decision in a computer. For this example, let us measure two things about each piece of fruit: its diameter and its color. The computer program

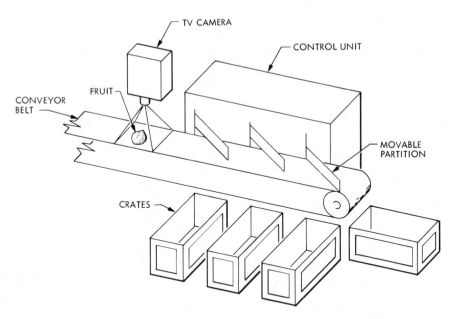

Figure 15-2   The fruit-sorter example

will process each digitized image and compute both the diameter of the fruit in milli-meters and a parameter indicative of color. Suppose the TV camera has switchable red and yellow color filters and the program computes the object contrast through each filter. It can then derive a parameter that takes on low values for yellow fruit and high values for red fruit. We shall call this parameter the *redness measure*.

Figure 15-3 shows the two-dimensional measurement space defined by the two features, diameter and redness, and the expected clusters produced by each of the four types of fruit. By placing appropriate decision lines in Figure 15-3, we can parti-tion the measurement space, and in so doing define a classification rule.

When any fruit approaches the TV camera, it is measured, and its measurements specify a point in the two-dimensional measurement space. Depending on where this point falls in the measurement space, the item of fruit is assigned to one of four classes. As soon as the classification decision is made, the mechanism drops the partition that will deflect the fruit into the appropriate shipping container.

While the preceding system may not find wide usage in the fruit-packing industry, it serves to illustrate statistical pattern recognition. The role of statistics will become clear in the next chapter. For the present, it suffices to say only that each type of fruit produces a probability density function in the measurement space. The decision lines can be determined from the interaction of the pdfs.

Pattern Recognition System Design. The design of a pattern recognition system is usually done in the five steps listed in Table 15-1. These are object locator design, feature selection, classifier design, classifier training, and performance evaluation.

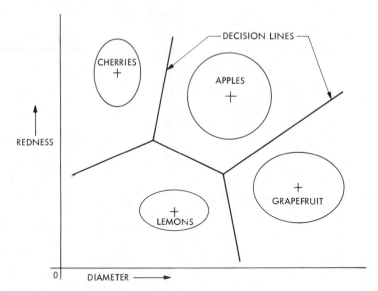

Figure 15-3 The measurement space

Table 15-1. Pattern recognition system design

Step	Function
1. Object locator design	Select the scene segmentation algorithm that will isolate the individual objects in the image.
2. Feature selection	Decide which properties of the objects best distinguish the object types and how to measure these.
3. Classifier design	Establish the mathematical basis of the classification algorithm and select the type of classifier structure to be used.
4. Classifier training	Fix the various adjustable parameters (decision boundaries, etc.) in the classifier to suit the objects being classified.
5. Performance evaluation	Estimate the expected rates of the various possible misclassification errors.

The object locator is the algorithm that isolates the images of the individual objects in the complex scene. This is scene segmentation, the topic addressed in this chapter. Feature selection involves deciding which properties of the object (size, shape, etc.) best distinguish among the various object types and thus should be measured. Classifier design consists of establishing a mathematical basis for the classification procedure. The various adjustable parameters of the classifier itself (decision thresholds, etc.) are pinned down in the training stage. Finally, it is usually desirable to estimate the error rates that can be expected with the system. This constitutes the performance evaluation step.

The Process of Image Segmentation

We can define the image segmentation process as one that partitions a digital image into disjoint (nonoverlapping) regions. For our purposes, a region is a connected set of pixels. A connected set is one in which all the pixels are adjacent or touching (Ref. 1). The formal definition of connectedness is as follows. Between any two pixels in a connected set, there exists a connected path wholly within the set. A connected path is one that always moves between neighboring pixels. Thus, in a connected set, you can trace a connected path between any two pixels without ever leaving the set.

When a human observer views a scene, the neurological processing that takes place in the retina and the optic cortex essentially segments the scene for him. This is done so effectively that he sees not a complex scene but rather something he thinks of as a collection of objects. With digital processing, however, we must isolate the objects in the image by breaking up that image into sets of pixels, each of which is the image of one object. While the task of image segmentation hardly has a counterpart in human visual experience, it is a nontrivial task in digital image analysis.

Image segmentation can be approached from two different philosophical perspectives. In the case we shall call the *region approach*, one assigns pixels to particular objects or regions. In the *boundary approach*, one attempts only to locate the boundaries that exist between the regions. Both approaches are useful for visualizing the problem.

In this chapter, we examine several techniques for isolating the objects in a digital image. Once isolated, these objects can be measured and classified. Techniques for measurement and classification are addressed in the next chapter.

IMAGE SEGMENTATION BY THRESHOLDING

Thresholding is a particularly useful region approach technique for scenes containing solid objects resting upon a contrasting background. It is computationally simple and never fails to define disjoint regions with closed connected boundaries. When using a threshold rule for image segmentation, one assigns all pixels at or above the threshold gray level to the object. All pixels with gray level below threshold fall outside the object. The boundary is then that set of interior points each of which has at least one neighbor outside the object.

Thresholding works well if the objects of interest have uniform interior gray level and rest upon a background of unequal but uniform gray level. If the object differs from its background by some property other than gray level (texture, etc.), one can first use an operation that converts that property to gray level. Then gray level thresholding can segment the processed image.

Global Thresholding. In the simplest implementation of boundary location by thresholding, the value of the threshold gray level is held constant throughout the image. If the background gray level is reasonably constant throughout the image, and if the objects all have approximately equal contrast above the background, then a fixed global threshold will usually work well, provided the threshold gray level is properly selected.

Adaptive Thresholding. In many cases, the background gray level is not constant and object contrast varies within the image. In such cases, a threshold that works well in one area might work poorly in other areas of the image. In these cases, it is convenient to use a threshold gray level that is a slowly varying function of position in the image.

Figure 15-4 shows a microscope image of the chromosomes from a single human blood cell. In this image, the background gray level varies due to nonuniform illumination, and contrast varies from one chromosome to the next. In Figure 15-4(a), a constant threshold gray level has been used throughout the image to isolate the chromosomes. Each chromosome was given a boundary and a sequence number. In Figure 15-4(b), the threshold was varied from one chromosome to the next commensurate with local background and chromosome contrast (Refs. 2, 3). This produced fewer segmentation errors—cases where multiple chromosomes were stuck together or individual chromosomes were broken up. A similar study showed that the accuracy of the area measurement for chromosomes was improved by adaptive thresholding. In Figure 15-4(b), the threshold for each chromosome was set approxi-

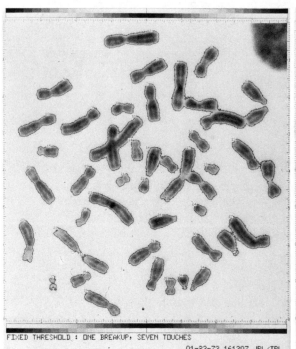

FIXED THRESHOLD : ONE BREAKUP, SEVEN TOUCHES

01-23-73 161307 JPL/IPL

(a)

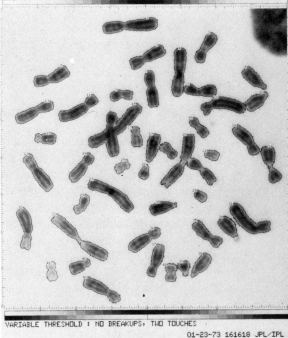

VARIABLE THRESHOLD : NO BREAKUPS, TWO TOUCHES

01-23-73 161618 JPL/IPL

(b)

Figure 15-4 Global and adaptive thresholding

mately midway between its mean interior gray level and the local background gray level (Refs. 2, 4).

OPTIMAL THRESHOLD SELECTION

Unless the object in the image has extremely steep sides, the exact value of threshold gray level can have considerable effect on the boundary position and overall size of the extracted object. This means that subsequent size measurements, particularly the area measurement, are sensitive to the threshold gray level. For this reason, we need an optimal, or at least consistent, method to establish the threshold.

Histogram Techniques. An image containing an object on a contrasting background has a bimodal gray level histogram (Figure 15-5). The two peaks correspond to the relatively large numbers of points inside and outside the object. The dip between the peaks corresponds to the relatively few points around the edge of the object and is commonly used to establish the threshold gray level (Refs. 5–8).

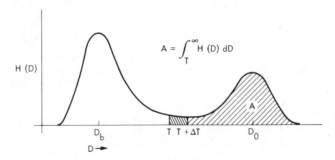

Figure 15-5   The bimodal histogram

Recall that the area of an object defined by a gray level threshold T is given by

$$A = \int_T^\infty H(D)\, dD \tag{1}$$

Notice that increasing the threshold from T to $T + \Delta T$ causes only a slight decrease in area if the threshold corresponds to the dip in the histogram. Thus placing the threshold at the dip in the histogram minimizes the sensitivity of the area measurement to small errors in threshold selection.

    If the image or the region of the image containing the object is noisy and not large, the histogram will be noisy. Unless the dip is extremely sharp, the noise will make its location obscure, or at least unreliable from one image to the next. This can be overcome to some extent by smoothing the histogram using either convolution or a curve fit. If the two peaks are of unequal size, the smoothing may tend to shift the position of the minimum. The peaks, however, are easy to locate and relatively stable under reasonable amounts of smoothing. A more reliable method is to place the threshold at some fixed position relative to the two peaks—perhaps midway (Ref. 4). The two peaks represent the mode (most commonly occurring) gray levels of the interior and exterior points of the objects. In general, these parameters can be estimated more reliably than the least commonly occurring gray level in the dip.

The adaptive segmentation technique of Figure 15-4(b) was implemented as a two-pass technique (Refs. 2, 4). Before the first pass, the image was divided into sectors of 100 by 100 pixels. From the gray level histogram of each sector, a threshold was determined midway between the background and data peaks. Sectors containing unimodal histograms were ignored. In the first pass, the object boundaries were defined using a gray level threshold that was constant within sectors but different for the various sectors. The objects so defined were not extracted from the image, but the interior mean gray level of each object was computed. On the second pass each object was given its own threshold that lay midway between its interior gray level and the background gray level of its principal sector. Examination of Figure 15-4 indicates that the number of touches dropped from 7 to 2, while breakups dropped from one to none.

The Analysis of Spots

In many important cases it is necessary to find objects that are roughly circular in shape. The following development is aimed primarily at circular objects. Restricting ourselves to circular objects allows us to pursue optimal threshold selection considerably further than we could otherwise. The concepts developed are nonetheless useful for more general cases as well.

Definitions. Suppose an image $B(x, y)$ contains a single *spot*. By definition, this image contains a point x_0, y_0 of maximum gray level. If we establish polar coordinates centered upon x_0, y_0, so that the image is given by $B_p(r, \theta)$, then

$$B_p(r_1, \theta) \geq B_p(r_2, \theta) \qquad \text{if } r_2 > r_1 \tag{2}$$

for all values of θ. We call $B(x, y)$ a *monotone spot* if equality is not allowed in Eq. (2). This means gray level strictly decreases along any line extending out from the center point x_0, y_0. For monotone spots, a flat top is not allowed and the center point x_0, y_0 is unique.

An important special case occurs if all contours of a monotone spot are circles centered on x_0, y_0. We call this special case a *concentric circular spot* (CCS). To a good approximation, this usually describes the noise-free images of stars in a telescope, certain cells in a microscope, and many other important cases. For a CCS, the function $B_p(r, \theta)$ is independent of θ, and we call this the *spot profile function*. This curve is useful for threshold selection. For example, we could locate the inflection point and select the gray level threshold to place the boundary at the point of maximum scope. This is approximately where the human eye places the boundary when viewing an image, and it is reasonably stable under smoothing and noise addition. Other unique points on the profile could be used as well.

If we threshold a monotone spot at a gray level T, we define an object having a certain area and perimeter. As we vary T throughout the range of gray levels, we generate the threshold area function $A(T)$ and the perimeter function $p(T)$. Both of these functions are unique for any spot. They are continuous for monotone spots, and either is sufficient to specify a CCS completely. As a matter of definition, two spots are *p-equivalent* if they have identical perimeter functions, and *H-equivalent* if they have identical histograms. It follows that *H*-equivalent spots have identical threshold area functions.

The Histogram and the Profile. Suppose a CCS image $B(x, y)$ is given by its profile function $B_p(r)$. We now seek an expression for the spot histogram in terms of its profile function. Suppose we threshold $B(x, y)$ at gray level D and again at gray level $D + \Delta D$. This defines two circular contours of radius r and $r + \Delta r$, respectively, as shown in Figure 15-6. The area of the annular ring between the contours is

$$\Delta A = \pi r^2 - \pi(r + \Delta r)^2 \approx -2\pi r \Delta r \qquad (3)$$

where the approximation was obtained by assuming Δr small and neglecting Δr^2.

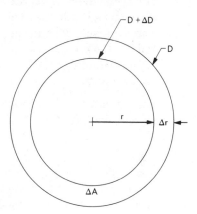

Figure 15-6 Thresholding a concentric circular spot

Equation (3) can be rearranged as

$$\frac{\Delta A}{\Delta r} \approx -2\pi r \qquad (4)$$

The histogram of the image is, by definition,

$$H_B(D) = \underset{\Delta D \to 0}{\mathcal{L}} \frac{\Delta A}{\Delta D} \qquad (5)$$

We can divide the numerator and denominator by Δr and substitute Eq. (4) to produce

$$H_B(D) = \underset{\Delta D \to 0}{\mathcal{L}} \frac{\Delta A / \Delta r}{\Delta D / \Delta r} = \frac{-2\pi r}{d/dr \, B_p(r)} \qquad (6)$$

In this equation, we noted that both Δr and ΔD approach zero and recognized the derivative of the profile function in the denominator.

We are not through yet, since the right side of Eq. (6) is a function of r instead of a function of D. Since $B(x, y)$ is the image of a monotone spot, $B_p(r)$ is a monotonically decreasing function and, hence, its inverse function

$$r(D) = B_p^{-1}(D) \tag{7}$$

exists. We can substitute this into the numerator and denominator of Eq. (6) to make the histogram a function of gray level, as desired. Notice also that, since the profile function $B_p(r)$ decreases monotonically with r, the denominator of Eq. (6) is negative. This cancels the minus sign in the numerator to make the histogram positive, as expected.

The Area Derived Profile. We now desire an expression for the profile of a CCS in terms of its histogram. The radius of the circular object obtained by thresholding a CCS at gray level T is

$$R(T) = \left[\frac{1}{\pi} A(T) \right]^{1/2} = \left[\frac{1}{\pi} \int_T^{\infty} H_B(D) \, dD \right]^{1/2} \tag{8}$$

For a monotone spot, the histogram $H_B(D)$ is nonzero between its minimum and maximum gray levels. This means that the area function $A(T)$ is monotonically increasing and, consequently, so is $R(T)$. Thus the inverse function of Eq. (8) exists, and it is the profile. We can compute the area-derived profile of a CCS by integrating the histogram to obtain the area function, taking first the square root and then the inverse function.

The Perimeter Derived Profile. Thresholding a CCS at gray level T produces a circular object of radius

$$R(T) = \frac{1}{2\pi} P(T) \tag{9}$$

where $p(T)$ is the perimeter function. As with the previous technique, the profile is merely the inverse function of Eq. (9). Thus if the perimeter function is known, the profile may be obtained by the inverse of Eq. (9).

Noncircular and Noisy Spots. For an image containing a noise-free CCS, we can most easily obtain the profile simply by taking the scan line that contains the peak. For noncircular spots and noisy spots, however, the foregoing techniques can be useful. For example, one could use the histogram of a noncircular spot to obtain the profile of the H-equivalent CCS and select the threshold gray level that maximizes the slope at the boundary. In other cases, it might be useful to measure the perimeter function and determine the profile of the p-equivalent CCS. Either of these techniques could produce thresholds suitable for the image at hand.

In digitized images of natural scenes, the noise level is frequently so high that differentiating a single scan line cannot reliably identify the inflection point on the profile. However, the area-derived and perimeter-derived profiles are computed using most or all of the pixels in the image. This process employs inherent noise reduction by averaging. Further noise reduction can be effected by smoothing the histogram or perimeter function before profile computation. The area-derived profile is the easier to compute, and it has superior noise discrimination properties.

Random noise in the image usually makes the threshold boundary jagged. While this may have little effect on the area function, it tends to make the perimeter function erroneously large. While this can be reduced by boundary smoothing built into the perimeter measurement routine, computational simplicity is still on the side of the area-derived profile.

Average Boundary Gradient

For highly noncircular spots, the *H*-equivalent and *p*-equivalent CCS profiles may not be acceptable for placing the gray level threshold. For objects of arbitrary shape, we can examine the average gradient around the boundary as a function of the threshold gray level that defines the boundary (Ref. 2). Suppose a noncircular monotone spot is thresholded at gray levels D and $D + \Delta D$, as shown in Figure 15-7. At some point

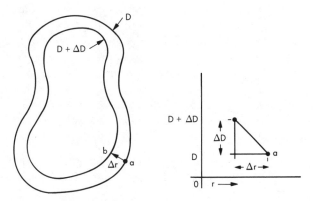

Figure 15-7 Thresholding a noncircular object

a on the outer boundary, Δr is the perpendicular distance to the inner boundary. Since Δr is perpendicular to a contour line, it lies in the direction of the gradient vector at point a. The magnitude of the gradient vector at point a on the outer boundary is given by

$$|\mathbf{VB}| = \underset{\Delta D \to 0}{\mathcal{L}} \frac{\Delta D}{\Delta r} \tag{10}$$

Since we are interested in the average gradient around the boundary, we can simply average $|\mathbf{VB}|$ around the outer boundary. If Δr is small with respect to the perimeter, the area between the two boundaries is merely

$$\Delta A = p(D)\, \overline{\Delta r} \tag{11}$$

where $\overline{\Delta r}$ is the average perpendicular distance from the outer to the inner boundary, and $p(D)$ is the perimeter function. To obtain the average gradient around the boundary, we need merely to substitute $\overline{\Delta r}$ for Δr in Eq. (10). This produces

$$|\overline{\mathbf{VB}}| = \underset{\Delta D \to 0}{\mathcal{L}} \frac{\Delta \mathbf{D}}{\Delta A} p(D) = \frac{p(D)}{H_B(D)} \tag{12}$$

which indicates that the average boundary gradient is merely the ratio of the perimeter function to the histogram.

The average boundary gradient function is not difficult to compute, and it readily identifies the threshold gray level that maximizes the slope at the boundary. For noisy images, the perimeter function and the histogram may require smoothing before their use.

Objects of General Shape

Although some of the foregoing results were developed primarily for restricted types of objects, they are useful for more general cases. Suppose an image contains objects of general shape on a low gray level background. While the objects may be relatively flat on top, nonmonotonic, and without a unique peak, they usually have sides that slope uniformly down toward the background. The psf of optical systems forbids sides of infinite slope in real images. On the sides of the objects, contour lines are closed and generally convex curves that may have local concavities. We can assume that each threshold gray level defines a single closed curve for each object. Under these conditions, we need consider only the range of gray levels corresponding to the sloping sides of the object. We now have four ways to establish the maximum slope threshold gray level T.

1. We can select T at a local minimum in the histogram. This is the easiest technique, and it minimizes the sensitivity of the area measurement to small variations in T.

2. We can select T corresponding to the inflection point in the H-equivalent CCS profile function. This is a simple computation, and it involves considerable averaging for noise reduction.

3. We can select T to maximize the average boundary gradient. This involves computing the perimeter function but requires no approximation regarding equivalent spot images.

4. We can select T corresponding to the inflection point in the p-equivalent CCS profile function.

Any one of the above methods can be implemented for routine use. For large-scale studies, one might use one of the above methods to characterize the objects under study. Then a shortcut method could be implemented for routine use. For example, suppose a profile analysis showed that the optimal threshold gray level for isolated star images in telescope pictures occurs midway between the peak and the background gray level. Then this simplified method could be employed for routine use.

GRADIENT BASED METHODS

The previously discussed region approach methods accomplish image segmentation by partitioning the image into sets of interior and exterior points. Boundary approach techniques attempt to find the edges directly by their high gradient magnitudes. In this section, we discuss two such methods.

Boundary Tracking

Suppose we start with the gradient magnitude image computed from an image that contains a single object on a contrasting background. We can start the boundary tracking process by identifying the pixel of highest gray level (i.e., highest gradient in the original image) as the first boundary point, since it certainly must be on the boundary. If several points have the maximum gray level, then we choose arbitrarily. Next we search the 3 by 3 neighborhood centered on the first boundary point and take the neighbor with maximum gray level as the second boundary point. If two neighbors have the same maximum gray level, we choose arbitrarily. At this point, we begin the iterative process of finding the next boundary point, given the current and last boundary points. Working in the 3 by 3 neighborhood centered on the current boundary point, we examine the neighbor diametrically opposite the last boundary point and the neighbors on each side of it (Figure 15-8). The next boundary point is the one of those three that has highest gray level. If all three or two adjacent boundary points share the highest gray level, then we choose the middle one. If the two nonadjacent points share the highest gray level, we choose arbitrarily.

In the noise-free image of a monotone spot, this algorithm will trace out the maximum gradient boundary; however, even small amounts of noise can send the tracking temporarily or hopelessly off the boundary. Noise effects can be reduced

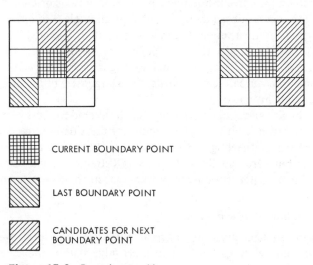

CURRENT BOUNDARY POINT

LAST BOUNDARY POINT

CANDIDATES FOR NEXT
BOUNDARY POINT

Figure 15-8   Boundary tracking

by smoothing the gradient image before tracking or by implementing a "tracking bug." Even so, boundary tracking does not guarantee closed boundaries, and the tracking algorithm can get lost and run off the border of the image.

A tracking bug works as follows. First we define a rectangular averaging window (the bug), usually having uniform weights (Figure 15-9). The last two or last few

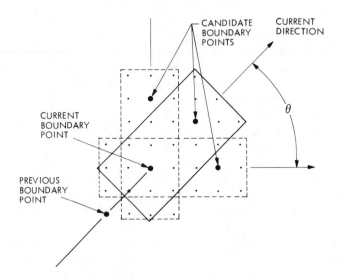

Figure 15-9   The boundary tracking bug

boundary points define the current boundary direction. The rear portion of the bug is centered on the current boundary point, with its axis oriented along the current direction. The bug is subsequently oriented at an angle θ to either side. In each position, the average gradient under the bug is computed. The next boundary point is taken as one of the pixels under the front portion of the bug when it is in the highest average gradient position. Clearly, the tracking bug is a larger implementation of the boundary tracking procedure descirbed earlier. The larger size of the bug implements smoothing of the gradient image.

The size and shape of the bug may be altered for best performance. The "inertia" of the bug can be increased by reducing the side-looking angle θ. In practice, the exact shape of the bug appears to have little effect on its performance. Gradient tracking bugs are usually useful only in extremely low-noise images or in situations where human intervention can prevent catastrophic derailments.

Gradient Image Thresholding

If we threshold a gradient image at moderate gray level, we find both object and background below threshold and most edge points above. Kirsch has advanced a method that makes use of this phenomenon (Ref. 9). In this technique, one first thresholds the gradient image at a moderately low level to identify the object and the background, which are separated by bands of edge points. Then the threshold is gradually increased. This causes both the object and the background to grow. When they touch, they are not allowed to merge, but rather the points of contact define the boundary. This method is computationally expensive, but it tends to produce maximum gradient boundaries while avoiding many of the problems of gradient tracking bugs. For multiple object images, the segmentation is correct if and only if it is done correctly by the initial thresholding step. Smoothing the gradient image beforehand produces smoother boundaries.

An Alternative to Gradient

Another local differential operator for edge finding has been advanced by Roberts (Ref. 10). It is given by

$$g(x, y) = \{[\sqrt{f(x, y)} - \sqrt{f(x + 1, y + 1)}]^2 + [\sqrt{f(x + 1, y)}$$
$$- \sqrt{f(x, y + 1)}]^2\}^{1/2} \tag{13}$$

where $f(x, y)$ is the input image in units of intensity and with integer pixel coordinates x and y. The inner square roots are used to make the operation resemble the processing that takes place in the human visual system.

    The edge operator advanced by Kirsch (Figure 15-10) also detects the presence of edges (Ref. 9). Each 3 by 3 neighborhood is convolved with eight kernels. The maximum value over each of the eight orientations is taken as the output value.

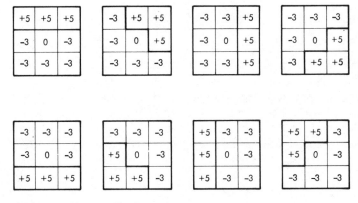

Figure 15-10   The Kirsch operator

REGION GROWING TECHNIQUES

Region growing (Refs. 11–14) is an approach to image segmentation that has received considerable attention in the computer vision segment of the artificial intelligence community. With this approach, one begins by dividing an image into many tiny regions. These initial regions may be small neighborhoods or even single pixels. In each region, the properties that reflect object membership are computed. The properties that discriminate among and identify the individual objects might include average gray level, texture, or color information. Thus the first step assigns to each region a set of parameters whose values reflect the object to which they belong.

Next, all boundaries between adjacent regions are examined. A measure of boundary strength is computed utilizing the differences of the averaged properties of the adjacent regions. A given boundary is strong if the properties differ significantly on either side of that boundary, and it is weak if they do not. Strong boundaries are allowed to stand, while weak boundaries are dissolved and the adjacent regions merged. The process is iterated by first recomputing the object membership properties for the enlarged regions and then dissolving weak boundaries. The region merging process is continued until a point is reached where no boundaries are weak enough to be dissolved. In this case, image segmentation is complete. Monitoring this procedure gives one the impression of regions in the interior of objects growing until their borders correspond with the edges of the object.

Region growing algorithms are computationally more expensive than the simpler techniques; however, region growing is able to utilize several image properties directly and simultaneously. Perhaps it shows greatest promise in the segmentation of natural scenes, where strong *a priori* knowledge is not available.

Figure 15-11 shows four stages in the region growing of one muscle fiber from a microscope slide. In this example low gradient was the sole region membership property. The lower right quadrant shows the final boundary.

SEGMENTED IMAGE STRUCTURE

If only gross measurements of each object are required, it is not necessary to extract the objects from the original image. In other cases, we may wish to compose a new image showing the objects somehow rearranged. Also, we may wish to do further measurements or other processing on the individual objects one at a time. In these cases, it is necessary to extract and store the individual objects in a convenient format. In general, each object should be assigned a sequence number as it is found. This object number can be used to identify the individual objects in the scene. In this section we discuss three ways to structure the segmented image.

The Object Membership Map

One way to store the segmentation information is to generate a separate image, the same size as the original, and encode object membership on a pixel-by-pixel basis. In the object membership map, the gray level of each pixel encodes the sequence number of the object to which the corresponding pixel in the original image belongs. For example, all pixels belonging to object 27 in the image will have a gray level of 27 in the membership map.

The membership map technique is perfectly general, but it is not a particularly

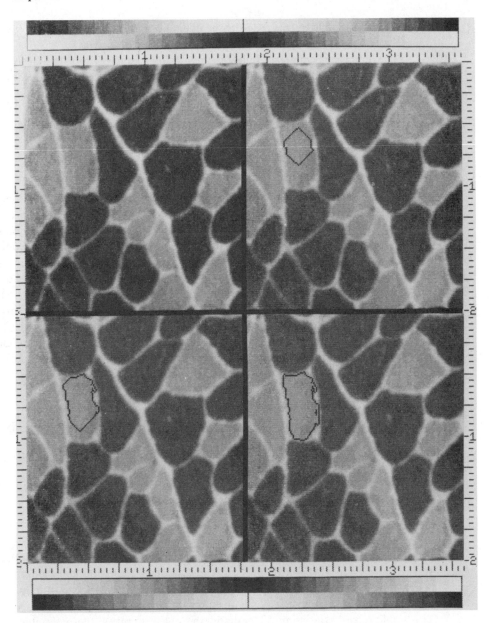

Figure 15-11　Region growing example

compact way to store the segmentation information. It requires an additional full-size digital image to describe a scene containing even one small object. If only object size and shape are of interest, however, the original image may be discarded after segmentation. Further data reduction results if there is only one object or if the objects need not be differentiated. In this case the membership map becomes a binary (two-level) image.

The data requirements of image segmentation frequently dictate that the process be done in several passes over the data. A binary or multilevel membership map is often useful as an intermediate step in a multiple-pass image segmentation procedure. Except in cases of small images, it is most practical to store both the input image and membership map on disk. In this case, the input image is processed and the membership map generated one line at a time.

The Boundary Chain Code

A more compact format for storing the image segmentation information is the boundary chain code (Refs. 15–18). Since it is the boundary that defines an object, it is not necessary to store the location of interior points. Furthermore, the boundary chain code utilizes the fact that boundaries are connected paths. The chain code starts by specifying the x, y coordinates of an arbitrarily selected starting point on the boundary of the object. That pixel has eight neighbors, and at least one of these must also be a boundary point. The chain code specifies the direction in which a step must be taken to go from the present to the next boundary point. Since there are eight possible directions, they can be numbered, say, from 0 through 7. Figure 15-12 shows a possible assignment of the eight direction codes. The boundary chain code then consists of the coordinates of the starting point, followed by the sequence of direction codes that specify the path around the boundary.

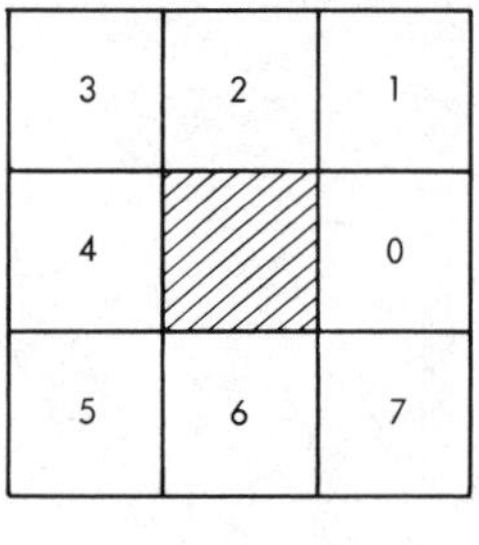

3	2	1
4	////	0
5	6	7

Figure 15-12   The boundary direction code

With the boundary chain code, the segmentation of an object requires, for storage, only one x, y coordinate and three bits for each boundary point. This is considerably less storage than that required for the object membership map. When a complex scene is segmented, it is convenient to store each object boundary on disk as a single record consisting of, perhaps, the object number, the perimeter (number of boundary points), and the chain code. The object shape (silhouette) can be reconstructed from the chain code alone. Furthermore, several size and shape measurements can be extracted directly from the boundary chain code, as shown in Chapter 16.

Generation of the boundary chain code usually requires random access to the input image since the boundary must be tracked through the image. With boundary tracking techniques of image segmentation, generation of the chain code is a natural adjunct. With boundary location by thresholding, the chain code usually must be generated in a subsequent step. Generation of the boundary chain code does not fit quite so well with line-by-line processing of images stored on disk. Since interior points are discarded, the chain code is less useful when further processing of the individual object images is required.

Line Segment Encoding

A line-by-line technique that stores extracted objects is line segment encoding. This process is best illustrated by the example in Figure 15-13. Suppose we wish to segment an image using a gray level threshold T. The program examines the image, line-by-line, working down from the top, looking for pixels having gray level greater than T. In Figure 15-13, the segment labeled 1-1 is a sequence of three adjacent pixels on line 100 having gray level at or above threshold. Thus segment 1-1 is the first line segment of the object encountered by the program. Upon examination of line 101, the program encounters two segments, 1-2 and 2-1, which are above threshold. Since it does not know at this time that both segments actually correspond to the same object, it assumes that the second segment on line 101 begins a second object, which it calls object number 2. Since segment 1-2 underlies segment 1-1, the program assumes that both segments correspond to object 1. The process continues through line 102, but at line 103 only a single segment is found, and it underlies segments of both objects 1 and 2. The program now recognizes that objects 1 and 2 are the same, and segment numbering continues for object 1.

On line 105 the program again finds two segments. Since they both underlie segment 1-5, however, they obviously belong to object 1. On line 107, no segments that underlie 1-8 or 1-9 are found, and the isolation of object 1 is complete. In this way, it is the line segments which, taken together, specify the object as it has been segmented.

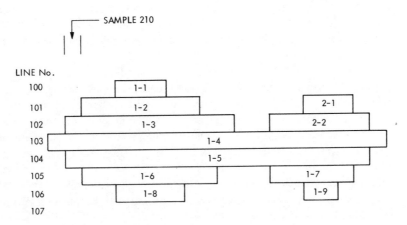

Figure 15-13   Object line segments

Figure 15-14 shows how the object segment information can be conveniently stored on disk. Each time a new object is located, the program generates an object file. This file begins with an object label containing the object number and the number of segments in that object. The latter entry cannot be made until object segmentation is complete. Following the object label, the individual line segments are stored. In Figure 15-14, they are stored with a segment label, followed by the gray level values of the pixels in that segment. The segment label contains the number of the line from which the segment was extracted, the coordinate of the first pixel on the line segment, and the number of pixels in the line segment.

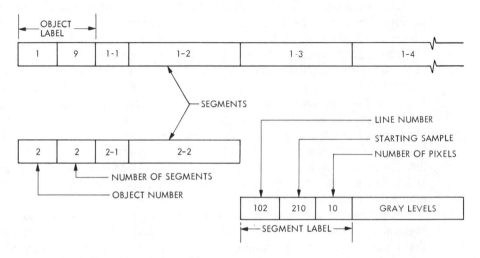

Figure 15-14 The object segment file

For the object in Figure 15-13, two object segment files would be opened. After only two segments have been stored in object file 2, however, the program discovers that objects 1 and 2 are the same and must be merged. Further construction of object file 2 is discontinued. After object segmentation is complete for this object, the two object segment files may be merged.

The net result of the single pass line segment encoding technique is a set of segment files, one for each object. If each segment file is stored as a single record, only one revolution of the disk is required to read or write an entire object. The object image can be easily reconstructed in main memory simply by unpacking the segment file. This is particularly useful when further processing of object images is desired.

Typically, the input image is read line by line from disk, and object segment files are assembled in main memory. As soon as an object file is completed, its label is finalized and that file is written to disk as one record. This technique minimizes the main memory requirements for this type of image segmentation. Another advantage of this method is that object area, perimeter, IOD, and horizontal and vertical extent measurements are easily built into the object extraction step. In this way, several important measurements are known by the time the object extraction step is complete.

SUMMARY OF IMPORTANT POINTS

1. Image segmentation is the process of partitioning a digital image into disjoint connected sets of pixels, each of which corresponds to an object or region.

2. Image segmentation can be approached either as the process of assigning pixels to objects or of finding boundaries between objects.

3. Gray level thresholding is a simple segmentation technique that always produces closed connected boundaries.

4. It is frequently necessary to allow the threshold gray level to vary within the image.

5. For images of simple objects on a contrasting background, placing the threshold at the dip of the bimodal histogram minimizes the sensitivity of measured area to threshold variations.

6. The profile function of a concentric circular spot may be derived from the histogram or the perimeter function of its image.

7. The average gradient around a contour line can be computed from the perimeter function and the histogram [Eq. (12)].

8. Object segmentation can be implemented by tracking the boundaries in the gradient image or by thresholding the gradient image.

9. Region growing techniques are useful for complex scenes and complex object definitions.

10. The segmentation of an image may be stored by a membership map, by a boundary chain code, or with line segment encoding.

REFERENCES

1. A. ROSENFELD, "Connectivity in Digital Pictures," *Journal of the ACM*, **17**, 146–160, 1970.

2. R. J. WALL, *The Gray Level Histogram for Threshold Boundary Determination in Image Processing with Applications to the Scene Segmentation Problem in Human Chromosome Analysis*, Ph.D. Thesis, University of California at Los Angeles, 1974.

3. K. CASTLEMAN and R. WALL, "Automatic Systems for Chromosome Identification," in T. Caspersson, ed., *Nobel Symposium* **23**—*Chromosome Identification*, Academic Press, New York, 1973.

4. K. CASTLEMAN and J. MELNYK, *An Automated System for Chromosome Analysis: Final Report,* Document No. 5040-30, Jet Propulsion Laboratory, Pasadena, California, July 4, 1976.

5. J. PREWITT and M. MENDELSOHN, "The Analysis of Cell Images," *Annals of the N.Y. Academy of Sciences,* **128**, 1035–1053, 1966.

6. J. PREWITT, "Object Enhancement and Extraction," in B. Lipkin and A. Rosenfeld, eds., *Picture Processing and Psychopictorics,* Academic Press, New York 1970.

7. R. J. WALL, A. KLINGER, and K. R. CASTLEMAN, "Analysis of Image Histograms," *Proc. 2nd. Int. Joint Conf. on Pattern Recognition* (IEEE Publ. 74CH-0885-4C), 341–344, Copenhagen, August 1974.

8. J. WESZKA, "A Survey of Threshold Selection Techniques," *Computer Graphics and Image Processing,* **7**, 259–265, 1978.

9. R. A. KIRSCH, "Computer Determination of the Constituent Structure of Biological Images," *Computers and Biomedical Research,* **4**, 315–328, 1971.

10. L. G. ROBERTS, "Machine Perception of Three-Dimensional Solids," in J. T. Tippett, *et al.,* ed., *Optical and Electro-Optical Information Processing,* 159–197, MIT Press, Cambridge, Massachusetts, 1965 (reprinted in Ref. 18).

11. C. R. BRICE and C. L. FENNEMA, "Scene Analysis Using Regions," *Artificial Intelligence,* **1**, 205–226, Fall, 1970 (reprinted in Ref. 18).

12. Y. YAKAMOVSKY and J. A. FELDMAN, "A Semantics-Based Decision Theory Region Analyzer," *Proc. 3rd Int. Joint Conf. on Artificial Intelligence,* 580–588, August 1973 (reprinted in Ref. 18).

13. S. L. HOROWITZ and T. PAVLIDIS, "Picture Segmentation by a Directed Split-and-Merge Procedure," *Proc. 2nd Int. Joint Conf. on Pattern Recognition* (IEEE Publ. 74CH-0885-4C), 424–433, August 1974 (reprinted in Ref. 18).

14. S. ZUCKER, "Region Growing: Childhood and Adolescence," *Computer Graphics and Image Processing,* **5**, 382–399, 1976.

15. H. FREEMAN, "On the Encoding of Arbitrary Geometric Configurations," *IRE Transactions on Electronic Computers,* **EC-10**, 260–268, June 1961 (reprinted in Ref. 18).

16. H. FREEMAN, "A Review of Relevant Problems in the Processing of Line-Drawing Data," in A. Grasselli, ed., *Automatic Interpretation and Classification of Images,* Academic Press, New York, 1969.

17. H. FREEMAN, "Boundary Encoding and Processing," in B. Lipkin and A. Rosenfeld, eds., *Picture Processing and Psychopictorics,* Academic Press, New York, 1970.

18. J. K. AGGARWAL, R. O. DUDA, and A. ROSENFELD, eds., *Computer Methods in Image Analysis,* IEEE Press, New York, 1977 (also available from John Wiley & Sons, New York).

MEASUREMENT AND CLASSIFICATION

━━

INTRODUCTION

In Chapter 15, we discussed the isolation and extraction of objects from a complex scene. In this chapter, we address the problem of measuring the objects and classifying them into groups. Much has been written on this subject, and we can only hope to introduce the basic concepts. For a more complete treatment, the reader is referred to a text on the subject (Refs. 1–6). We begin with a summary of the steps involved in pattern recognition system design.

Feature Selection

If we desire a system to distinguish objects of different types, we must first decide which parameters, descriptive of the objects, will be measured. The particular parameters that are measured are called the *features*. Proper selection of the features is important, since only they will be used to distinguish the objects.

There are few analytical means to guide the selection of features. Frequently intuition guides the listing of potentially useful features. Feature ordering techniques compute the relative power of the various features. This allows the list to be pared to the best few features.

Features should have four characteristics:

1. *Discrimination*. Features should take on significantly different values for objects belonging to different classes. Diameter is a good feature in the fruit-sorting example of Chapter 15, since it takes on significantly different values for cherries and grapefruit.

2. *Reliability*. Features should take on similar values for all objects of the same class. Color might be a poor feature for apples if they occur in varying degrees of ripeness. A green apple and a ripe apple might differ significantly in color even though they both belong to the class of apples.

3. *Independence*. The various features used should be uncorrelated with each other. The diameter and the weight of the fruit form highly correlated features, since weight is approximately proportional to the cube of the diameter. Both diameter and weight essentially reflect the same property, the size of the fruit. While highly correlated features might be combined to reduce noise sensitivity, they generally should not be used as separate features.

4. *Small Numbers*. The complexity of a pattern recognition system increases rapidly with the number of features used. More importantly, the amount of data required to train the classifier and to measure its performance increases exponentially with the number of features (Ref. 5). In some cases, it may be impractical to acquire enough data to train the classifier adequately. Finally, adding more features that are either noisy or correlated with existing features can actually degrade the performance of the classifier.

In practice, the feature selection process usually involves testing a set of intuitively reasonable features and reducing this set to an acceptable number of the best ones. Frequently the features are less than ideal in terms of the above characteristics.

Classifier Design

Many different classifier structures have been investigated (Refs. 1–6). Most classification decisions reduce to a threshold rule. If the values of the features fall within specified ranges, then the object is assigned to a particular group. Each range of feature values corresponds to a single group. In some cases, one or more such ranges may correspond to a group called *unknown*.

Classifier Training

Once the basic decision rules of the classifier have been established, one must determine the particular threshold values that separate the classes. This is generally done by training the classifier on a group of known objects. A number of objects from each class, previously identified by some accurate method, constitutes the training set. Objects in the training set are measured, and the measurement space is partitioned by decision lines that maximize the accuracy of the classifier when operating on the training set. When training a classifier, one might use a simple rule such as minimizing the total number of classification errors. If some misclassifications are more undesirable than others, one might establish a cost function that weights the errors appropriately and place the decision lines to minimize the overall cost.

If the training set is representative of the objects as a whole, then the classifier should perform about as well on new objects as it did on the training set. Obtaining a large enough training set is frequently a laborious task. In order to be representative, the training set should include examples of all types of objects that might be encountered, including those rarely seen. If the training set excludes certain uncommon objects, or if it contains classification errors, it is a biased training set.

Performance Measurement

A classifier's performance can be estimated directly by tabulating it on a known test set of objects. If the test set is large enough to be representative of the objects at large, and if it is free of errors, the resulting performance estimate can be useful.

An alternative is to use a test set of known objects to estimate the pdfs of the features for objects belonging to each group. Given the underlying pdfs, one can use the classification parameters to calculate the expected error rates. If the general form of the underlying pdfs is known, this technique can be superior.

One is tempted to take classifier performance on the training set as a measure of overall performance, but this estimate is usually biased optimistically. A better approach is to use a separate test set for performance evaluation. This, however, increases significantly the requirement for preclassified data. If previously classified objects are at a premium, one can use a round-robin procedure in which the classifier is trained on all but one of the available objects, and that object is then classified. When this is done for all objects, one has an estimate of overall performance.

SIZE MEASUREMENTS

In this section, we consider several commonly used features that reflect object size. These features have come into common usage because they are important in a variety of pattern recognition problems, and they lend themselves well to computation by digital image processing techniques.

Area and Integrated Optical Density

These two measures are easily obtained and can be computed during the extraction of an object from the image. The area measurement is simply the number of pixels inside (and including) the boundary, multiplied by the area of a single pixel. The integrated optical density (IOD) is the sum of the gray levels of all pixels inside the object. Computation of the IOD was covered in Chapter 5.

Area is a convenient measure of overall size. It is dependent only on the boundary of the object and disregards gray level variations inside. The IOD reflects the "mass" or "weight" of the object. It is numerically equal to the area multiplied by the mean interior gray level.

Length and Width

It is easy to compute the horizontal and vertical extent of an object while it is being extracted from an image. One needs only the minimum and maximum line number and sample number for this computation. For objects of random orientation, however, horizontal and vertical may not be the directions of interest. In this case, it is necessary to locate the major axis of the object and measure length and width relative to it.

There are several ways to establish the principal axis of an object once its boundary is known. One can compute a best-fit straight (or curved) line through the points in the object (Refs. 7, 8). The principal axis can be computed from moments, as discussed in the following section. A third way uses the minimum enclosing rectangle (Ref. 7). With this technique, the boundary of the object is rotated through 90° in steps of 3° or so. After each incremental rotation, a horizontally oriented minimum enclosing rectangle (MER) is fit to the boundary. Computationally, this merely involves keeping track of the minimum and maximum x and y values of the rotated boundary points. At some angle of rotation, the area of the MER goes through a minimum. The dimensions of the MER at that point can be taken as the length and width of the object. The angle at which the MER is minimized gives the principal axis of the object by this method. This technique is particularly useful for rectangular objects, but it gives satisfactory results for more general object shapes.

Perimeter

Frequently the circumferential distance around the boundary is useful for classification purposes. The perimeter is easily obtained from the boundary chain code. It is also simple to compute from the object segment file, provided one is careful to compute accurately the center-to-center distance between adjacent pixels on the boundary. Image noise usually produces artifactual jaggedness in the object boundary. This generally combines with sampling grid effects to make perimeter measurements artificially large. The perimeter can be reduced by judicious use of boundary smoothing built into its measurement. Without boundary smoothing, the perimeter of an object is given by

$$p = N_e + \sqrt{2}\, N_o \tag{1}$$

where N_e is the number of even and N_o is the number of odd steps in the boundary chain code, when the convention of Figure 15-11 is used.

SHAPE MEASUREMENTS

Frequently the objects of one class can be distinguished from other objects by their shape. Shape features can be used independently of, or in combination with, size measurements. In this section, we consider some commonly used shape parameters.

Rectangularity

A measurement that reflects the rectangularity of an object is the rectangle fit factor,

$$R = \frac{A_o}{A_R} \tag{2}$$

This is simply the ratio of A_o the object's area to A_R, the area of its minimum enclosing rectangle. It represents how well an object fills its minimum enclosing rectangle. This parameter takes on a maximum value of 1.0 for rectangular objects. It assumes the value $\pi/4$ for circular objects and becomes small for slender, curved objects. The rectangle fit factor is bounded between 0 and 1.

Another related shape feature is the aspect ratio

$$A = \frac{W}{L} \tag{3}$$

which is the ratio of width to length of the minimum enclosing rectangle. This feature can distinguish slender objects from roughly square or circular objects.

Circularity

A group of shape features are called *circularity measures* because they are minimized by the circular shape. Their magnitude tends to reflect the complexity of the boundary. The most commonly used circularity measure is

$$C = \frac{P^2}{A} \tag{4}$$

the ratio of perimeter squared to area. This feature takes on its minimum value of 4π for a circular shape. More complex shapes yield higher values. This measurement is roughly correlated with the subjective concept of complexity of the boundary.

A related circularity measurement is the boundary energy (Ref. 9). Suppose an object has perimeter P and we measure distance around the boundary from some starting point with the variable p. At any point, the boundary has an instantaneous radius of curvature $r(p)$. That is the radius of the circle tangent to the boundary at that point (Figure 16-1). The curvature function at point p is given by

$$K(p) = \frac{1}{r(p)} \tag{5}$$

The function $K(p)$ is periodic with period P. We can compute the average energy per unit length of boundary as

$$E = \frac{1}{P} \int_0^P |K(p)|^2 \, dp \tag{6}$$

The circle has, for fixed area, minimum boundary energy given by

$$E_o = \left(\frac{2\pi}{P}\right)^2 = \left(\frac{1}{R}\right)^2 \tag{7}$$

where R is the radius of the circle. Curvature and, hence, boundary energy are easily computed from the chain code (Ref. 9). Young has shown that the boundary energy

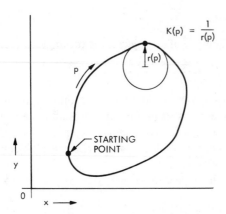

Figure 16-1   Radius of curvature

reflects the perceptual concept of boundary complexity better than the circularity measure of Eq. (4) (Ref. 9).

A third circularity measure makes use of the average distance from an interior point to the boundary (Ref. 10). This distance is given by

$$\bar{d} = \frac{1}{N} \sum_{i=1}^{N} x_i \tag{8}$$

where x_i is the distance from the ith pixel to the nearest boundary point in an object of N points. The shape measure is

$$g = \frac{A}{\bar{d}^2} = \frac{N^3}{\left(\sum\limits_{i=1}^{N} x_i\right)} \tag{9}$$

Figure 16-2 shows an object image and its distance transform. Gray level values in the transformed image reflect their distance from the boundary. The sum in the denominator of Eq. (9) is the IOD of the distance transformed image. For circles and

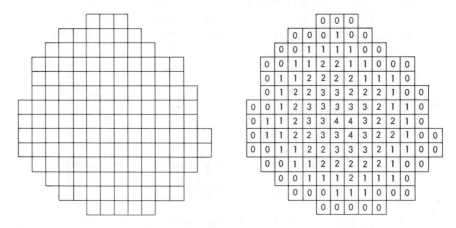

Figure 16-2   The distance transform

regular polygons, Eq. (9) gives the same value as Eq. (4); however, the discriminating power of the feature in Eq. (9) may be superior in some cases.

Invariant Moments

The moments of a function are commonly used in probability theory (Ref. 11, 12). There is a class of shape features having several desirable properties that can be derived from moments (Refs. 13, 14).

Definition. The set of moments of a bounded function $f(x, y)$ of two variables is defined by

$$M_{jk} = \int_{-\infty}^{\infty} \int_{-\infty}^{\infty} x^j y^k f(x, y) \, dx \, dy \tag{10}$$

where j and k take on all nonnegative integer values. The moments of pdfs are widely used in probability theory.

As j and k take on all nonnegative values, they generate an infinite set of moments. Furthermore, this infinite set of moments is sufficient to specify the function $f(x, y)$ completely. In other words, the set $[M_{jk}]$ is unique for the function $f(x, y)$ and only one function has that particular set of moments.

For shape descriptive purposes, suppose $f(x, y)$ takes on the value 1 inside the object and 0 elsewhere. This silhouette function reflects only the shape of the object and ignores internal gray level detail. Every unique shape corresponds to a unique silhouette and, furthermore, to a unique set of moments.

The parameter $j + k$ is called the *order of the moment*. There is only one zero-order moment

$$M_{00} = \int_{-\infty}^{\infty} \int_{-\infty}^{\infty} f(x, y) \, dx \, dy \tag{11}$$

and it is clearly the area of the object. There are two first-order moments and correspondingly more moments of higher orders. We can make all first- and higher-order moments invariant to object size by dividing them by M_{00}.

Central Moments. The coordinates of the center of gravity of the object are given by

$$\bar{x} = \frac{M_{10}}{M_{00}} \qquad \bar{y} = \frac{M_{01}}{M_{00}} \tag{12}$$

The so-called *central moments* are computed using the center of gravity as the origin.

$$\mu_{jk} = \int_{-\infty}^{\infty} \int_{-\infty}^{\infty} (x - \bar{x})^j (y - \bar{y})^k f(x, y) \, dx \, dy \tag{13}$$

The central moments are position invariant.

Principal Axes. The angle of rotation θ that causes the second-order central moment μ_{11} to vanish may be obtained from

$$\tan 2\theta = \frac{2\mu_{11}}{\mu_{20} - \mu_{02}} \tag{14}$$

The coordinate axes x', y' at an angle θ from the x, y axes are called the *principal axes* of the object. The 90° ambiguity in Eq. (14) can be eliminated if we specify that

$$\mu_{20} < \mu_{02} \qquad \mu_{30} > 0 \tag{15}$$

If the object is rotated through the angle θ before moments are computed, or if the moments are computed relative to the x', y' axes, then the moments are rotation invariant.

Invariant Moments. The area normalized central moments computed relative to the principal axis are invariant under magnification, translation, and rotation of the object. Only moments of third order and higher are nontrivial after such normalization. The magnitudes of these moments reflect the shape of the object and can be used as shape features. Invariant moments and combinations thereof have been applied to shape recognition of printed letters (Refs. 13, 14) and to chromosome analysis (Ref. 15).

While invariant moments definitely have some of the properties that good shape features must have, they may or may not have all of them in any particular problem. The uniqueness of the shape of a particular object is spread out over an infinite set of moments. Thus, a large set of features may be required to distinguish similar shapes. The resulting high-dimensional classifier may become quite sensitive to noise and to intraclass variations. In some cases, a few relatively low-order moments may reflect the distinguishing shape characteristics. Usually some experimentation will suggest which, if any, of the invariant moments are reliable and discriminating shape features.

Gray Level Images. If we let $f(x, y)$ be the gray level image of an object, rather than a binary-valued silhouette function, we can compute invariant moments as before. The zero-order moment [Eq. (11)] becomes the integrated optical density rather than the area as before. However, the preceding development applies in a similar manner. For gray level images, the invariant moments reflect not just the shape of the object but also the density distribution within it. As in the previous case, it must be shown, for each object recognition problem, that a reasonably small number of invariant moments can reliably distinguish among the different objects.

Shape Descriptors

The Differential Chain Code. Sometimes it is useful to describe the shape of an object in more detail than that offered by a single parameter but more compactly than in the object image itself. A shape descriptor is a compact representation of object shape. One such descriptor is the boundary chain code (BCC) discussed in the previous chapter. Figure 16-3 shows a simple object with its chain code and the derivative of the chain code. The differential chain code (DCC) reflects the curvature of the boundary, and convexities and concavities show up as peaks. The chain code itself shows the boundary tangent angle as a function of distance around the object. Both functions can be further analyzed to obtain shape measures. Polygonal shapes have one sharp convexity per vertex and are thus separable in the DCC. For example, a

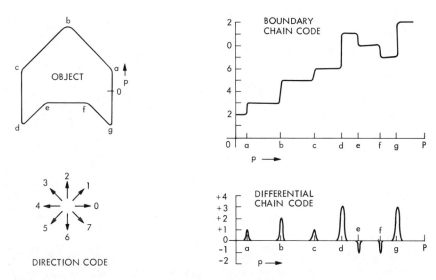

Figure 16-3 The chain code and its derivative

triangularity measure might be the amplitude of the third harmonic of a Fourier series expansion of the DCC. One might distinguish between triangles and squares by using the ratio of the third to the fourth harmonic amplitude. Smoothing of the BCC is usually required before differentiation.

Fourier Descriptors. Recall that the BCC is a periodic function that completely describes the object shape. Hence its Fourier transform is an alternate representation of object shape (Ref. 16). Since the BCC is periodic, it has a discrete (sampled) spectrum. The strengths of the impulses in the BCC spectrum correspond to the coefficients of the Fourier series expansion of the BCC. In many cases, one can lowpass filter the BCC spectrum without destroying the characteristic shape of the object. This means that only the low-frequency impulses in the spectrum (low-order Fourier coefficients) are required to characterize the basic shape.

Parametric Boundary Representation. Suppose we let p represent distance measured along the boundary from some starting point. The coordinates of any boundary point can then be written as $x(p)$, $y(p)$, as shown in Figure 16-4. We can define a complex-valued boundary function

$$B(p) = x(p) + jy(p) \tag{16}$$

Like the BCC, this function is periodic with a discrete spectrum. It thus forms the basis for another set of Fourier descriptors and is amenable to data reduction by lowpass filtering as before.

The Medial Axis Transform. Another data reduction technique that retains shape information is the medial axis transformation (MAT) (Refs. 17, 18). A point inside

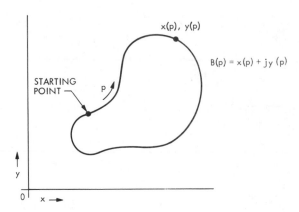

Figure 16-4 The parametric boundary function

the object is on the medial axis if and only if it is the center of a circle that is tangent to the boundary at two nonadjacent points. There is a value associated with each point on the medial axis. This value is the radius of the previously described circle. It represents the minimum distance to the boundary from that point.

One way to compute the MAT is to consider all interior points in some order, expanding a circle about each point. If the circle touches the boundary in two non-adjacent points as it grows, its center is a point on the medial axis. The value associated with that point is the radius of the circle when it first touched the boundary. If the expanding circle touches the boundary at only one point, then its center is not on the medial axis.

Another way to find the medial axis is by boundary peeling. Here one successively removes the outer perimeter of points in a manner similar to peeling an onion. If one point is encountered twice on the peeling excursion around the boundary, then removing that point would disconnect the object. Such a point is on the medial axis, and its value is simply the number of layers that have been previously peeled.

This approach is sometimes called the "grass-fire technique" because it is analogous to setting fire around the periphery of a grassy field and letting the fire burn inward toward the middle. The points where the flame fronts meet are on the medial axis.

For binary images, the medial axis transform retains the shape of the original object. This means that the transformation is invertable and the object can be reconstructed from its MAT. When programmed on digital images using a rectangular sampling grid, the inversion may differ slightly from the original (Ref. 19). Figure 16-5(b) shows the MAT of the chromosome in Figure 16-5(a) computed by an algorithm of R. J. Wall (Ref. 19). Figure 16-5(c) shows how the MAT depends on object orientation with respect to the sampling grid. A definition of an method for computing the MAT of gray level images has been proposed (Ref. 20).

The MAT is useful for finding the central axis of long, narrow, curved objects such as bent chromosomes (Refs. 19, 21). Frequently the medial axis is useful as a

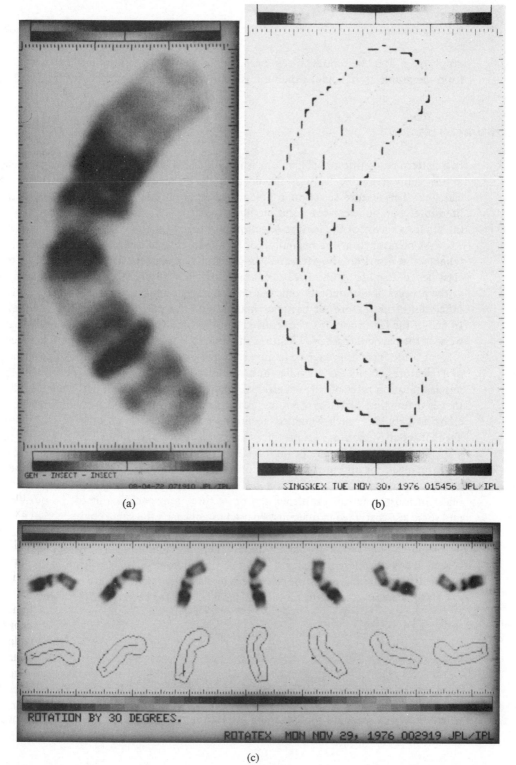

Figure 16-5 The medial axis transform: (a) digital image, (b) MAT, (c) the effect of orientation

graph only, and the values are ignored. Other shape descriptors can be computed from the graph itself, such as the number of branches and the total length (Ref. 22).

FEATURE SELECTION

In a pattern recognition problem, one is usually faced with the task of selecting which of the many available features should actually be measured and presented to the classifier. The feature selection problem has received considerable attention in the literature, but no clear-cut solution has emerged. This section attempts to give the reader a flavor of the feature selection problem.

As mentioned at the beginning of this chapter, one seeks a small set of highly reliable and discriminating features. In general, one expects the performance of the classifier to degrade as fewer features are used. Thus feature selection may be viewed as the process of eliminating some features, starting with the poorest, and combining others until the feature set becomes manageable but performance is still adequate. In fact, if the feature set is to be reduced from M features to some smaller number N, we seek the particular set of N features that maximizes overall classifier performance.

A brute force approach to feature selection is as follows. For all possible subsets of N features, train the classifier and quantify its performance by tabulating the misclassification rates of various groups. Then generate an overall performance index that is a function of the error rates. An example of this would be a linear sum of error probabilities, each weighted according to how serious an error it is. Finally, use that set of N features that produces the best performance index.

The problem with the brute force approach, of course, is the huge amount of work involved for all but the simplest of pattern recognition problems. In fact, there are frequently available only resources enough to train and evaluate the classifier once. In most practical problems, the brute force approach is out of the question, and some less costly technique must be used to approach the same goal.

In the following discussion, we consider the simple case of reducing a two-feature problem to a one-feature problem. Suppose there is available a training set containing objects from M different classes. Let N_j be the number of objects from class j. The two features obtained when the ith object in class j is measured are x_{ij} and y_{ij}. We can start by computing the mean value of each feature for each class. These are

$$\hat{\mu}_{xj} = \frac{1}{N_j} \sum_{i=1}^{N_j} x_{ij} \qquad (17)$$

and

$$\hat{\mu}_{yj} = \frac{1}{N_j} \sum_{i=1}^{N_j} y_{ij} \qquad (18)$$

where the "hats" remind us that these are estimates of the class means based upon the training set rather than the true class means.

Feature Variance. Ideally, the features should take on similar values for all objects within the same class. The estimated variance of the feature x within class j is given by

$$\hat{\sigma}^2_{xj} = \frac{1}{N_j} \sum_{i=1}^{N_j} (x_{ij} - \hat{\mu}_{xj})^2 \tag{19}$$

and, for y, by

$$\hat{\sigma}^2_{yj} = \frac{1}{N_j} \sum_{i=1}^{N_j} (y_{ij} - \hat{\mu}_{yj})^2 \tag{20}$$

Feature Correlation. The covariance of the features x and y in class j can be estimated by

$$\hat{\sigma}_{xyj} = \frac{\frac{1}{N_j} \sum_{i=1}^{N_j} (x_{ij} - \hat{\mu}_{xj})(y_{ij} - \hat{\mu}_{yj})}{\hat{\sigma}_{xj} \hat{\sigma}_{yj}} \tag{21}$$

This quantity is bounded by -1 and $+1$. A value of zero indicates that the two features are uncorrelated, while a value near $+1$ implies a high degree of correlation. A covariance of -1 implies that each variable is proportional to the negative of the other. If the magnitude of the covariance is not low, the two features might well be combined into one, or one of them discarded.

Class Separation. A relevant measure of the ability of a feature to distinguish between two classes is the variance normalized distance between class means. For feature x, this is given by

$$\hat{D}_{xjk} = \frac{|\hat{\mu}_{xj} - \hat{\mu}_{xk}|}{\sqrt{\hat{\sigma}^2_{xj} + \hat{\sigma}^2_{xk}}} \tag{22}$$

where the two classes are j and k. Clearly, the superior feature is the one producing the widest class separation.

Dimension Reduction. There are many ways to combine the two features x and y into a single feature z. A simple way is to use a linear function

$$z = ax + by \tag{23}$$

Since classifier performance is not affected by magnitude scaling of the features, we can impose a magnitude restriction such as

$$a^2 + b^2 = 1 \tag{24}$$

This can be incorporated into Eq. (23) by writing

$$z = x \cos \theta + y \sin \theta \tag{25}$$

where θ is a new variable designating the proportions of x and y in the mixture.

If each object in the training set corresponds to a point in two-dimensional feature space (i.e., the x-y plane), then Eq. (25) describes the projection of those points onto the z-axis, which makes an angle θ with the x-axis. This is shown in Figure 16-6. Clearly, θ should be selected to maximize the class separation or some

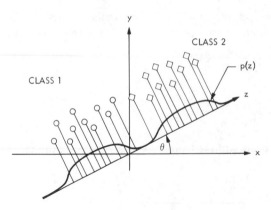

Figure 16-6 Dimension reduction by projection

other quality criterion. For further discussion of dimension reduction, the reader should consult a pattern recognition textbook (Refs. 1–6),

CLASSIFICATION

In this section, we consider some of the commonly used mathematical tools for classification.

Statistical Decision Theory

As an example, suppose we have a simplified fruit sorting problem as in Chapter 15, but with only two classes and a single feature. This means that the objects that present themselves belong either to class 1 (cherries) or to class 2 (apples). On each object, we measure one property, diameter, and this is the feature we call x.

It may be that the pdf of the diameter measurement x is known for one or both classes of objects. For example, the Cherry Farmers' Association may issue a report stating that mean cherry diameter is 20 mm and the pdf is approximately Gaussian with a standard deviation of 4 mm. If the diameter pdf for apples is unknown, we might estimate it by measuring a large number of apples and plotting a histogram of their diameter values. After normalization to unit area, and perhaps some smoothing, this histogram can be taken as an estimate of the corresponding pdf.

A Priori **Probabilities.** It may be that one class is, in general, more likely to occur than the other. For example, suppose that the conveyor belt of the fruit sorting example is known to transport twice as many cherries as apples over any extended period. Thus we can say that the *a priori* probabilities of the two classes are

$$P(C_1) = \tfrac{2}{3} \quad \text{and} \quad P(C_2) = \tfrac{1}{3} \tag{26}$$

These equations merely state that class 1 is twice as likely to occur as class 2. The *a priori* probabilities represent our knowledge about an object before it has been measured. In this example, we know that an unmeasured object is twice as likely to be a cherry as an apple.

Conditional Probabilities. Figure 16-7 shows what the two pdfs might look like. We denote the conditional pdf for cherry diameter as $p(x|C_1)$, which can be read *the probability that diameter* x *will occur, given that the object belongs to class* 1. Similarly, $p(x|C_2)$ is the probability of diameter x, given class 2 (apples).

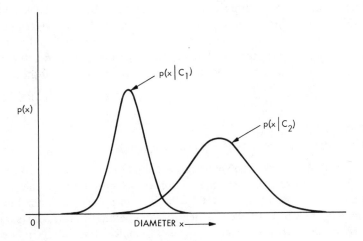

Figure 16-7 Conditional pdfs

Bayes' Theorem. Before an object has been measured, our knowledge of it consists merely of the *a priori* probabilities of Eq. (26). After measurement, however, we should be able to use the measurement and the conditional pdfs to improve our knowledge of class membership. After measurement, the so-called *a posteriori* probability that the object belongs to class i is given by Bayes' theorem:

$$P(C_i|x) = \frac{p(x|C_i)P(C_i)}{p(x)} \tag{27}$$

where

$$p(x) = \sum_{i=1}^{2} p(x|C_i)P(C_i) \tag{28}$$

is the normalization factor required to make the set of *a posteriori* probabilities sum to unity.

Bayes' theorem allows us to combine the *a priori* probabilities of class membership, the conditional pdf, and the measurement to compute, for each class, the probability that the measured object belongs to that class. Given this information, we might choose to assign each object to its most likely class. In our fruit sorting example, we would assign the object to class 1 (call it a cherry) if

$$P(C_1|x) \geq P(C_2|x) \tag{29}$$

and assign it to class 2 (apples) otherwise. Substituting Bayes' theorem (Eq. 27) into Eq. (29) and multiplying out the common denominator produces

$$p(x \mid C_1)P(C_1) \geq p(x \mid C_2)P(C_2) \tag{30}$$

as the condition for assignment to class 1 of a fruit having diameter x. At the decision threshold, where equality holds in Eq. (30), we may assign arbitrarily. The classifier defined by this decision rule is a *maximum-likelihood classifier.*

The General Case. Suppose we make not one but n measurements on each object. Rather than a single feature value, we now have a feature vector $\{x_1, x_2, \ldots, x_n\}$, and each measured object corresponds to a point in n-dimensional feature space. Suppose also that there are not two but m classes of objects. Under these conditions, the *a posteriori* probability of membership in class i is, by Bayes' theorem,

$$p(C_i \mid x_1, x_2, \ldots, x_n) = \frac{p(x_1, x_2, \ldots, x_n \mid C_i)P(C_i)}{\sum_{i=1}^{m} p(x_1, x_2, \ldots, x_n \mid C_i)P(C_i)} \tag{31}$$

where the conditional pdfs are n-dimensional.

Bayes' Risk. Every time we assign an object to a class, we risk making an error. In multiclass problems, some misclassifications may be more harmful than others. A quantitative way to account for this is with a cost function. Let l_{ij} be the cost of assigning an object to class i when it really belongs in class j. Usually, l_{ij} will take on zero value for correct decisions ($i = j$), small values for harmless errors, and large values for costly mistakes. The Bayes' risk is the expected long-term cost of operating the classifier. The risk is evaluated by integrating the probability-weighted cost function.

Suppose we measure an object and assign it to class i. The expected loss resulting from this assignment is the conditional risk

$$R(C_i \mid x_1, x_2, \ldots, x_n) = \sum_{j=1}^{m} l_{ij} p(C_j \mid x_1, x_2, \ldots, x_n) \tag{32}$$

which is just the cost averaged over all m of the groups to which the object might actually belong. Thus, given the feature vector, there is a certain risk involved in assigning the object to any group.

Bayes' Rule. The Bayes' decision rule states that each object should be assigned to the class producing the minimum conditional risk. If we do this, we can then let $R_m(x_1, x_2, \ldots, x_n)$ be the resulting minimum risk corresponding to the feature vector $[x_1, x_2, \ldots, x_n]$. The overall long-term risk of operating the classifier with the Bayes' decision rule is called *Bayes' risk* and is obtained by integrating the risk function over the entire feature space,

$$R = \int_{-\infty}^{\infty} R_m(x_1, x_2, \ldots, x_n) p(x_1, x_2, \ldots, x_n) \, dx_1, dx_2, \ldots, dx_n \tag{33}$$

Clearly, no other decision rule can reduce $R_m(x_1, x_2, \ldots, x_n)$ at any point, and the overall risk is minimized by using Bayes' decision rule.

Classifier Types

It is useful to distinguish among different types of classifiers based upon what is known about the underlying statistics and what must be estimated.

Parametric and Nonparametric Classifiers. If the functional form of the conditional pdfs is known but some parameters of the density function (mean value, variances, etc.) are unknown, then the classifier is called *parametric*. Since the *a priori* probabilities are also parameters, they may be unknown. With parametric classifiers, the functional form of the conditional pdfs is assumed, based upon some fundamental knowledge about the objects themselves. Frequently, functional forms are assumed for mathematical expediency as well as for more intrinsic reasons.

If the functional form of some or all of the conditional pdfs is unknown, the classifier is termed *nonparametric*. This means that all conditional pdfs must be estimated from training data. To do so requires considerably more training data than merely estimating a few parameters in a pdf of known functional form. Thus nonparametric techniques are used only when suitable parametric models are unavailable.

Classifier Training

Supervised and Unsupervised Training. The process of estimating the conditional pdfs or their parameters using object measurements is referred to as training. If the objects have been previously classified by some error-free entity, the training is referred to as supervised. With unsupervised training, the conditional pdfs are estimated using samples of unknown class. Unsupervised learning is ordinarily used only when it is inconvenient or impossible to obtain a preclassified training set.

We shall concern ourselves with two commonly used approaches to supervised training. These are the maximum-likelihood and Bayesian techniques. While the two techniques exhibit considerable philosophical difference in approach, they usually produce similar results.

Maximum-Likelihood Estimation. This approach assumes that the parameters to be estimated are fixed but unknown. A given sample (the training set) is drawn, and the parameter estimate is taken as that value which makes the occurrence of the observed training set most likely. For example, suppose that 100 samples are drawn from a normal distribution of unknown mean but with a standard deviation of 2.0. Suppose further that the mean value of the 100 samples is 12. Now it is much more likely that the 100 samples came from a population having a mean value of 12 than from a population with a mean of zero. Although the latter situation is possible, it requires a conspiracy of highly unlikely events. It can be shown that the population mean that makes the observed sample mean most likely is 12.

Maximum-likelihood estimation is a well-developed subject and considerably beyond our scope. We shall be content here merely to introduce the concept and quote the well-known result that the maximum-likelihood estimate of the mean and standard deviation of a normal distribution are the sample mean and sample standard deviation, respectively.

Bayesian Estimation. Unlike maximum-likelihood estimation, the Bayesian approach treats the unknown parameter as a random variable. Furthermore, it assumes that something is known about the unknown parameter in advance. Bayesian estimation assumes that the unknown parameter has a known, or assumed, *a priori* pdf before any samples are taken. After the training set has been measured, Bayes' theorem is used to allow the sample values to update, or refine, the *a priori* pdf. This results in an *a posteriori* pdf of the unknown parameter value. We hope that this pdf has a single narrow peak, centered on the true value.

As an example of Bayesian estimation, suppose we wish to estimate the mean of a normal distribution with known variance. Before measuring the training set, we can use whatever knowledge is available to establish an *a priori* pdf on the unknown mean value. We shall call this *a priori* density function $p(\mu)$. We shall denote the known functional form of the pdf of unknown mean by $p(x \mid \mu)$. This states that, given a value for μ, we then know $p(x)$. If we let X represent the set of sample values obtained by measuring the training set, Bayes' theorem gives the *a posteriori* pdf of μ after the training set has been measured,

$$p(\mu \mid X) = \frac{p(X \mid \mu)p(\mu)}{\int p(X \mid \mu)p(\mu) \, d\mu} \tag{34}$$

What we really want is $p(x \mid X)$, the best estimate of the density $p(x)$, given the training set measurements X. One way to do this is to set up the joint (two-dimensional) pdf of both x and μ, and then integrate out the μ-component; that is,

$$p(x \mid X) = \int_{-\infty}^{\infty} p(x, \mu \mid X) \, d\mu \tag{35}$$

The joint density in the integrand can be written as a product of two independent one-dimensional pdfs. Then Eq. (35) becomes

$$p(x \mid X) = \int_{-\infty}^{\infty} p(x \mid \mu)p(\mu \mid X) \, d\mu \tag{36}$$

This is the desired result, since $p(x \mid \mu)$ is the assumed functional form and $p(\mu \mid X)$ is the *a posteriori* pdf of the unknown mean from Eq. (34).

To see how $p(\mu \mid X)$ affects $p(x \mid X)$, suppose that $p(\mu \mid X)$ has a single sharp peak at $\mu = \mu_o$. This means that our *a priori* knowledge has combined with the training set to specify μ within narrow limits around the value μ_o. If the peak is sufficiently sharp, we can approximate $p(\mu \mid X)$ by an impulse at μ_o.

$$p(\mu \mid X) \approx \delta(\mu - \mu_o) \tag{37}$$

Then Eq. (36) becomes

$$p(x \mid X) = \int_{-\infty}^{\infty} p(x \mid \mu)\delta(\mu - \mu_o) \, d\mu \tag{38}$$

which, by the sifting property of the impulse, is

$$p(x \mid X) = p(x \mid \mu_o) \tag{39}$$

This says that μ_o is the best estimate of the unknown mean.

Suppose, on the other hand, that the *a posteriori* distribution of the unknown mean $p(\mu \mid X)$ has a relatively broad peak about μ_o. In this case, $p(x \mid X)$ becomes a weighted average of many pdfs, all having different means in the neighborhood of μ_o. This has the effect of smearing or broadening $p(x \mid X)$ to reflect our uncertainty about the mean value.

As mentioned earlier, maximum-likelihood and Bayesian estimation produce similar if not identical results in many common cases. For example, both approaches tend to establish the unknown mean at the mean of a large training set. Bayesian estimation allows us to combine any *a priori* knowledge we have with the quantitative data of the training set to estimate the unknown parameter. Furthermore, the width of $p(\mu \mid X)$ is an indication of how confidently we have estimated the unknown parameter.

In summary, the steps involved in Bayesian estimation are as follows. First, we assume an *a priori* pdf for the unknown parameter or parameters. Secondly, we collect sample values from the population by measuring the training set. Thirdly, we use Bayes' theorem to refine the *a priori* pdf into the *a posteriori* pdf using the sample values. Finally, we form the joint density of x and the unknown parameter and integrate out the latter to leave the desired pdf estimate.

If we have strong ideas about the probable values of the unknown parameter, we may assume a narrow *a priori* pdf. If, on the other hand, we know little about the parameter, we should assume a relatively broad pdf.

The effect of using sample values to refine the *a priori* pdf is shown from

$$p(\mu \mid X) = \frac{1}{c} p(X \mid \mu) p(\mu) = \frac{1}{c} \prod_{i=1}^{n} p(x_i \mid \mu) p(\mu) \tag{40}$$

where c is the denominator of Eq. (34) and Π indicates an n-term product. Since the n samples are all taken independently, the probability of pulling out the entire training set is merely the product of the individual probabilities of pulling out each sample.

If the samples are tightly clustered about the sample mean μ_s, then $p(X \mid \mu)$ has a sharp peak at or near $\mu = \mu_s$. If the assumed *a priori* density $p(\mu)$ is relatively flat in that area, then

$$p(x \mid X) = \int_{-\infty}^{\infty} p(x \mid \mu) p(X \mid \mu) p(\mu) \, d\mu \tag{41}$$

The function $p(x \mid \mu)$ is the assumed form of the pdf with μ as a parameter. As far as the integral in Eq. (41) is concerned, $p(x \mid \mu)$ is a function of x and μ. The function $p(X \mid \mu)$ is the probability that the sample set X would be drawn if the pdf indeed has mean value μ. It is given by Eq. (40), is a function μ, and becomes increasingly sharp as n increases. Our prior knowledge of the unknown parameter μ is given by $p(\mu)$, which is the assumed *a priori* pdf of the unknown mean.

Let us consider two different cases that illustrate the role of prior knowledge and the training set in Bayesian estimation. For case 1, we have strong feelings about the value of μ, and we take a relatively small number of samples in the training set. This means that we would assume $p(\mu)$ is narrow about μ_o, our preconceived idea of the mean value. If n is small, $p(X|\mu)$ is broad about the sample mean μ_s. Then Eq. (41) can be approximated by

$$p(x|X) \approx \int p(x|\mu)\delta(\mu - \mu_o)\, d\mu = p(x|\mu_o) \tag{42}$$

This indicates that the Bayesian estimate of the unknown pdf is basically the assumed parametric form with our preconceived value μ_o substituted for the mean.

In the second case, suppose we do not have strong feelings about the mean value and that we employ a large training set. Thus we assume a $p(\mu)$ that is broad about μ_o. Furthermore, if n is large, $p(X|\mu)$ becomes relatively sharp about μ_s, the sample mean. Then the Bayesian estimate of the unknown pdf is

$$p(x|X) \approx \int p(x|\mu)\delta(\mu - \mu_s)\, d\mu = p(x|\mu_s) \tag{43}$$

In this case, the large training set has overpowered our timid *a priori* estimate and substituted the sample mean into the assumed form of the unknown pdf. Thus, as the number of samples increases, the final estimate of the mean moves from our initial estimate μ_o toward the sample mean μ_s. Our *a priori* confidence is represented by the sharpness of $p(\mu)$. The sharper this function is, the more slowly the estimate moves toward μ_s with increasing n.

Maximum-likelihood estimation allowed us to use the training set to estimate the unknown mean. Bayesian estimation allows us to combine our prior knowledge with the training set to estimate the unknown mean. If our prior knowledge is small compared to that of the training set, then both methods tend to converge toward the sample mean.

An Example. We conclude this discussion with an example that illustrates training a classifier. The objects to be classified are the human chromosomes. Under the light microscope, the 46 chromosomes from the nucleus of a human lymphocyte (white blood cell) appear in scattered disarray [Figure 16-8(a)]. The 46-chromosome complement is known to consist of 22 pairs of morphologically similar homologous chromosomes and 2 sex determinant chromosomes (XX for the female and XY for the male). The two long arms and the two short arms are all connected at the "centromere."

For diagnostic purposes, it is customary to arrange the chromosome images into groups of similar morphology. This arrangement produces the "karyotype" of Figure 16-8(b). The groups are designated by the letters A through G, as indicated. This display format facilitates the visual examination for abnormal or missing chromosomes.

As a pattern recognition problem, our task is merely to assign each incoming chromosome to one of the seven groups, A through G. We shall measure two features of each chromosome, total length and the arm length ratio. The latter feature is called *centromeric index* and is the ratio of long arm length to total length.

Figure 16-9(a) shows a two-dimensional histogram of the measurements from the 2300 chromosomes in 50 normal cells. In the two-dimensional feature space, the abscissa is chromosome length, while the ordinate is centromeric index. Gray level is indicated by a combination of derivative shading and contour lines. The histogram

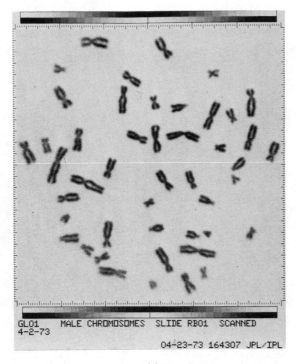

(a)

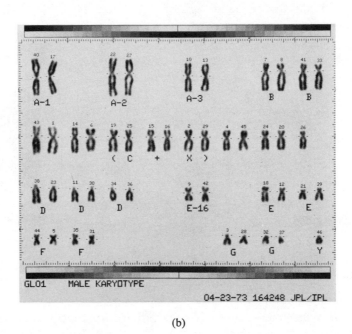

(b)

Figure 16-8 Human chromosomes: (a) digitized microscope image, (b) karyotype

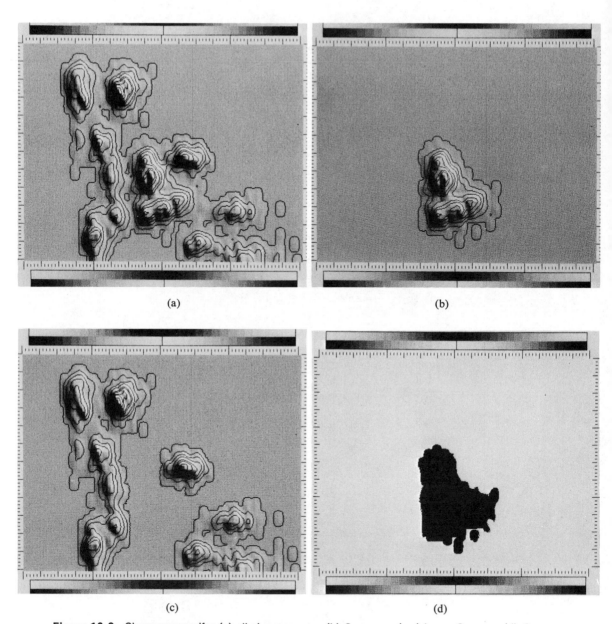

(a)

(b)

(c)

(d)

Figure 16-9 Chromosome pdfs: (a) all chromosomes, (b) C group only, (c) non-C group, (d) C group decision region

has been smoothed slightly by convolution with a lowpass filter. Multiple clusters are clearly evident, indicating the morphological differences of the homologous pairs.

Figure 16-9(b) shows a similar histogram for only those chromosomes that belong to the C group. This subset of chromosomes was identified by an expert cytogeneticist. The histogram of all non-C group chromosomes appears in Figure 16-9(c).

Training the classifier in this case consists of partitioning the feature space into disjoint regions, one for each karyotype group. The histogram in Figure 16-9(b) can be viewed as an unnormalized estimate of the pdf for C group chromosomes. It can be written as

$$f_c(x, y) = Np(C)p(x, y \,|\, C) \tag{44}$$

where N is number of cells in the training set (50), $p(C)$ is the *a priori* probability that an unmeasured chromosome belongs to the C group, and $p(x, y \,|\, C)$ is the pdf for C group chromosomes. The normal male karyotype has 15 and the female karyotype 16 chromosomes in the C group, which includes the X chromosomes. Thus, if males and females are equally likely, the *a priori* probability is

$$p(C) = \frac{15.5}{46} \tag{45}$$

Equation (30) gives the decision rule for the maximum-likelihood classifier. This means that we should assign a chromosome with feature values x, y to the C group if the histogram of Figure 16-9(b) is greater at x, y than the histogram of Figure 16-9(c). We can identify this region by subtracting the digital image of Figure 16-9(c) from that of Figure 16-9(b). The area of positive difference is shown in Figure 16-9(d). This establishes the maximum-likelihood region for the C group. A similar procedure for the other groups produces the classifier presented in Figure 16-10.

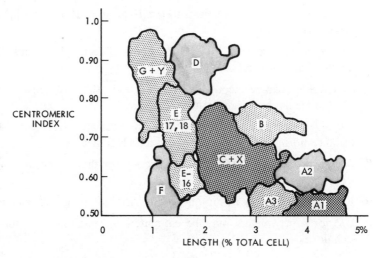

Figure 16-10   Chromosome classifier

Classifier Performance

There are several ways to estimate the performance of a classifier after it has been designed and trained. If the pdfs are known or have been estimated, one can compute the probability of error as the area under the tails. Alternatively, one can run the classifier on a known test set. The test set should be different from the training set. Computing the area under the design pdfs is analogous to using the training set as a test set.

As a general rule in pattern recognition applied to digital images, the quality of the image limits the reliability of the measurements, and this in turn limits the accuracy of classification by causing overlap in the pdfs. Image quality is degraded by optics, noise, and distortion. This combines with the in-class variability of the objects to broaden the pdfs. The classifier should be appropriate to the problem, but a highly sophisticated classifier may not produce commensurate improvements in classification accuracy when compared with a simple one.

SUMMARY OF IMPORTANT POINTS

1. Features used for classification should be discriminating, reliable, independent, and few in number.

2. A training set used to establish classifier parameters should be representative and unbiased.

3. Classifier performance (error rates) can be estimated by classifying a known test set.

4. Object size is reflected in measurements of area, IOD, length, width, and perimeter, among others.

5. Object shape is reflected in measurements of rectangle fit and circularity and in the invariant moments.

6. Object shape can be encoded in the chain code, the parametric boundary function, and the medial axis transform.

7. Effective features have small within-class variance, low correlation, and large variance-normalized separation between class means.

8. Bayes' theorem [Eq. (27)] gives the probability that a measured object belongs to a particular class.

9. The Bayes' decision rule minimizes the "risk" of operating a classifier.

10. Unknown parameters may be estimated by maximum-likelihood or Bayesian techniques.

REFERENCES

1. R. O. DUDA and P. E. HART, *Pattern Classification and Scene Analysis*, John Wiley & Sons, Inc., New York, 1973.

2. K. FUKUNAGA, *Introduction to Statistical Pattern Recognition*, Academic Press, New York, 1972.

3. K. S. FU, *Sequential Methods in Pattern Recognition and Machine Learning*, Academic Press, New York, 1968.

4. E. A. PATRICK, *Fundamentals of Pattern Recognition*, Prentice-Hall, Inc., Englewood Cliffs, New Jersey, 1972.

5. W. MEISEL, *Computer-Oriented Approaches to Pattern Recognition*, Academic Press, New York, 1972.

6. H. C. ANDREWS, *Introduction to Mathematical Techniques in Pattern Recognition*, John Wiley & Sons, Inc., New York, 1972.

7. J. HILDITCH and D. RUTOVITZ, "Chromosome Recognition," *Annals of the New York Academy of Science*, **157**, 339–364, 1969.

8. H. A. LUBBS and R. S. LEDLEY, "Automated Analysis of Differentially Stained Human Chromosomes," in T. Caspersson and L. Zech, eds., *Nobel 23—Chromosome Identification*, Academic Press, New York, 1973.

9. I. T. YOUNG, J. E. WALKER, and J. E. BOWIE, "An Analysis Technique for Biological Shape. I," *Information and Control*, **25**, 357–370, 1974.

10. P. E. DANIELSON, "A New Shape Factor," *Computer Graphics and Image Processing*, **7**, 292–299, 1978.

11. A. PAPOULIS, *Probability, Random Variables, and Stochastic Processes*, McGraw-Hill Book Company, New York, 1965.

12. E. KREYSZIG, *Introductory Mathematical Statistics*, John Wiley & Sons, Inc., New York, 1970.

13. M. K. HU, "Visual Pattern Recognition by Moment Invariants," *IRE Trans. Info. Theory*, February 1962, 179–187.

14. F. L. ALT, "Digital Pattern Recognition by Moments," *JACM*, **9**, 240–258, 1962.

15. J. W. BUTLER, M. K. BUTLER, and A. STROUD, "Automatic Classification of Chromosomes," in K. Enslein, ed., *Data Acquisition and Processing in Biology and Medicine*, **3**, Pergamon Press, New York, 1964.

16. C. T. ZAHN and R. Z. ROSKIES, "Fourier Descriptors for Plane Closed Curves," *IEEE Trans. Computers*, **C-21**, 269–281, 1972.

17. H. BLUM, "Biological Science and Visual Shape (Part I)," *J. Theor. Biol.*, **38**, 205–287, 1973.

18. H. BLUM, "A Transformation for Extracting New Descriptors of Shape," in W. Wathen-Dunn, ed., *Models for the Perception of Speech and Visual Form*, MIT Press, Cambridge, Massachusetts, 1967.

19. R. J. WALL, A. KLINGER, and S. HARAMI, "An Algorithm for Computing the Medial Axis Transform and Its Inverse," *Proceedings of the 1977 Workshop on Picture Data Description and Management*, 121–122, Proceedings 77CH1187-4C, IEEE Computer Society, Piscataway, New Jersey, 1977.

20. G. LEVI and U. MONTANARI, "A Gray-Weighted Skeleton," *Information and Control*, **17**, 62–91, 1970.

21. K. CASTLEMAN and R. WALL, "Automatic Systems for Chromosome Identification," in T. Caspersson, ed., *Nobel Symposium 23—Chromosome Identification*, Academic Press, New York, 1973.

22. T. PAVLIDIS, "A Review of Algorithms for Shape Analysis," *Computer Graphics and Image Processing*, **7**, 243–258, 1978.

Chapter 17

THREE-DIMENSIONAL
IMAGE PROCESSING

IIIIIIIIIIIIIIIIIIIIIIIIIIIIIIIIIIIIIIIIIIIIIIIIIIIIIIIIIIIIIIIIIIIIIIIIIIIIIIIIIIIIIIIIIIIIIIIIIIIIIIIIIIIIIIIIIIIIIIIIIIIIIIIIIIIIIIIIIIIIIIIIIIIIIIIIIII

INTRODUCTION

In previous chapters, we have discussed two-dimensional digital image processing. Such images can be thought of as having gray level that is a function of two spatial variables. The most straightforward generalization to three dimensions would have us deal with images having gray level that is a function of three spatial variables. These we call *spatially three-dimensional images*. One can conceive of several examples of these, such as ocean water temperature as a function of x, y, and depth; atmospheric pollution levels as a function of x, y, and altitude; and gravity field strength as a function of three dimensions in outer space. Perhaps more common examples are three-dimensional images of transparent microscope specimens or of larger objects viewed with X-ray illumination. In such images, the gray level represents some local property such as optical density per millimeter of path length.

More common in human experience is the ordinary three-dimensional world in which we live. Indeed, most of the two-dimensional images we see have been derived from this three-dimensional world by camera systems that employ a perspective projection to reduce the dimensionality from three to two. By modeling this projection, one can implement the inverse projection to learn more about the three-dimensional object that produced a given image in the first place. Similarly, given a mathematical description of a three-dimensional object, one can compute the image that would be obtained by a camera at a specified location. Thus another topic deserving of the name three-dimensional image processing concerns the simulation of image-forming projections and their inverses.

We shall address five topics in three-dimensional image processing in this chapter. These topics were chosen primarily because they are appropriate for treatment using hardware and software oriented toward two-dimensional digital image processing. Thus these applications logically build upon the techniques discussed in previous chapters. By contrast, three-dimensional computer graphics is a different discipline, which dictates a different hardware and software emphasis. For an introduction to this fascinating field, the interested reader should consult a textbook on that subject (Ref. 1). The following paragraphs introduce the five topics treated in this chapter.

Multispectral Analysis

A three-dimensional image can be formed by sampling not only the two spatial coordinates of an optical image but also the wavelength spectrum of the light at each point. Thus, instead of quantizing the total light intensity falling upon each pixel, one would sample and quantize the electromagnetic spectrum of that illumination. This forms a three-dimensional image in which gray level is a function of two spatial variables and a third variable, optical wavelength.

The discipline concerned with processing such images is commonly called *multispectral analysis*. The resulting images are often referred to as *multidigital images*. They are usually organized as a series of two-dimensional digital images, each of which was obtained by digitizing the original image in a narrow spectral band.

Spatially Three-Dimensional Images

Consider a three-dimensional object that is not perfectly transparent but does allow light to pass through. We can think of a local property that is distributed throughout the object in three dimensions. This property is local optical density. It might be specified in units of optical density per millimeter of path length. For example, if the object were a slab of uniform local property oriented perpendicular to the illuminating beam, the measured optical density of the slab would be proportional to both the value of the local property and the slab thickness.

Thin specimens of biological tissue appear transparent under a microscope. In the X-ray portion of the electromagnetic spectrum, the entire human body is transparent. In a later section, we discuss how three-dimensional imaging can be performed using light microscopy and the X-ray technique called *tomography*.

Stereometry

When a camera forms an image of a three-dimensional scene, it necessarily discards certain information about that scene. This loss of information is a direct result of the perspective projection that reduces the dimensionality from three to two. For example, a feature of a certain size in the image could result from either a large distant object or a small nearby object. This "range ambiguity" is a result of the information loss in the imaging projection.

When a three-dimensional scene is photographed by a pair of cameras located at slightly different positions, the range ambiguity can be resolved. The two images produced are called a *stereoscopic image pair*. A *range image* is an image in which gray level represents not brightness but rather the distance from the camera to the reflecting surface in the scene that gives rise to the corresponding pixel brightness. Each pixel in a digitized image can be viewed as projecting a slender cone out through the imaging lens (Figure 17-1). In the brightness image, the gray level of a particular pixel indicates the amount of light reflected off the first surface intersected by the pixel cone. In the range image, the gray level represents the length of the pixel cone.

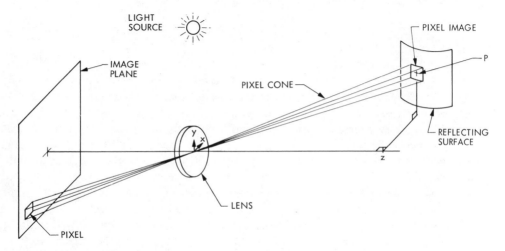

Figure 17-1   Imaging geometry

The combination of a brightness image and a range image restores much of the information lost in the imaging projection. It is not, however, a complete description of the original scene, since surfaces may be obscured in the image. For many purposes, however, the range image is a useful adjunct to the brightness image. Stereometry is the technique of deriving a range image from a stereo pair of brightness images. It has long been used as a manual technique for creating elevation maps of the earth's surface. Later in this chapter, we discuss computer-implemented stereometry.

Stereoscopic Display

If it is possible to compute a range image from a stereo pair, then it should be possible to generate a stereo pair given a single brightness image and a range image. This technique makes it possible to generate stereoscopic displays that give the viewer a sensation of depth. If a stereoscopic image pair is presented to a viewer in such a way that each eye sees one of the two images, the resulting visual sensation of depth can duplicate that of viewing the original scene. Stereo display techniques can increase the available information in a computer-driven display.

Shaded Surface Display

It is frequently desirable to generate either monocular or stereo pair images of a three-dimensional object that exists only as a mathematical description. By modeling the imaging system, one can compute the digital image that would result if the object existed and if it were digitized by conventional means. Shaded surface display grew out of the domain of computer graphics and has developed rapidly in the past few years. It is commonly done on hardware systems designed for two-dimensional digital image processing and is thus appropriate for discussion here.

MULTISPECTRAL ANALYSIS

The most common examples of multispectral imaging are color photography and color television (Ref. 2). In both cases, the visible spectrum is divided into three bands: red, green, and blue. In color photography, three images are formed on three sandwiched photographic emulsions. In color television, three separate camera tubes are employed, one each behind red, green, and blue optical filters. Color photography and color television are more concerned with display of the image than with its analysis. For display purposes, red, green, and blue images are superimposed, either on the color print or on the color television monitor.

Perhaps the greatest effort has been devoted to multispectral analysis in the field of remote sensing (Ref. 3). Multispectral images are obtained from aircraft or spacecraft that overfly a region of interest on the earth's surface. Each pixel of the image is sensed by a battery of narrow band light measuring devices. Thus the image is digitized with multi-valued pixels. Twenty-four or more spectral channels are commonly used. The resulting image data is processed as a set of 24 or so two-dimensional digital images. Each two-dimensional image shows the object as it would appear through a narrow band optical filter. The spectral range covered by multispectral analysis need not be limited to the visible spectrum. Commonly, the range of interest extends from the infrared through the visible spectrum and into the ultraviolet.

A considerable portion of multispectral analysis is devoted to "pixel classification." The image is partitioned into regions that correspond to different types of surfaces, such as lakes, fields, forests, and industrial areas. Each multi-valued pixel is classified as to surface type using its set of spectral intensity measurements. The classification is accomplished with techniques similar to those discussed in Chapter 16. Frequently, algebraic operations such as subtraction and ratioing are performed on the set of images to enhance surface differences. The interested reader should consult the remote sensing literature for an introduction to this subject (Ref. 3).

OPTICAL SECTIONING

The light microscope is a commonly used tool in histology and microanatomy. These disciplines are concerned with the structure and function of physiological specimens on a microscopic scale. The specimens, however, are three-dimensional, and this presents problems for analysis with a conventional light microscope. First, only those structures in or near the plane of focus are visible. Furthermore, although structures just outside the focal plane are visible, they appear blurred. This effect can be overcome by serial sectioning, a technique that involves slicing the specimen to produce a series of thin sections that may be studied individually to develop an understanding of the three-dimensional structure. Serial sectioning has two major disadvantages. One is the loss of registration occurring when the sections become separated after slicing. Also, there is unavoidable geometric distortion as the slices are processed. These artifacts include stretching, curling, folding, and tearing of the thin sections.

In many applications, it would be advantageous to obtain a three-dimensional display of the biological specimen. The three-dimensional display is important because improper interpretation of two-dimensional section images has led to a variety of structural misunderstandings (Ref. 4). A three-dimensional display can be produced by digitizing the specimen with the focal plane situated at various levels along the optical axis (optical sectioning) and then processing each resulting image to remove the defocused information from structures in neighboring planes. In this section, we address the use of digital image processing for deblurring optical section images and for three-dimensional display of the optically sectioned specimen.

Thick Specimen Imaging

Figure 17-2 diagrams the optical system of a microscope imaging a specimen of thickness T. The three-dimensional coordinate system has its origin at the bottom of the specimen, and the z-axis coincides with the optical axis of the microscope. The lens-to-image-plane distance d_i is fixed (commonly 160 mm), and the in-focus plane falls at $z = z'$, a distance d_f below the lens center. The image plane has its own coordinate system x', y', with its origin on the z-axis.

The focal length of the objective lens determines the distance d_f to the focal plane from the lens equation

$$\frac{1}{d_i} + \frac{1}{d_f} = \frac{1}{f} \tag{1}$$

This, in turn, determines the magnification or power of the obejctive

$$M = \frac{d_i}{d_f} \tag{2}$$

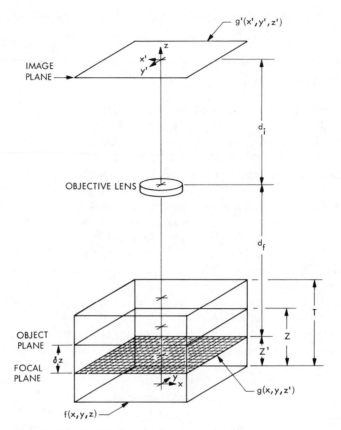

Figure 17-2 Thick specimen imaging

Since the image distance d_i and the focal length f are fixed, the focal plane may be moved through the specimen simply by moving the objective lens and image plane up and down as a unit. Thus we can place the focal plane at any desired level z'. The focal length of the objective is related to the other microscope parameters by

$$f = \frac{d_i}{M+1} = \frac{M}{M+1}d_f = \frac{d_i d_f}{d_i + d_f} \tag{3}$$

and the distance from lens center to focal plane is given by

$$d_f = \frac{d_i}{M} = \frac{M+1}{M}f = \frac{fd_i}{d_i - f} \tag{4}$$

We can describe the three-dimensional distribution of optical density within the specimen by the function $f(x, y, z)$. We denote the image that results with the focal plane located at level z' by $g'(x', y', z')$. Since the dimensions of interest are those of the specimen and not those of the magnified image, we can simplify the notation somewhat. We define an ideal (distortion-free) projection from the image plane back into the focal plane. This projection of $g'(x', y', z')$ to form $g(x, y, z')$ counteracts the magnification and 180° rotation induced by the imaging projection, and it places the image back into the specimen coordinate system. Thus a point at x, y, z images to a point at x, y, z' in the focal plane. We are ignoring the slight change in magnification produced by defocus.

Since we are processing a digital image anyway, it is more convenient to refer all scale factors (pixel spacing, spatial frequency, etc.) to the specimen coordinate system. We now wish to establish the relationship between the image $g(x, y, z')$ and the specimen function $f(x, y, z)$.

Figure 17-3 illustrates the simplified case, where the specimen has zero density except in the object plane located at $z = z_1$; that is,

$$f(x, y, z) = f_1(x, y)\delta(z - z_1) \tag{5}$$

This corresponds to two-dimensional imaging with the object out of focus by the amount $z_1 - z'$. Since a defocused lens is still a linear system, we can write the convolution relation

$$g_1(x, y, z') = f(x, y, z_1) * h(x, y, z_1 - z') \tag{6}$$

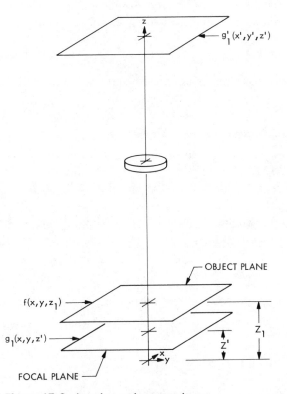

Figure 17-3 Imaging a planar specimen

where $h(x, y, z_1 - z')$ is the point spread function of the optical system defocused by the amount $z_1 - z'$.

We can model the three-dimensional specimen as a stack of object planes located at small intervals Δz along the z-axis; that is,

$$\sum_{i=1}^{N} f(x, y, i\Delta z) \, \Delta z \tag{7}$$

where

$$N = \frac{T}{\Delta z} \tag{8}$$

The image of this stack obtained with the focal plane located at z' is the sum of the individual plane images; that is,

$$g(x, y, z') = \sum_{i=1}^{N} f(x, y, i\Delta z) * h(x, y, z' - i\,\Delta z)\Delta z \tag{9}$$

If we substitute $z = i\Delta z$ and take the limit as Δz approaches zero (and N approaches infinity), the summation becomes an integral, and Eq. (9) reduces to

$$g(x, y, z') = \int_0^T f(x, y, z) * h(x, y, z' - z) \, dz \tag{10}$$

If we specify that $f(x, y, z)$ is zero outside the field of view and outside the range $0 \le z \le T$ and write out the two-dimensional convolution, we are left with

$$g(x, y, z') = \int_{-\infty}^{\infty} \int_{-\infty}^{\infty} \int_{-\infty}^{\infty} f(x', y', z)h(x - x', y - y', z' - z) \, dx' \, dy' \, dz \tag{11}$$

Thus microscope imaging of a thick specimen involves a three-dimensional convolution.

Deblurring Optical Section Images

We now seek a means to remove the defocused information from optical section images. In other words, we wish to recover the function $f(x, y, z)$ from a series of images $g(x, y, z')$ taken at different focal plane levels z'.

Deconvolution. We could recover the specimen function $f(x, y, z)$ by three-dimensional deconvolution, subject to the restrictions imposed by zeros in the transfer function. Transforming Eq. (11) yields the frequency domain relation

$$G(u, v, w) = F(u, v, w)H(u, v, w) \tag{12}$$

where u, v, and w are frequency variables in the x-, y-, and z-directions, respectively. The spectrum of the specimen function is

$$F(u, v, w) = G(u, v, w)H'(u, v, w) \tag{13}$$

where

$$H'(u, v, w) = \frac{1}{H(u, v, w)} \tag{14}$$

is the inverse three-dimensional OTF.

Transforming back to the spatial domain yields

$$f(x, y, z) = g(x, y, z) * h'(x, y, z) \tag{15}$$

Writing out the z-component of the convolution integral produces

$$f(x, y, z) = \int_{-\infty}^{\infty} g(x, y, z') * h'(x, y, z - z') \, dz' \tag{16}$$

where z' is now a dummy variable of integration.

If we discretize the z-axis into intervals Δz by letting $z = j\Delta z$, $z' = i\,\Delta z$, and $dz' = \Delta z$, Eq. (16) becomes

$$f(x, y, j\,\Delta z) = \sum_{i=-\infty}^{\infty} g(x, y, i\,\Delta z) * h'(x, y, j\,\Delta z - i\,\Delta z)\,\Delta z \tag{17}$$

When the focal plane moves outside the specimen ($i < 0$ or $i > N$), the information content of the resulting image becomes rather small. Thus we can approximate Eq. (17) by the finite summation

$$f(x, y, j\,\Delta z) \approx \sum_{i=-M}^{N+M} g(x, y, i\,\Delta z) * h'(x, y, j\,\Delta z - i\,\Delta z)\,\Delta z \tag{18}$$

where M is some positive integer. This reduces the restoration of each object plane to a finite summation of two-dimensional convolutions.

While three-dimensional deconvolution might result in restoration of the specimen function $f(x, y, z)$, it is fraught with difficulties. First is the complexity of computing the spectrum of the three-dimensional psf. Second is the computation of $h'(x, y, z)$, the inverse three-dimensional transform of Eq. (14). Finally, Eq. (18) represents considerable computational effort, especially if Δz is small and if $N + 2M$ must be large to attain a reasonable approximation.

Simultaneous Equations. For a second approach, let us again approximate the specimen with a stack of object planes separated at equal intervals Δz along the z-axis. We generate a series of optical section images by digitizing the specimen repeatedly while moving the focal plane up the z-axis in the same increments Δz. We make the substitutions

$$z' = j\,\Delta z \qquad 1 \leq j \leq N \qquad dz = \Delta z \tag{19}$$

and the jth section image is obtained from Eq. (9).

$$g(x, y, j\,\Delta z) = \sum_{i=1}^{N} f(x, y, i\,\Delta z) * h(x, y, i\,\Delta z - j\,\Delta z)\,\Delta z \tag{20}$$

where $h(x, y, z)$ is assumed symmetric in z.

We can simplify the notation temporarily by dropping x, y, and the constant Δz as understood and writing i and j as subscripts. With these changes, Eq. (20) becomes

$$g_j = \sum_{i=1}^{N} f_i * h_{i-j} = \sum_{i=1-j}^{N-j} f_{i+j} * h_i \tag{21}$$

This states simply that the jth image is a sum of convolutions of the various specimen planes with the appropriate defocus psfs. Recall that $(i - j)\,\Delta z$ is the defocus distance.

We can simplify the situation by taking the two-dimensional Fourier transform of Eq. (21). This moves us from the spatial to the frequency domain, where convolution reduces to multiplication. By definition,

$$G_j = \mathcal{F}\{g(x, y, j\,\Delta z)\} \qquad F_i = \mathcal{F}\{f(x, y, i\,\Delta z)\} \qquad H_i = \mathcal{F}\{h(x, y, i\,\Delta z)\} \quad (22)$$

and Eq. (21) becomes

$$G_j = \sum_{i=1-j}^{N-j} F_{i+j} H_i \qquad (23)$$

Given a set of optical section images, G_j for $1 \le j \le N$, Eq. (23) represents a set of N simultaneous linear equations in N unknowns. Thus we have a second possibility for recovering the specimen function $f(x, y, z)$. We could use Cramer's rule, or some such technique, to solve the system of equations indicated by Eq. (23) for the F_j's. The computational complexity of this task is formidable. In reality, F_j, G_j, and H_j are two-dimensional functions of frequency. Thus the system of equations would have to be solved for every sample point in the (two-dimensional) frequency domain. While this could be done, provided a solution exists, it is doubtful that the results would justify the computational expense.

An Approximate Method. Rather than an exact solution, which recovers the specimen function completely, what may be of more practical use is an approximate method that significantly improves the situation, but at reasonable expense (Refs. 5, 6). If we abandon the notion of an exact (and consequently simultaneous) solution, we can perhaps develop a reasonable technique that yields good performance.

Let us pull the $i = 0$ term out of Eq. (21), leaving two summations, one for positive i and one for negative i,

$$g_j = f_j * h_0 + \sum_{i=1-j}^{-1} f_{i+j} * h_i + \sum_{i=1}^{N-j} f_{i+j} * h_i \qquad (24)$$

which may be rearranged to yield

$$f_j * h_0 = g_j - \sum_{i=1-j}^{-1} f_{i+j} * h_i - \sum_{i=1}^{N-j} f_{i+j} * h_i \qquad (25)$$

The h_0 is the in-focus point spread function of the microscope. This equation states that the specimen at level j, convolved with the in-focus psf, is given by the image at level j minus a sum of adjacent specimen planes that have been blurred by out-of-focus psfs h_i. In this summation, i represents the distance between the focal plane and the object plane.

Equation (25) suggests that we can recover the specimen at level j by subtracting from the image at level j a series of adjacent specimen planes blurred by the defocus transfer function. If we abandon a simultaneous solution approach, we do not have available the adjacent specimen planes f_{i+j}. However, we do have access to adjacent plane images g_{i+j}. We see from Eq. (24) that each image contains the corresponding specimen plane plus a sum of defocused adjacent specimen planes. Since the defocus transfer function tends to discriminate against high spatial frequencies (fine detail), but passes low frequency information, we can make the general statement that the image spectrum G_j contains the specimen spectrum F_j plus excess low-frequency information from adjacent planes. We can approximate the specimen f_j by a highpass-filtered version of the image g_j; that is,

$$f_j \approx g_j * k_0 \tag{26}$$

where k_0 is some heuristically determined highpass filter with a transfer function that takes on the value zero at zero frequency and unity at the high frequencies of interest. This will remove the large amount of excess low frequency information and make the approximation reasonable. If we furthermore ignore the effect of the in-focus psf, we can write an approximation to Eq. (25) as

$$f_j \approx g_j - \sum_{i=1-j}^{-1} g_{i+j} * k_0 * h_i - \sum_{i=1}^{n-j} g_{i+j} * k_0 * h_i \tag{27}$$

It may be necessary to use only some small number M of adjacent planes to remove most of the troublesome defocused information. Equation (27) then becomes

$$f_j \approx \tilde{f}_j = g_j - \sum_{i=1}^{M} (g_{j-i} * h_{-i} + g_{j+i} * h_i) * k_0 \tag{28}$$

This suggests that we can partially remove the defocused structures by subtracting $2M$ adjacent plane images that have been convolved with the appropriate defocus point spread function and a highpass filter k_0. The highpass filter and the number M of adjacent planes must be selected to give reasonable results. While we cannot expect this technique to recover the specimen function exactly, it does promise to improve optical section images at reasonable expense.

Figure 17-4 illustrates the results of a simple deblurring algorithm for optical sections. It involves only the two adjacent plane images ($M = 1$) and is given by (Ref. 5)

$$\hat{f}_j = 5g_j - 2(g_{j-1} + g_{j+1}) * h_1 \tag{29}$$

where h_1 is a psf that approximates the blurring due to defocus by the amount Δz. Figure 17-4(a) through (c) shows three digitized optical section images of a Golgi stained (silver impregnated) horizontal cell in the catfish retina ($\Delta z = 5\mu$). The blurred upper and lower plane images appear in Figure 17-4(d) and (f). The result of deblurring Figure 17-4(b) with Eq. (29) is shown in Figure 17-4(e). Notice that structures that appear only in Figure 17-4(b) are preserved at full contrast, while defocused structures from adjacent planes are removed. Structures visible in all three planes lose contrast because the excess low frequency information was not removed from adjacent plane images [Eq. (26) was not used].

The Defocus OTF

The deblurring technique presented in Eq. (28) requires a set of defocus point spread functions. We now investigate the transfer function of a defocused optical system.

Square Aperture. Recall from Chapter 13 that the OTF of an optical system under incoherent illumination is the autocorrelation of its pupil function. For a square aperture of width l, the pupil function with defocus becomes (Ref. 7)

$$P(x, y) = \Pi\left(\frac{x}{l}\right)\Pi\left(\frac{y}{l}\right)e^{j\pi(\epsilon/\lambda)(x^2+y^2)} \tag{30}$$

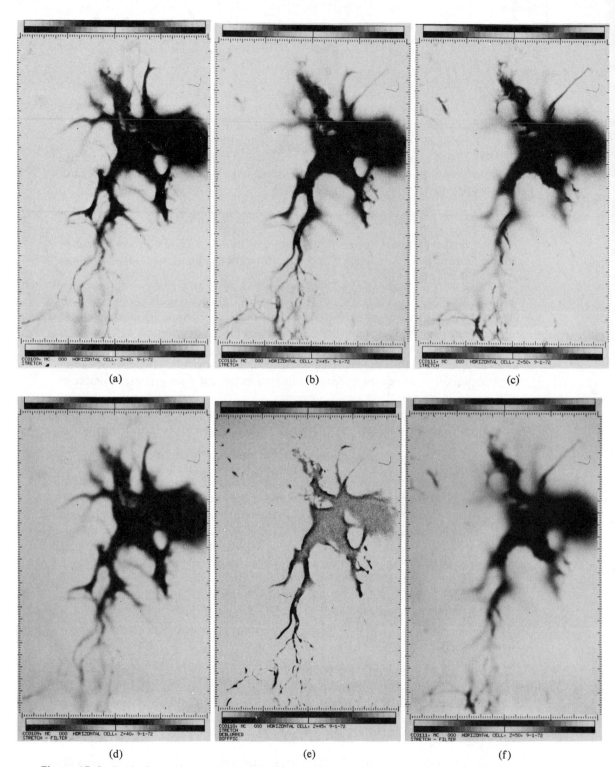

Figure 17-4 Deblurring optical sections: (a), (b), (c) digitized optical section images, (d), (f) blurred adjacent plane images, (e) deblurred image

358

where the complex exponential represents the phase disturbance due to the optical path length error that results from defocus. The defocus error is (Ref. 7)

$$\epsilon = \frac{1}{d_i} + \frac{1}{d_o} - \frac{1}{f} = \frac{\delta z}{d_f(d_f - \delta z)} \tag{31}$$

The u-axis component of the image plane OTF is (Ref. 7)

$$T(u, 0) = \Lambda\left(\frac{u}{f_c}\right) \operatorname{sinc}\left[l^2 \frac{\epsilon}{\lambda}\left(1 - \frac{|u|}{f_c}\right)\frac{u}{f_c}\right] \qquad f_c = \frac{l}{\lambda d_i} \tag{32}$$

where

$$\operatorname{sinc}(x) = \frac{\sin(x)}{x} \tag{33}$$

The object plane OTF results if we substitute d_f for d_i in Eq. (32).

    Notice that this is a sinc function in an envelope that is the in-focus OTF. For $\epsilon = 0$ (no defocus), the argument of the sinc is zero and we are left with the in-focus OTF. Notice also that the argument is quadratic in the frequency variable u. This effects frequency modulation of the sinc. The "frequency" of the sinc decreases linearly to zero as u goes from zero to f_c.

Circular Aperture. For an optical system with a circular aperture of radius A, the pupil function with defocus becomes

$$P(r) = \Pi\left(\frac{r}{2A}\right)e^{jkwr^2/A^2} \qquad k = \frac{2\pi}{\lambda} \qquad r^2 = x^2 + y^2 \tag{34}$$

where the maximum defocus path length error is (Ref. 8)

$$w = -d_i - \delta z \cos\alpha + (d_i^2 + 2d_i\delta z + \delta z^2 \cos^2\alpha)^{1/2} \qquad \alpha = \arctan\frac{A}{d_i} \tag{35}$$

assuming the image is recorded on a plane located $d_i + \delta z$ behind the lens. Hopkins (Ref. 9) showed that the recording plane OTF of a defocused optical system is given by

$$
\begin{aligned}
T_H(s) = &\frac{4}{\pi a}\cos\left(\frac{1}{2}as\right)\left\{\beta J_1(a) + \sum_{n=1}^{\infty}(-1)^{n+1}\frac{\sin(2n\beta)}{2n}[J_{2n-1}(a) - J_{2n+1}(a)]\right\} \\
&- \frac{4}{\pi a}\sin\left(\frac{1}{2}as\right)\sum_{n=0}^{\infty}(-1)^n\frac{\sin[(2n+1)\beta]}{2n+1}[J_{2n}(a) - J_{2n+2}(a)]
\end{aligned}
\tag{36}
$$

where

$$a = 2kws \qquad \beta = \cos^{-1}\frac{q}{f_c} \qquad s = \frac{2q}{f_c} \qquad q^2 = u^2 + v^2 \qquad f_c = \frac{2A}{\lambda d_i} \tag{37}$$

Stokseth (Ref. 8) derived an approximation of the form

$$T_s(s) = (1 - 0.69s + 0.0076s^2 + 0.043s^3)\operatorname{jinc}\left[4kw\left(1 - \frac{|s|}{2}\right)\frac{s}{2}\right] \tag{38}$$

where

$$\operatorname{jinc}(x) = 2\frac{J_1(x)}{x} \qquad |s| < 2 \tag{39}$$

The coefficients of the third-order polynomial in Eq. (38) were selected to make the approximation accurate at large values of defocus ($w \geq 5\lambda$). At zero defocus ($w = 0$), the jinc term is unity and the polynomial differs only slightly from the in-focus OTF. When deblurring optical sections, we are primarily interested in the adjacent planes, where defocus is relatively small. We can make the approximation more accurate for small defocus by substituting the in-focus OTF for the polynomial. This produces

$$H(w, q) \approx \frac{1}{\pi}(2\beta - \sin 2\beta)\, \mathrm{jinc}\left[4kw\left(1 - \frac{|q|}{f_c}\right)\frac{q}{f_c}\right] \tag{40}$$

Note the similarity between the approximate OTF for a circular aperture in Eq. (40) and the square aperture OTF in Eq. (32).

Equation (40) allows us to calculate the defocus microscope OTF. Figure 17-5 illustrates the effect of defocus on the OTF. The circularly symmetric OTFs may be inverse Fourier transformed to produce the defocus psfs required for deblurring by Eq. (28). Shantz (Ref. 6) has demonstrated agreement between theoretical and experimental psfs in the light microscope.

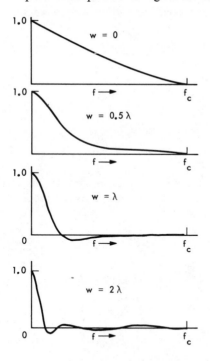

Figure 17-5 The defocus OTF (after Stokseth. Ref. 6)

COMPUTERIZED AXIAL TOMOGRAPHY

Biological tissue, including the human body, is opaque to light in the visible spectrum, except in very thin sections. However, biological tissue does transmit X rays. Some structures in the body, bones for example, absorb more heavily than other structures. Conventional radiography (Figure 17-6) produces an image in which the three-dimen-

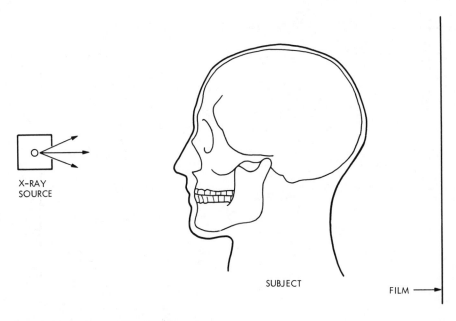

Figure 17-6 Conventional radiography

sional structures in the body are projected onto a plane and superimposed. In radiography, no lens is involved, but rather the subject stands between a point source and the recording film. The structures in the body cast superimposed shadows on the film. This creates difficulty in interpreting the multiple overlapping images of different structures. Radiologists frequently use multiple views (X rays taken at different angles) to resolve ambiguities.

Tomography

Conventional tomography is an X-ray technique that isolates objects in a particular plane of interest (Figure 17-7). Tomography employs a source and film that move during the exposure. In Figure 17-7, the source moves down while the film moves up in such a way that any point P in the plane of interest always lies on a line connecting the source with the corresponding point P' on the film. Structures outside the plane of interest become blurred because their images on the film move during the exposure. Objects near the plane of interest are blurred less than remote objects. The technique is useful where image detail is required in deeply imbedded structures, such as those of the middle ear. One disadvantage is that the X-ray dosage is usually higher than in normal radiography.

Axial Tomography

Computerized axial tomography (CAT) is a relatively recent technique, which incorporates digital image processing (Refs. 10–15). The devices, commonly called CAT

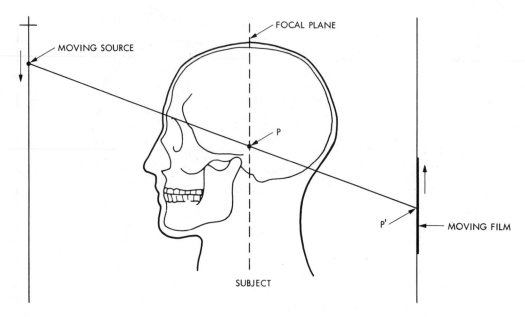

Figure 17-7 Conventional tomography

scanners, reconstruct the three-dimensional image of the X-ray absorbing object. The technique is illustrated in Figure 17-8. A planar X-ray beam penetrates the object, and transmitted beam intensity is measured by a linear array of X-ray detectors. This produces the transmitted intensity function shown in Figure 17-8. A series of these intensity functions are recorded as the apparatus rotates about the object through a small angle between each exposure. A complete series would cover 180° of rotation in steps of from 2 to 6°. The resulting set of one-dimensional intensity functions are used to compute a two-dimensional cross-section image of the object at the level of the beam. This process is repeated as the beam/detector unit is moved down the object in small steps, producing a set of cross-section images that can be "stacked" to form a three-dimensional image of the object. CAT scanners have become important in the diagnosis of many diseases, including brain tumors in particular.

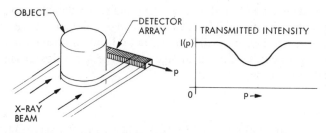

Figure 17-8 Axial tomography

Image Reconstruction

Figure 17-9 illustrates the Fourier transform technique for CAT image reconstruction. Suppose we define a three-dimensional coordinate system having its origin inside the object. The two-dimensional function that describes the density of the object in the

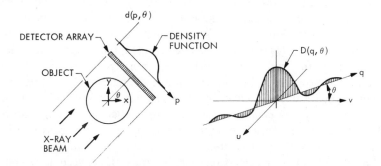

Figure 17-9   CAT reconstruction

plane at level z is $d(x, y)$. The beam direction forms an angle θ with the x-axis. The transmitted intensity function is used to compute the projected density function $d(p, \theta)$ using

$$d(p, \theta) = \log \left[\frac{I_0}{I(p, \theta)} \right] \tag{41}$$

where I_0 is the incident beam intensity and $I(p, \theta)$ is the transmitted intensity at position p along the linear detector array.

The projected density function represents the result of collapsing (projecting) $d(x, y)$ onto a line that makes an angle θ with the y-axis. Under the similarity properties of the two-dimensional Fourier transform (see Chapter 10) we can write

$$\mathcal{F}\{d(p, \theta)\} = D_r(q, \theta) \tag{42}$$

where θ is an angle measured with respect to the v-axis,

$$q = \sqrt{u^2 + v^2} \tag{43}$$

and

$$D(u, v) = \mathcal{F}\{d(x, y)\} \tag{44}$$

Thus each projected density function $d(p, \theta)$ yields a function $D_r(q, \theta)$ that is a radial slice through the two-dimensional Fourier transform of the object. A set of $D_r(q, \theta)$, where θ covers 180° in small steps, can be interpolated to determine $D(u, v)$ approximately. This, in turn, can be inverse transformed, yielding $d(x, y)$. Performed over a range of z, this technique produces $d(x, y, z)$, the three-dimensional X-ray image of the object.

Some CAT scanners use simpler, though less exact, reconstruction algorithms to reduce the computational load. The simplest such algorithm is known as *back projection*. With this technique, each projected density function $d(\theta, p)$ is expanded (projected back) along the beam axis to form a two-dimensional image containing bars parallel to the beam axis. When all such images for one *z*-level are superimposed, one obtains an approximate reconstruction of that cross section. While this technique was used in early CAT scanners, the current trend seems to be toward more exact, but more expensive, methods.

The accuracy or resolution obtained by a CAT scanner depends on several parameters. These include (1) how finely the projected density function is sampled, (2) how finely it is quantized, (3) the reconstruction algorithm used, (4) the interpolation method used, (5) the beam thickness, and (6) the sample spacing in the *z*-direction.

As in other radiography techniques, noise presents a problem in axial tomography. The principal noise source is due to the random distribution of photons in the illuminating beam. This effect is called *quantum mottle* in radiology and is a result of the necessarily low exposure dosage to the patient. This noise source is essentially what we called photoelectronic noise in Chapter 14. Lowpass filtering of the reconstructed cross-section image discriminates against the random noise, but at the expense of resolution. Thus, in each case, there is a tradeoff between noise and resolution. The techniques discussed in Chapter 14 are generally applicable. The noise situation also can be improved by higher beam energy. Thus there is a clinical tradeoff between image noise and X-ray dosage to the patient.

STEREOMETRIC RANGING

Stereometry is a technique by which one can deduce the three-dimensional shape of an object from a stereoscopic image pair. To do this, one must model the geometry of image formation. Figure 17-1 diagrams an object, a light source, and a camera system. We establish a three-dimensional coordinate system centered upon the optical center of the lens system. The optical axis of the camera coincides with the *z*-axis.

The object of interest consists of an opaque surface in front of the camera. A portion of the light striking the surface is reflected, scattering in all directions. Some portion of the scattered light passes through the lens aperture and forms an image of the object at the image plane of the camera.

If the image is to be digitized, we can think of the image plane as being covered with an array of pixels. In Figure 17-1, one of the pixels is projected back through the lens to form an image of the pixel on the object. The projection of the pixel forms a "pixel cone," extending out from its apex at the lens center until it intersects the object.

The pixel image defines that region of the object to which the pixel corresponds. A portion of the light incident upon the pixel image is scattered back into the lens aperture. All this light is converged to fall upon the pixel and determine its brightness.

In addition to brightness, we can associate another value with this pixel. The distance from lens center to the point *p* defines the range of this pixel. Notice that if other surfaces lie behind the object, they are obscured. Thus the range of a pixel is the distance along its pixel cone from lens center to the first opaque surface encountered. We can generate a "range image" by assigning each pixel a gray level proportional not to pixel brightness but to the length of its pixel cone.

Stereoscopic Imaging

Figure 17-10 diagrams a dual camera configuration suitable for stereoscopic imaging. A three-dimensional coordinate system has its origin at the lens center of the left camera. The optical axes of both cameras are parallel and lie in the *x-z* plane. The *z*-axis coincides with the optical axis of the left camera. Both cameras have a focal length *f* and are separated by the distance *d*. The positive *y*-axis direction is into the page.

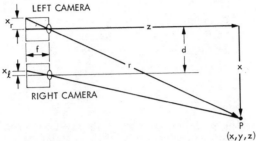

Figure 17-10 Stereoscopic imaging

   Suppose the point *P* having coordinates (x, y, z) is situated in front of the cameras. The image of point *P* will have horizontal coordinates x_l in the left and x_r in the right camera image. Making use of similar triangles, we can write

$$x = \frac{zx_l}{f} \tag{45}$$

for the left camera and

$$x - d = \frac{zx_r}{f} \tag{46}$$

for the right camera. The latter equation can be rearranged to yield

$$x = \frac{zx_r + fd}{f} \tag{47}$$

Equating the right-hand sides of Eqs. (45) and (47) and solving for *z* produces

$$z = \frac{fd}{x_l - x_r} \tag{48}$$

the "normal range" equation. It relates the normal component *z* of range to the amount of pixel shift between the two images. Notice that in Eq. (48) *z* is a function only of the difference between x_l and x_r and not their magnitudes.

   Again using similar triangles, we can write

$$\frac{r}{z} = \frac{\sqrt{f^2 + x_l^2 + y_l^2}}{f} \tag{49}$$

Rearranging and substituting Eq. (48) for *z* produces

$$r = \frac{d\sqrt{f^2 + x_l^2 + y_l^2}}{x_l - x_r} \tag{50}$$

which is the "true range" equation. This gives the total distance from the origin to the point P. For narrow-angle (telephoto) systems, x_l is small compared to f, and Eq. (50) can be approximated by Eq. (48).

Given corresponding pixels in left and right images, one can calculate either normal or true range from Eqs. (48) or (50), respectively. However, it is a nontrivial task to find the value of x_r that corresponds to each x_l.

Stereometric ranging can be done in the following way. First, for each pixel in the left image, find the pixel in the right image that corresponds to the same point on the object. For the system diagrammed in Figure 17-10, this can be accomplished on a line-by-line basis, since any point on the object images to the same vertical position on both images. Secondly, using the pixel shift, calculate z by Eq. (48). Next, calculate the x-coordinate of the point by Eq. (45) and its y-coordinate by

$$y = \frac{z}{f} y_l \qquad (51)$$

This procedure allows us to calculate the x, y, z-coordinates of every point on the object that images to a pixel. If we express z as a function of x and y, we have a normal range image. Using Eq. (50) produces a true range image. In either case, we have mapped the visible surface of the object in three dimensions.

In Figure 17-10, the cameras are boresighted. Except in cases where z is much larger than d, it may be necessary to "converge" the cameras to ensure that their fields of view overlap for objects in the near field. In converged systems, the camera axes are not parallel but converge to some point in the xz plane. In this case, the same techniques apply, but the range equations are slightly more complex. If the camera axes do not lie in a plane, the situation is still more complex (Ref. 16).

Pixel Correlation

Figure 17-11 illustrates a technique to locate the right image pixel that corresponds to a particular left image pixel. Suppose the given pixel in the left image has coordinates x_l, y. We fit imaginary windows around that pixel and the pixel having the same coordinates in the right image. Next, we compute a measure of the "agreement" between the images inside the two windows. We repeat this process as the window in

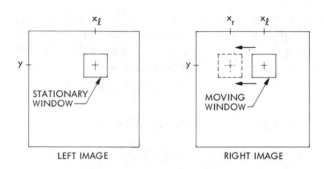

Figure 17-11 Pixel shift calculation

the right image moves toward the left. At some point, the moving window will be centered at x_r, y and will contain the same features as the fixed window in the left image. When this happens, the image content in the two windows is approximately the same, and the measure of image agreement is maximized. Image agreement within the two windows can be calculated using cross-correlation, a sum of squared differences, or a similar technique. In any case, the image agreement measure should reach a maximum when the two windows contain the same features.

Noise in the images tends to corrupt the image agreement measure. This situation can be improved somewhat by increasing the size of the correlation window. Doing so reduces the resolution of the resulting range image, however, since large windows tend to smear over any abrupt changes in range. Thus the window size should be as small as possible, consistent with maintaining a low probability of miscalculating the pixel shift. The pixel shift calculation is also more reliable if the surface of the object exhibits considerable texture or high-frequency detail. Smooth surfaces are very difficult to range properly. Sometimes it is helpful to project a random texture pattern onto a smooth surface to achieve accurate ranging.

Stereometry with Angle Scanning Cameras

The Viking Mars Lander spacecraft employed stereoscopic imaging. Each Lander had two digitizing cameras spaced 1 meter apart. In these angle scanning cameras, however, the pixels were equally spaced in azimuth and elevation angle rather than being equally spaced in the image plane. Thus the coordinates of a pixel are given by the azimuth and elevation angles of the center line of its pixel cone. As illustrated in Figure 17-12, the azimuth is the angle between the y-z plane and a vertical plane containing the pixel cone axis. The elevation angle is the angle between the x-z plane and a plane containing the x-axis and the pixel cone axis. The reference axes (zero azimuth, zero elevation) of the two cameras lie parallel in the x-z plane.

Using the geometry in Figure 17-12, we can write the normal range component in terms of the two camera azimuth coordinates θ_l and θ_r as

$$z = \frac{d}{\tan \theta_l - \tan \theta_r} \tag{52}$$

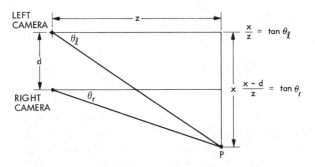

Figure 17-12　Stereoscopic angle scanning camera

Applications
Part III

The two remaining coordinates of the point P are given by

$$x = z \tan \theta_l \tag{53}$$

and

$$y = z \tan \phi_l \tag{54}$$

where ϕ_l is the elevation coordinate and is the same for both cameras.

Figure 17-13 shows a stereo pair from the Viking Lander. Stereometry was used in the near field to establish a set of grid lines on the surface.

Figure 17-13 Viking Lander example

STEREOSCOPIC IMAGE DISPLAY

Display Geometry

Figure 17-14 illustrates the viewing geometry for stereoscopic display. The stereoscopic image pair is positioned a distance D in front of the viewer's eyes, which are separated by the interocular distance S. A small feature located at coordinates x_l, y_l in the left image and x_r, y_l in the right image will appear to the observer as if it were located at

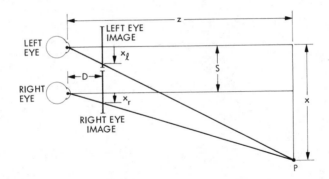

Figure 17-14   Stereoscopic display

point *P*. A geometric development similar to that surrounding Figure 17-10 produces a range relation

$$z = \frac{DS}{x_l - x_r} \tag{55}$$

which is reminiscent of Eq. (48). The *x*-coordinates of corresponding points in the two images are related by

$$x_r = x_l - \frac{DS}{z} \tag{56}$$

This implies that, for objects located at $z = \infty$, the right and left eye coordinates are identical. As an object is shifted left in the right eye image, its apparent position moves toward the observer.

Stereoscopic photography is a technique that uses a camera configuration similar to that of Figure 17-10 and a viewing apparatus similar to that shown in Figure 17-14 to reproduce three-dimensional scenes. Suppose that the two cameras in Figure 17-10 produce positive transparencies at the image plane. These transparencies can be rotated 180° about the *z*-axis and positioned in front of the observer as in Figure 17-14. If the relationship

$$DS = fd \tag{57}$$

is satisfied, the scene will appear as if the observer had viewed it firsthand.

Stereo Display Generation

Suppose the left eye image and the normal range image are given, and it is desired to produce the right eye image for stereoscopic display. This requires only a geometric transformation of the form

$$x_r = x_l - \frac{DS}{z} \qquad y_r = y_l \tag{58}$$

which is merely a copying operation with variable horizontal shift. The transformation for a single line of the image is illustrated in Figure 17-15. The right eye image is

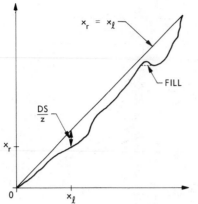

Figure 17-15 The pixel shift transformation

generated by copying the gray level at coordinate x_l into the pixel located at x_r. At each point x_l, y, the amount of shift is a reciprocal function of range.

It is desirable that x_r be a nondecreasing function of x_l. If this function has negative slope over some interval, it produces a right for left reversal in the generated image. As illustrated in Figure 17-15, we can use a horizontal fill technique to remove any local areas of negative slope. In areas of zero slope, features of finite size in the left image become compressed to a point in the right image. This occurs, for example, when the right eye is looking directly along a surface that is visible to the left eye. Negative slope in the pixel shift transformation corresponds to the case in which the eyes are looking at opposite sides of the same surface. In normal scenes, both of these conditions are rare. For proper stereoscopic effect and comfortable viewing, one should keep the maximum pixel shift to no more than approximately 5% of the image width. In this case, the occurrence of zero or negative slope in the pixel shift function is quite rare.

Given a monocular brightness image and a range image of a particular scene, one might obtain a more pleasing stereo display by generating both right eye and left eye images. The right eye image is generated by Eq. (58), but using only half the prescribed amount of shift. The left eye image is generated by shifting an equal amount in the opposite direction. This technique can produce superior displays for images containing nearby objects with considerable shape detail.

Figure 17-16 illustrates a stereoscopic image pair produced by a transformation of the form of Eq. (58). The left eye image is a grid pattern, and the normal range image was a Gaussian function. The right eye image was produced by Eq. (58).

Display Quality

The surfaces displayed by stereoscopic techniques should have an abundance of fine detail or texture to assist the viewer's eye in the matching process. The human visual system executes a process that is apparently similar to that described for stereometric ranging. Thus smooth (untextured) surfaces are usually difficult to view properly in a stereoscopic display. The introduction of surface texture can be helpful.

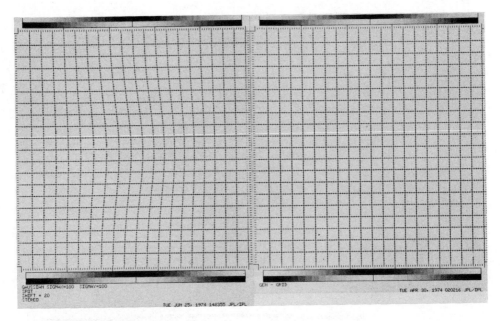

Figure 17-16 Stereo pair of a grid

In stereo work, it is important that the display pixels not be visible. If they are, the two eyes will be confused in an attempt to cross-correlate the pixel patterns in the two images. This creates visual discomfort and can destroy depth perception in the display. Also reproduction by a "screening" process can introduce disturbing texture. For further discussions of the human visual system in this regard, the interested reader should consult a text on the subject (Refs. 17, 18).

Display of Optical Section Stacks

A series of optical section images can be displayed as shown in Figure 17-17. The stack of transparent sections can be observed from any viewpoint specified by an azimuth, elevation, and range. The section images are projected onto an imaginary viewing screen, where they are superimposed by summation. The projection is accomplished by a geometric operation, as illustrated in Figure 17-18. A computer-generated rectangular grid image was projected with azimuth 60°, elevation 45°.

Figure 17-19 shows two stereo pairs, each generated by projecting a stack of retinal cell images to two viewpoints.

SHADED SURFACE DISPLAY

The basic idea of shaded surface display is to generate an image of a three-dimensional object that exists only as a mathematical description. This discipline is usually considered to be a branch of computer graphics (Ref. 1). It cannot be performed on

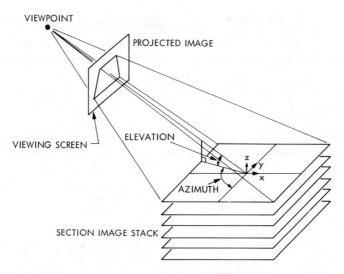

Figure 17-17   Image stack projection geometry

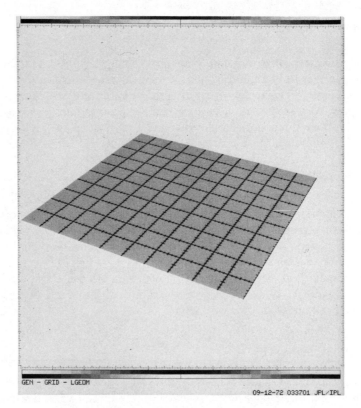

GEN - GRID - LGEOM

09-12-72 033701 JPL/IPL

Figure 17-18   Projection of a grid image

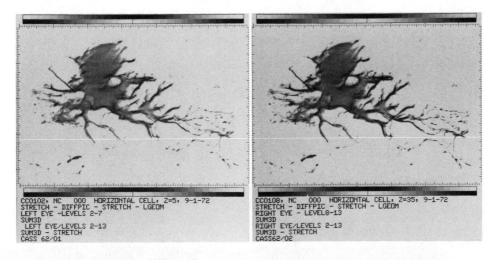

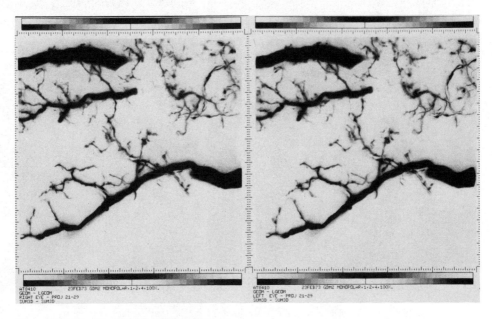

Figure 17-19   Stereo pairs of image stacks

conventional computer graphics systems, however, but requires a system designed for digital image processing. Therefore, we shall introduce the subject here.

The object of interest is given by a mathematical description of its opaque outer surface in a three-dimensional coordinate system. The user specifies the location of all light sources and of the imaginary camera that is to generate the image. The latter position is called the *viewpoint*. The display algorithm then computes the image that the imaginary camera would make of the object.

Surface display requires modeling of three things: (1) the spatial surface description, (2) the light-reflecting phenomenon at the surface, and (3) the geometry of the light sources and the imaging projection.

Surface Description

The three-dimensional surface of the object is ordinarily described by a polyhedral approximation (Ref. 19). Points on the object's surface form the vertices of the polygonal faces of the polyhedron. Triangles are commonly used for the faces, since it is always possible to pass a plane through three points. A planar quadrilateral cannot always connect four points on a surface. Figure 17-20 shows an image generated using a polyhedral approximation employing rectangles.

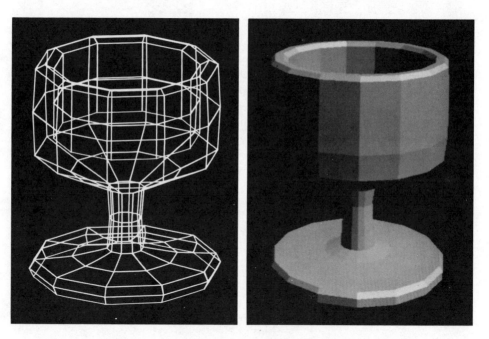

Figure 17-20   (a) Graphic (wire grid) display; (b) Shaded surface display

The surface description may consist of a polygon list describing each polygon by the three-dimensional coordinates of its vertex points. This description is somewhat redundant, since each point is actually the vertex of several adjacent polygons and thus will appear in the list more than once. The actual format of the file containing the surface description (the "polygon file") involves a tradeoff between compact storage and ease of access while computing the projected image. Clearly, the more polygons used to define the surface, the more accurate the representation will be.

Surface Reflection Phenomena

Figure 17-21 illustrates the reflection of light from a flat surface. A point source at distance r provides incident light that makes an angle θ with the normal to the surface. A camera is located on a line that makes an angle ϕ with the normal. The light intensity falling upon the surface is proportional to $\cos(\theta)/r^2$.

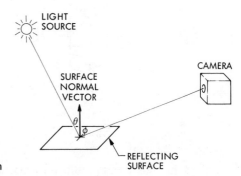

Figure 17-21 Surface reflection

There are two important types of reflection—diffuse scattering and specular reflection. Diffuse scattering is characteristic of "matte" or "chalky" surfaces, and reflected intensity can be modeled as proportional to $\cos\phi$. Specular reflection is characteristic of "shiny" or "metallic" surfaces. The intensity due to specular reflection can be modeled as proportional to $[\cos(\theta - \phi)]^n$, where n is between 0.5 and 10. The larger values of n make the surface appear more shiny.

The apparent brightness of a uniformly radiating surface varies as $1/\cos(\phi)$ because, as the viewer moves away from the normal (increasing ϕ), the same amount of energy from the surface projects into a smaller area of his retina. We can now write the reflected intensity equation as

$$I = A\frac{\cos(\theta)}{r^2}\frac{1}{\cos(\phi)}\{B\cos(\phi) + (1 - B)[\cos(\theta + \phi)]^n\} \qquad 0 \leq B \leq 1 \quad (59)$$

where B and n are surface reflectance parameters and A is a constant of proportionality.

The parameter B determines how the incident light is divided between diffuse and specular reflection. For a purely diffusing surface, we can let $B = 1$. If r is very large, it can be assumed constant over the extent of the object and can be absorbed into the proportionality constant. Then Eq. (59) reduces to

$$I = C\cos(\theta) \qquad\qquad\qquad (60)$$

for a computationally simple surface brightness rule.

Imaging Geometry

Figure 17-22 shows the model for computing the image of the object's surface. The cone from any pixel p projects through the lens, and its axis intersects the surface at

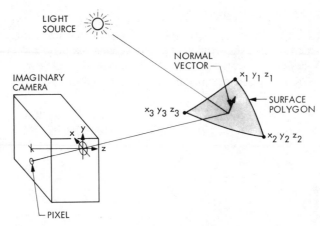

Figure 17-22   Surface display geometry

some point P that falls on a particular polygon. Thus the gray level (brightness) of pixel P can be computed from Eq. (59) if the surface normal vector of that polygon is known. The image is generated, pixel by pixel, by first determining upon which polygon the pixel cone axis falls and then computing the light intensity reflected into the camera using the geometry of Figure 17-22.

Perhaps the most challenging aspect of shaded surface display is the organization of, and search algorithm for, the polygon file. The generation of such images can be quite slow and expensive, particularly if file management is not handled efficiently. Image generation for real-time display usually requires special-purpose high-speed image processing hardware.

The image is ordinarily generated in the conventional line-by-line digital image processing fashion. Any image line intersects only a limited number of polygons. Thus the search algorithm need work only with a few "active" polygons at a time. If one pixel on the line does not fall on an active polygon, the previously unused polygons are searched to find which one should be "activated." If no pixel on the line falls on a particular active polygon, it becomes inactive. Some pixels may fall on no polygon at all (outside the object), and these can be set to black or some other background gray level. The design of search algorithms is beyond our scope, and the interested reader is encouraged to consult a textbook (Ref. 1) and the computer graphics literature.

Smooth Shading

The polyhedral approximation to a curved surface produces an artificial appearance in the computed image (Figure 17-20). The polygon edges represent highly visible discontinuities in brightness. Using more polygons is an expensive remedy. Goroud (Ref. 20) has advanced a computationally simple method to achieve a smooth surface approximation. Each vertex on the surface is actually the vertex of several adjacent polygons. The surface normal vector at each vertex point is defined as the average of the normal vectors of the surrounding polygons. When a pixel cone axis intersects a surface polygon, the local normal surface vector is obtained by interpolation from the surrounding vertices, as illustrated in Figure 17-23. This technique causes the

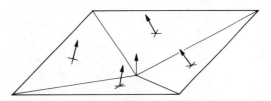

Figure 17-23 Normal vector interpolation

surface normal vector to vary smoothly rather than abruptly over the surface, and it produces a smooth surface appearance, as shown in Figure 17-24.

SUMMARY OF IMPORTANT POINTS

1. Multispectral images are digitized functions of x, y, and optical wavelength that show the reflectance spectrum of the object at each pixel.

2. Optical sectioning produces images contaminated by defocused images of structures outside the focal plane.

3. Thick specimen imaging involves a three-dimensional convolution of the specimen function with the defocus psf.

4. Theoretically, optical section images can be deblurred exactly by three-dimensional deconvolution or by a simultaneous linear equation approach.

5. Practically, optical section images can be deblurred approximately by subtraction of blurred neighboring plane images [Eq. (28)].

6. The defocus OTF of a circular lens is given by Eq. (40).

7. Computerized axial tomography uses the projection property of the two-dimensional Fourier transform to reconstruct an image from a set of its projections.

8. A range image can be computed from a stereo pair [Eq. (48)].

9. A stereo pair can be generated from a brightness image and a range image [Eq. (58)].

10. Surface texture is helpful in stereometry and stereoscopic display.

11. Shaded surface display techniques produce images of objects that exist only as a mathematical surface description.

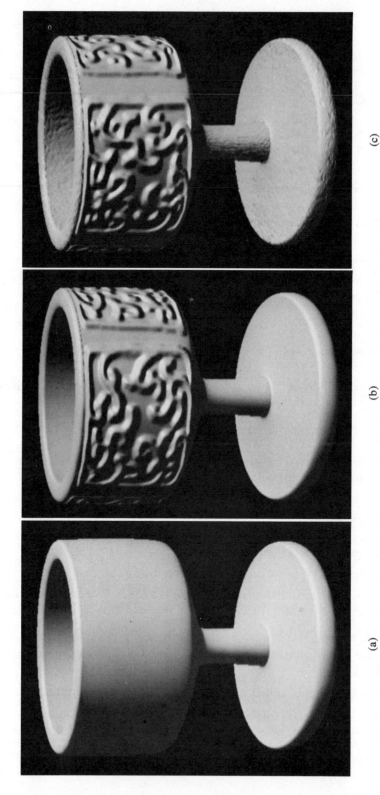

(a)

(b)

(c)

Figure 17-24 (a) Surface display with smooth shading, (b) relief pattern added, (c) random surface texture added

REFERENCES

1. W. M. NEWMAN and R. F. SPROULL, *Principles of Interactive Computer Graphics*, McGraw-Hill Book Company, New York, 1973.

2. W. K. PRATT, *Digital Image Processing*, John Wiley & Sons, Inc., New York, 1978.

3. R. G. GREEVES, A. ANSON, and D. LANDEN, eds., *Manual of Remote Sensing (Vol. I)*, American Society of Photogrammetry, Falls Church, Virginia, 22046, 1975.

4. H. ELIAS, "Three-Dimensional Structure Identified from Single Sections," *Science*, **174**, 993–1000, December 1971.

5. M. WEINSTEIN and K. R. CASTLEMAN, "Reconstructing 3-D Specimens from 2-D Section Images," *Proceedings of the SPIE*, **26**, 131–138, May 1971.

6. M. J. SHANTZ, *Description and Classification of Neuronal Structure in the Frog Retina*, Ph.D. Thesis, California Institute of Technology, Pasadena, California, 1976.

7. J. W. GOODMAN, *Introduction to Fourier Optics*, McGraw-Hill Book Company, New York, 1968.

8. P. A. STOKSETH, "Properties of a Defocused Optical System," *J. Optical Soc. Amer.*, **59**, No. 10, 1314–1321, October 1969.

9. H. H. HOPKINS, "The Frequency Response of a Defocused Optical System," *Proc. Royal Soc.*, (London) **A 231**, 91–103, 1955.

10. R. GORDON and G. T. HERMAN, "Reconstruction of Pictures from Their Projections," *Comm. ACM*, **14**, No. 12, 759–768, December 1971.

11. R. GORDON and G. T. HERMAN, "Three-Dimensional Reconstruction from Projections," *International Review of Cytology*, **38**, 111–151, 1974.

12. Z. H. CHO, ed., "Special Issue on Three-Dimensional Image Reconstruction," *IEEE Transactions on Nuclear Science*, **NS-21**, No. 3, June 1974.

13. R. GORDON, G. T. HERMAN, and S. A. JOHNSON, "Image Reconstruction from Projections," *Scientific American*, **233**, No. 4, 56–68, October 1975.

14. R. N. BRACEWELL and S. J. WERNECKE, "Image Reconstruction over a Finite Field of View," *J. Opt. Soc. Amer.*, **65**, No. 11, 1342–1346, November 1975.

15. J. T. PAYNE and E. C. McCULLOUGH, "Basic Principles of Computer-Assisted Tomography," *Applied Radiology*, **103**, 53–60, March/April 1976.

16. Y. YAKIMOVSKY and R. CUNNINGHAM, "A System for Extracting Three-Dimensional Measurements from a Stereo Pair of TV Cameras," *Computer Graphics and Image Processing*, **7**, 195–210, 1978.

17. T. N. CORNSWEET, *Visual Perception*, Academic Press, New York, 1970.

18. B. JULESZ, *Foundations of Cyclopean Perception*, University of Chicago Press, Chicago, Illinois, 1971.

19. H. FUCHS, Z. KEDEM, and S. USELTON, "Optimal Surface Reconstruction from Planar Contours," *Comm. ACM*, **20**, 693–702, 1977.

20. H. GOROUD, "Continuous Shading of Curved Surfaces," *IEEE Trans. on Computers*, **C-20**, No. 6, 623–629, June 1971.

APPENDICES

APPENDICES

A HISTORY OF DIGITAL
IMAGE PROCESSING AT JPL

This discussion of the development of image processing at the Jet Propulsion Laboratory (JPL) is included for several reasons. First, it pays tribute to the people whose foresight and efforts conceived and evolved a facility that made a significant contribution to the field treated in this book and to the nation's space program. Second, it serves to reconstruct an interesting story, in which all Americans can take pride, before the facts slip away into obscurity. Third, this section gives the interested reader an insight into the work that went into and the experience that resulted from the first 20 years of digital image processing at JPL. It should give an appreciation for the amount of work that led to the present state of image processing at this one facility.

Digital image processing was begun at JPL for a specific purpose—to obtain the maximum information from images sent back by unmanned lunar and planetary exploration spacecraft. In the course of this work, new hardware and software technology was developed, and considerable experience was accumulated in this relatively new field. This experience manifests itself in a unique insight into imagery and image processing, and a certain polish apparent in the work done.

The reader is cautioned against reading into this section any claims that digital image processing was discovered or invented at JPL. Several such efforts—for example, optical character recognition (Ref. 1)—predate the work at JPL (see also

Ref. 2). The task of tracing the origin of each technique is beyond our scope and is left to other authors. The reader interested in comparing the JPL development with the general movement of the field may use existing literature surveys (Refs. 3–8).

THE BEGINNING

Robert Nathan came to JPL in 1959 to establish a digital capability for science ground data handling for the upcoming Ranger missions to the moon. He came from Caltech, where he received a Ph.D. in 1956 and stayed on through 1959 to manage the digital computer center there. Since the Ranger missions would return primarily image data, Nathan decided to concentrate on image data handling. He had at his disposal a small NCR 102D computer. When Nathan saw the 1961 Soviet pictures of the far side of the moon, he realized the Ranger pictures could be at least that good (Ref. 9). He noticed that the Soviets had used photographic techniques to clean up areas contaminated by severe telemetry noise and felt the computer could be a much more powerful tool for removing defects from spacecraft TV images.

Fred Billingsley worked in the central computing facility, which housed an IBM 7094 computer. In 1962 he became involved with Nathan and the image data handling. The Ranger program would need a high-quality device to record the analog FM video data from the spacecraft on film to make pictures, but no suitable hardware existed. Billingsley and Roger Brandt wrote the specifications for a device called the *video film converter* (VFC). Billingsley convinced NASA to fund both the construction of the VFC and the development of Nathan's image processing software on the IBM 7094 for the Ranger program. The VFC was built by LINK General Precision and installed at JPL in May 1963. In their design, Billingsley and Brandt included not only film recording of analog video but also the capability to digitize film and to record digital images on film. The VFC was years ahead of its time, only being equalled by commercial equipment in the mid-1970s. Much modified and improved, this instrument was used through the late 1970s for digital image playback for the unmanned planetary missions.

Ranger

The Ranger spacecraft used television cameras with vidicon image sensor tubes. Even though these were the best available at the time, they were subject to geometric distortion, photometric nonlinearity, and noise contamination. Using preflight calibration images, Nathan designed, and Robert Selzer and Howard Frieden implemented, contrast enhancement, photometric correction, geometric correction (Figure I-1), and line streak noise removal techniques on the IBM 7094 computer in JPL's Central Computing Facility (Ref. 10). In mid-1963, John Morecroft implemented several image processing programs on the NCR 102D. The Fourier transform and a

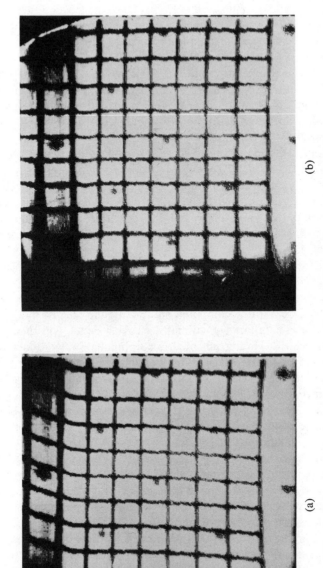

(b)

(a)

Figure I-1 Geometric decalibration of an early Ranger camera: (a) before, (b) after

smear removal technique were programmed on this tiny machine and tested on synthesized images of craters.

The early Ranger program was plagued with disappointments. In 1962, Rangers III, IV, and V, the first three Rangers to carry cameras, failed to return pictures of the moon (Refs 11, 12). The following year saw the mission objectives revised from lunar science to Apollo support, and all nonvisual science experiments removed. In January 1964, Ranger VI went to the moon, but its cameras failed to turn on. Throughout this period, Nathan was preparing to process images that never came. Knowing the limitations of the cameras, he used preflight test images and simulated images to develop restoration techniques.

On July 31, 1964, Ranger VII sent back 4316 pictures before making its mark on the moon (Refs. 11, 12). Many of its images were corrected for photometric and geometric distortion (decalibrated) on the IBM 7094. Ranger VII also developed a periodic noise pattern, for which Nathan developed a removal technique (Ref. 10). This worked so well that on May 12, 1965, the decision was made to use digital image processing for coherent noise removal on the images for the Ranger atlases of the moon (Ref. 11).

While doing contrast enhancement on Ranger VII data, Nathan realized that he could perform more contrast enhancement on small features without introducing gray scale saturation if the images had less dynamic range. He conceived the idea of blurring the image in the computer, subtracting the blurred image from the original to reduce the dynamic range without destroying detail, and then performing a contrast enhancement. He implemented this technique by two-dimensional convolution with the help of Selzer and Frieden. Familiar with the Fourier transform from his background in X-ray crystallography, Nathan soon couched this operation in terms of two-dimensional linear filtering. He discovered that modulation transfer function (MTF) data was being taken by the Ranger television team. Using this data, he designed deconvolution filters for Ranger images and produced an impressive increase in apparent resolution (Ref. 10).

After the success of Ranger VII, Tom Rindfleisch developed a photoclinometry technique to produce topographical maps of the lunar surface from Ranger pictures (Refs. 13, 14). This technique used the location of the sun and the reflective properties of the lunar surface to decode the shadow pattern and produce elevation maps (Figure I-2). In those days, there was concern that a three-legged Surveyor or Apollo spacecraft might not be able to land on the lunar surface without tipping over. Rindfleisch used the computer to simulate thousands of landings on the Ranger-derived lunar surface maps and concluded that the probability of tipping over was only about 0.03 (Ref. 15).

On February 20, 1965, Ranger VIII sent back 7137 pictures (Refs. 11, 12). This spacecraft impacted the lunar surface at an angle, thereby covering a larger area with its cameras but also introducing smear (motion blur) into the images. Nathan extended his deconvolution to cover smear removal and applied it to Ranger VIII pictures. One month later, Ranger IX sent back 5814 pictures to end the Ranger program.

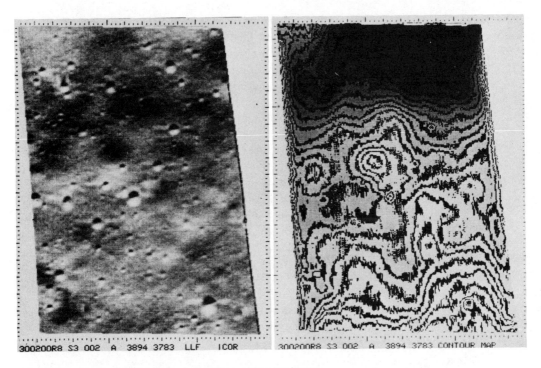

300200R8 S3 002   A 3894 3783 LLF   ICOR   300200R8 S3 002   A 3894 3783 CONTOUR MAP

Figure I-2   Lunar elevation map derived from a Ranger image

Mariner Mars 1964

Mariner 4 was launched on November 28, 1964, and flew by Mars on July 14, 1965, sending back twenty-two 200 by 200 pixel images. Mariner 4 was the first of the Mariner series to carry a camera. Unlike the Rangers, Mariner sent back its video data in digital form at a data rate of $8\frac{1}{3}$ bits per second. Digital transmission was used because of the great distances over which the images were sent. John Morecroft obtained a PDP-7 computer in early 1965 and implemented a real-time display for Mariner 4 images. The pictures were accumulated in core memory, displayed on a CRT, and photographed on Polaroid film. This system was also used for line drop replacement, reseau removal, and shading correction. Reseau position and camera shading had been carefully measured before launch, but all this changed during the trauma of blastoff. Morecroft lowpass filtered frames 2 through 20 and averaged them together to determine the modified shading pattern so it could be removed. Mariner 4's images mapped about 1 % of the low-contrast Martian surface.

The Image Processing Laboratory

As the Ranger program was winding down, the Surveyor program was coming up to speed. Nathan and Billingsley approached the Surveyor Project Office with numerous suggestions for camera design changes to facilitate digital processing, but other spacecraft design constraints prevented their suggestions from being implemented. The success of Ranger and Mariner digital processing did not go unnoticed, however. In 1965, Billingsley did an economic analysis showing that the Surveyor project would save money by buying their own dedicated computer for image processing rather than using the Central Computing Facility. The SDS-930 computer won the bidding competition but, after delivery, developed a myriad of problems. Finally JPL refused acceptance of the machine, and IBM, the next lowest bidder, inherited the sale. Eager to participate in image processing for the space program, IBM hurriedly assembled components from all over the country and installed a working 360/44 system within six weeks. This machine became the workhorse for digital image processing at JPL for a decade until it was replaced, in late 1975, with an IBM 360/65 for the Viking mission.

During the time between Ranger and Surveyor, Billingsley and Nathan also convinced JPL management that digital image processing and science data analysis should be centralized. An administrative reorganization brought Billingsley and his group into the Space Sciences Division to join Nathan. The new IBM 360/44 was installed to replace Nathan's NCR computer. The Video Film Converter had been sent back to LINK for an upgrade and, upon its return to JPL, was installed adjacent to the IBM 360/44. These activities constituted the establishment of the Image Processing Laboratory (IPL) as a facility within JPL.

VICAR

Billingsley's group was to be responsible for processing the image data from the upcoming Surveyor missions. The software on the IBM 7094 had been somewhat cumbersome to use in production. Minor changes in the processing protocol required recompilation of the FORTRAN-language image processing programs. Rather than a straight transfer of software from the IBM 7094 to the 360/44, Billingsley wanted a software system that would be easy for insiders to use in production and simple for outsiders to learn. Stan Bressler and Howard Frieden defined an image processing language called VICAR (Video Image Communication And Retrieval). Under Bressler's direction, IBM employees John Campbell, Tom King, and Ed Efron, under contract to JPL, implemented VICAR in 1966.

THE YEARS OF GROWTH

Surveyor

The Surveyor spacecraft were lunar soft-landers, each carrying a TV camera, a soil-sampling scoop, and a variety of Lunar surface experiments (Ref. 16). There was concern that the Surveyor program would experience problems like those that had plagued Ranger. To the surprise of many, Surveyor I, launched on May 30, 1966, performed flawlessly. Between then and January 1968, a total of five successful Surveyors soft-landed on the moon, scooping up soil and sending back 87,674 images

(Refs 16, 17). The Surveyor cameras sent back analog FM video, which was recorded at the antenna site and then digitized and processed in the IPL. Figure I-3 shows the image improvement realized with deconvolution. Surveyor IV had a glare problem because of lunar dust on its TV camera. As a "spin-off" from the space program, Robert Selzer had developed a filter for medical X-ray enhancement (Ref. 18). In perhaps the first case of image processing "spin-back" into the space program, Selzer's X-ray enhancement filter was used to reduce the glare from Surveyor (Ref. 19).

The Surveyor images were used to produce lunar panorama mosaics, color pictures, polarization pictures, stereoscopic range images, and topographic maps that contributed greatly to our knowledge of the lunar surface (Ref. 16). The most important of the Surveyor images were decalibrated and enhanced by the IPL.

Figure I-3 MTF deconvolution of a Surveyor image: (a) before, (b) after

Mariner Mars 1969

During the Surveyor period, the Mariner program flew only one mission, Mariner Venus 1967. This spacecraft had no cameras aboard, and the IPL was not involved. The Mariner Mars 1969 (MM'69) mission, however, launched two spacecraft (Mariners 6 and 7) carrying two cameras each to Mars. As Surveyor was winding down, MM'69 was gaining momentum. With the success of digital image processing on the Surveyor mission, the pendulum of Project Office acceptance had swung far in favor of digital image processing. The Mariner Mars 1969 mission marked its farthest excursion. Whereas digital image processing had been a post-mission tack-on to both Ranger and Surveyor, the MM'69 television system design involved digital image processing at a very fundamental level.

The two Mariner 1969 spacecraft had to send back large amounts of image data from a distant, low-contrast planet. A complex on-board encoding scheme was designed to obtain the most from those images while reducing the amount of data to be transmitted (Refs. 20, 21). The required image decoding gave the mission a built-in dependence on digital image processing. The MM'69 TV system used an automatic exposure meter, an automatic gain control (AGC), a signal cuber, clipping of the two most significant bits (MSBs), analog tape recording, a time-out for telemetry data in the middle of each line, and transmission of every pixel, every seventh pixel, and every twenty-eighth pixel in three separate video data streams. A 2400-Hz power supply coupled into the video circuits, leaving a superimposed basketweave pattern on the images. The interested reader can explore the complexity of the MM'69 TV system and the difficulty of unraveling the pictures in Ref. 21.

During the mission, the camera noise characteristics changed, and Mars proved to have more contrast than the images from Mariner 4 had indicated. These developments made the MSB restoration very difficult. The AGC returned no data regarding its own behavior, and this had to be inferred from the images. Automatic restoration of the two most significant bits worked properly a large percentage of the time. The MSB restoration program failed primarily in areas where Mars changed brightness rapidly—at the limb (edge of the planet), the terminator (twilight zone), and the polar cap. Unfortunately, these were precisely the features of primary interest to the experimenters, and thousands of important pixels had to have their MSBs restored by hand.

Tom Rindfleisch and his group turned this difficult situation into a successful mission. Rindfleisch developed a nonlinear filtering technique for removing the power supply noise pattern (Figure I-4), while Reuben Ruiz and Howard Frieden restored missing MSBs and photometric integrity to the images. Mariner 1969 mapped about 10% of the Martian surface, showing much about the varied topography of the planet and its CO_2 ice polar cap.

Mariner Mars 1971

Mariner Mars 1971 was a Mars orbiter mission. For the first time, the data restriction was on the ground rather than in space. The spacecraft could take pictures and send them back faster than they could be processed. Mariner 9 returned more than 7300 images from Mars orbit, all of which required geometric and photometric decalibration (Ref. 22). These were images of a low-contrast planet, taken by a sensitive camera under severe lighting conditions. Before processing, the images showed almost no detail. Furthermore, the spacecraft arrived during a severe dust storm that covered the entire surface of the planet. Again digital image processing was necessary for mission success.

The Mariner 1971 mission was the first ever to map an entire planet (Refs. 22, 23). Figure I-5 shows a 4-foot-diameter Martian globe covered with 2000 orthographically projected Mariner 9 images. The mission also produced an atlas of Phobos and Deimos, the two moons of Mars.

During the preparation for MM'71, Joel Seidman and John Kreznar devised a geometric decalibration technique that used as references the reseau marks on the image tube faceplate. This was necessary because the geometric distortion in a vidicon

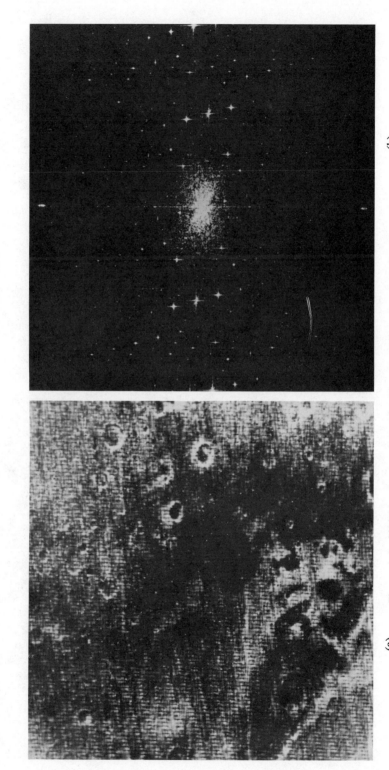

(a)   (b)

Figure I-4 MM'69 noise removal: (a) origonal, (b) Fourier amplitude specturum

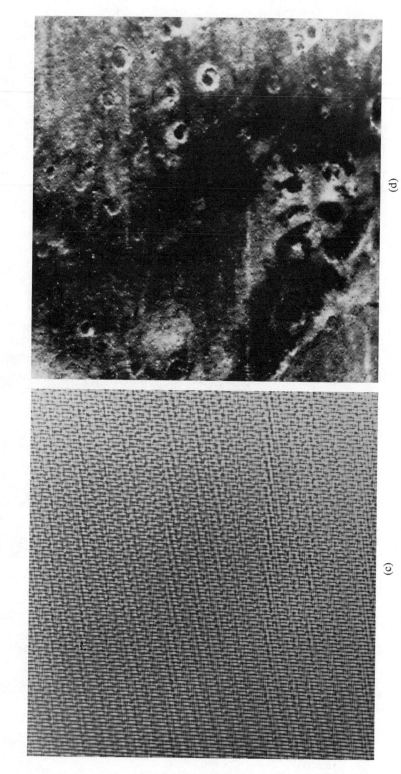

(c)   (d)

Figure I-4 (*cont.*) (c) periodic noise pattern, isolated from (b). (d) processed image

Figure I-5   Mariner 9 photomosaic globe of Mars

tube is image-dependent and preflight calibration is insufficient. Bill Green and Reuben Ruiz developed photometric decalibration software, while Paul Jepsen implemented a residual image removal technique. Alan Gillespie and Arnold Schwartz wrote cartographic projection programs for making photographic maps from MM'71 images.

Mariner Venus Mercury 1973

The Mariner Venus Mercury 1973 mission profited from the experience of its predecessors. Well-managed and highly successful, it operated on a short time scale and a fixed budget that was small in comparison to previous missions. MVM'73, the first multiplanet photography mission, used a gravity assist from Venus to reach Mercury. Mariner 10 returned over 16,000 images of the earth, the moon, Venus, and, on three separate encounters, the planet Mercury. This mission produced the first closeup pictures of the Venusian cloud cover (Ref. 24) and of the sun's nearest neighbor (Refs. 25, 26). The image quality was so good that residual image removal and coherent noise removal were unnecessary. The telecommunications link allowed each image to be sent over the 90-million-mile distance in only 42 seconds at a data rate of 117,600 bits per second (Ref. 26). Exacting preflight calibration permitted photometric and geometric decalibration to be performed more accurately than ever before.

Don Lynn directed the development of IPL's role in the mission. James Soha developed a scene-dependent filtering technique to bring out details near the terminator (Figure I-6). John Kreznar and David Haas advanced the geometric decalibration software, and Arnold Schwartz and Denis Elliott developed new map projection techniques. Jean Lorre devised and implemented techniques for producing color ratio and polarization images. Joel Mosher developed software for producing large computer mosaics of map projected images, and Paul Jepsen improved the bit error removal software. These efforts produced a very successful mission and a highly refined package of image processing software.

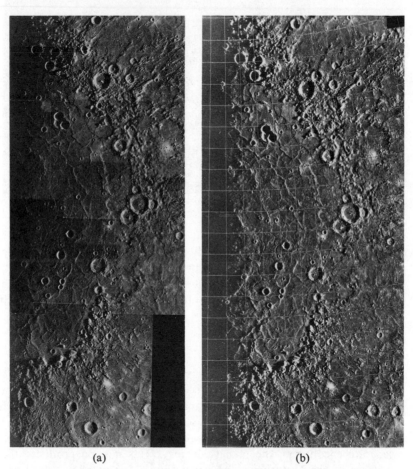

          (a)   (b)

Figure I-6   Mariner 10 scene-dependent filter: (a) before, (b) after

The Viking Mission

In terms of image processing the Viking mission to Mars was the most ambitious to date. In preparation for Viking the IBM 360/44 in the IPL was replaced with an IBM 360/65 with 1 megabyte of memory and 900 megabytes of disk storage, producing a threefold increase in throughput (Ref. 27). A PDP-11/40 was attached to support five interactive image processing terminals. Joel Seidman designed and Mike Girard implemented LIBEXEC, an interactive version of VICAR. As a result of this upgrade, in fiscal year 1977, the IPL ran 110,000 jobs and produced 86,000 photographic prints.

The Viking mission sent two orbiter and two lander spacecraft to Mars, with each of the four carring two cameras. The orbiter images were used for high resolution surface mapping and for stereometry using software developed by Jean Lorre. The lander mission was the first with a designed-in color objective and the first to use spherical pixel coordinates (azimuth/elevation scanning) (Refs. 28–30). Viking Orbiter image size was 1024 by 1240 pixels while Viking Lander images had up to 3600 512-pixel columns. Arnold Schwartz and Denis Elliott developed interactive stereometry with one-millimeter accuracy, and this was used to select the digging sites in front of the two landers (Ref. 31). Special stereoscopic viewing hardware was developed at Stanford University by Dr. Sidney Liebes. The hardware was transferred to the IPL and used in stereoscopic analysis of the area around both landing sites. Lander Imaging Flight Team members Robert Tucker of Stanford University, Dr. Ken Jones of Brown University, and Steve Wall of the NASA Langley Research Center were directly involved in the production of Viking Lander images. Viking image processing consumed 100 hours per week of IPL computer time for over a year. An example of the product appears in Figure 17-13.

Voyager

In 1977, the Voyager mission sent two Mariner-type spacecraft on a course passing by both Jupiter and Saturn. Each spacecraft carried one wide-angle and one narrow-angle vidicon camera to return close-up images of those distant planets and their satellites. This five-year mission was the first designed for routine production of large color mosaics and for the analysis of cloud motion during the 90-day planetary encounters. At these great distances radio navigation is not sufficiently accurate to steer the spacecraft around one planet and on to the next. Voyager '77 was the first mission to rely on optical navigation using its television cameras.

NEW APPLICATIONS

Robert Nathan, a crystallographer at heart, always had an interest in biomedical problems. During the Ranger program, he worked in his spare time on such biomedical problems as electron microscope images of catalase (Ref. 32). Robert Selzer also did early work in the enhancement of X-ray images (Refs. 10, 18, 33). In 1968, Nathan went to full-time biomedical work, leaving the space program in the hands of his followers. He obtained support from the National Institutes of Health to apply the digital image processing hardware and software at JPL to relevant problems in medicine and biology. Beginning under this grant, Robert Selzer's X-ray work led to a quantitative diagnostic tool for atherosclerosis that can detect small changes in disease state (Refs. 34, 35); the light microscope work, under Ken Castleman, led to a prototype clinical system for chromosome analysis (Refs. 36–42); and Nathan's own EM work produced electron microscope pictures with atomic resolution (Ref. 43).

The image processing technology has been applied to a number of other fields as well. Billingsley and Goetz performed lunar surface analysis using Apollo multispectral photographs (Refs. 44–46). Richard Blackwell used frequency domain filtering to improve the readability of fingerprints (Refs. 47, 48). Don Williams, Yoram Yakamovski, and Martin Levine developed scene analysis techniques for future use by a mobile Mars rover spacecraft (Refs. 49, 50). Peter Paluzzi developed digital sonar image enhancement techniques, and Jean Lorre, Don Lynn, William Benton, and Denis Elliott developed a variety of astronomical applications of digital image processing (Refs. 51–53).

The JPL image processing facilities have been applied to multispectral images from Landsat and from aircraft (Refs. 54, 55). Richard Blackwell developed a lake eutrophication technique for remote determination of water quality (Ref. 56). Nevin Bryant and Albert Zobrist developed an information system that combines nonvisual geographically referenced data (census data, etc.) with remotely sensed images (Ref. 57). William Stromberg and James Soha developed synthetic aperture radar image processing techniques to monitor ocean wave patterns from orbiting satellites, and Alan Gillespie developed a technique to produce thermal inertia images (Ref. 58).

Much of the success and growth of digital image processing at JPL is due to the efforts of Benn Martin, Bill Spuck, and Bill Green. Each served as Manager of the Science Data Analysis Section, which contains the IPL. Martin served from 1964 through 1969, Spuck from 1970 through 1975, and Green from 1975.

OBSERVATIONS

Throughout the previous discussion, several trends are evident. First, digital image processing progressed from an untested idea in early Ranger days to a fortuitous late addition to Ranger and Surveyor and then to an integral component of the success of the Mariner missions, all within the span of 15 years. In the absence of digital image processing, the recent history of planetary exploration would be much different. A second trend is the explosion of returned data, from Mariner 4's twenty-two 200×200 pixel images, requiring 8 hours each for transmission, to Mariner 10's 16,000 700×832 pixel images, arriving one every 42 seconds.

A third trend is the increasing complexity and scientific success of the missions themselves. The Ranger series, with its early disappointments, paved the way for later successes by serving as the proving ground for ambitious techniques never before attempted in human history. The overall success of the Mariner and Viking series speaks well of man's ability to send his creations over vast distances in search of knowledge.

Finally, it is noteworthy how a technology that was developed for the rather restricted field of planetary exploration has been so readily applicable in other fields.

JPL's activities in nonspace areas illustrate the versatility of general purpose image processing hardware and software.

REFERENCES

1. G. L. Fischer, *et al.*, eds., *Optical Character Recognition*, Spartan, Baltimore, 1962.

2. R. A. Kirsch, L. Cahn, L. C. Ray, and G. H. Urban, "Experiments in Processing Pictorial Information with a Digital Computer," *Proceedings of the Eastern Joint Computer Conference*, December 9, 1957 (published by the Association for Computing Machinery).

3. A. Rosenfeld, *Picture Processing by Computer*, Academic Press, New York, 1969.

4. A. Rosenfeld, "Progress in Picture Processing: 1969–1971 (A Bibliography)," *Computing Surveys*, **5**, 81–108, June 1973.

5. A. Rosenfeld, "Picture Processing: 1972," *Computer Graphics and Image Processing*, **1**, 394–416, 1972.

6. A. Rosenfeld, "Picture Processing: 1973," *ibid.*, **3**, 178–194, 1974.

7. A. Rosenfeld, "Picture Processing: 1975," *ibid.*, **5**, 215–237, 1976.

8. A. Rosenfeld, "Picture Processing: 1977," *ibid.*, **7**, 211–242, 1978.

9. *Christian Science Monitor*, Wed., March 24, 1965.

10. R. Nathan, *Digital Video Handling*, Jet Propulsion Laboratory Technical Report 32-877, January 5, 1966.

11. R. Cargill Hall, *The Ranger Chronology*, JPL Historical Report No. 2.

12. R. Cargill Hall, *Lunar Impact—A History of Project Ranger*, U.S. Government Printing Office, Washington, D.C., 20402, Stock No. 033-000-00699-3, Catalogue No. 1.21:4210, 1977.

13. T. C. Rindfleisch, *A Photometric Method for Deriving Lunar Topographic Information*, Technical Report No. 32-786, Jet Propulsion Laboratory, Pasadena, California, September 15, 1965.

14. T. C. Rindfleisch, "Photometric Method for Lunar Topography," *Photogrammetric Engineering*, **32**, 262–276, March 1966.

15. L. Jaffe, ed., *Lunar Scientific Model*, Surveyor Project Document 54, Jet Propulsion Laboratory, April 1, 1966.

16. *Surveyor Project Final Report, Part II—Science Results*, JPL Technical Report 32-1265, June 15, 1968.

17. D. H. Le Croissette, "The Scientific Instruments on Surveyor," *IEEE Transactions on Aerospace and Electronic Systems*, **AES-5**, No. 1, 2–21, Jan. 1969.

18. R. H. Selzer, "The Use of Computers to Improve Biomedical Image Quality," *Proceedings of the Fall Joint Computer Conference*, **33**, 817–834, 1968.

19. M. I. Smokler, "Calibration of the Surveyor Television System," *Journal of the SMPTE*, **77**, No. 4, 317–323, April 1968.

20. G. E. DANIELSON and D. R. MONTGOMERY, "Calibration of the Mariner Mars 1969 Television Cameras," *J. Geophys. Res.*, **76**, 418–431, 1971.

21. T. C. RINDFLEISCH, J. A. DUNNE, H. J. FRIEDEN, W. D. STROMBERG, and R. M. RUIZ, "Digital Processing of the Mariner 6 and 7 Pictures," *J. Geophys. Res.*, **76**, 394, 1971.

22. W. B. GREEN, P. L. JEPSEN, J. E. KREZNAR, R. M. RUIZ, A. A. SCHWARTZ, and J. B. SEIDMAN, "Removal of Instrument Signature from Mariner 9 Television Images of Mars," *Applied Optics*, **14**, 105, January 1975.

23. E. C. LEVINTHAL, *et al.*, "Mariner 9—Image Processing and Products," *Icarus*, **18**, 75–101, 1973.

24. *Science*, **183**, 1289–1321, 29 March 1974, Mariner 10 Venus Encounter Issue.

25. *Science*, **185**, 12 July 1974, Mariner 10 Mercury Encounter Issue.

26. *J. Geophys. Res.*, **80**, No. 17, June 10, 1975, Mariner 10 Mercury Encounter Issue.

27. W. B. GREEN, "Applications of Interactive Digital Image Processing to Problems of Data Registration and Correlation," *Proceedings of the National Computer Conference*, June 1978.

28. W. B. GREEN, "Viking Image Processing," *Proceedings of the SPIE*, **119**, 2–9, 1977.

29. W. B. GREEN, "Computer Image Processing—The Viking Experience," *Proceedings of the 18th Annual IEEE Chicago Spring Conference on Consumer Electronics*, June 1977.

30. E. C. LEVINTHAL, W. B. GREEN, K. L. JONES, and R. TUCKER, "Processing the Viking Lander Camera Data," *J. Geophys. Res.*, **82**, No. 28, 4412–4420, September 30, 1977.

31. S. LIEBES and A. A. SCHWARTZ, "Viking 75 Mars Lander Interactive Video Computerized Stereophotogrammetry," *J. Geophys. Res.*, **82**, 4421–4429, September 5, 1977.

32. F. C. BILLINGSLEY, "Applications of Digital Image Processing," *J. of Applied Optics*, **9**, No. 2, 289–299, February 1970.

33. R. H. SELZER, "Recent Progress in Computer Processing of X-Ray and Radioisotope Scanner Images," *Biomedical Sciences Instrumentation*, **6**, Instrument Society of America, Pittsburgh, Pennsylvania, 1969.

34. D. W. CRAWFORD, S. H. BROOKS, R. H. SELZER, R. BARNDT, E. S. BECKENBACH, and D. H. BLANKENHORN, "Computer Densitometry for Angiographic Assessment of Arterial Cholesterol Content and Gross Pathology in Human Atherosclerosis," *J. Lab. & Clin. Medicine*, **89**, 378–392, 1977.

35. D. W. CRAWFORD, E. S. BECKENBACH, and D. H. BLANKENHORN, "Grading of Coronary Atherosclerosis: Comparison of a Modified IAP Visual Grading Method and a New Quantitative Angiographic Technique," *Atherosclerosis*, **19**, 231–241, 1974.

36. K. R. CASTLEMAN, "Pictorial Output for Computerized Karyotyping" in *Perspectives in Cryogenetics*, 316–323, S. W. Wright, B. F. Crandall, L. Boyer, eds., C. C. Thomas, Springfield, Illinois, 1970.

37. K. R. CASTLEMAN and R. J. WALL, "Automatic Systems for Chromosome Identification," *Proc. XXIII Nobel Symposium*, Stockholm, Sweden, Academic Press, 1972.

38. T. CASPERSSON, K. R. CASTLEMAN, *et al.*, "Automatic Karyotyping of Quinacrine Mustard Stained Human Chromosomes," *Experimental Cell Res.* **67**, 233–235, 1971.

39. D. A. O'HANDLEY, E. S. BECKENBACH, K. R. CASTLEMAN, R. H. SELZER, and R. J. WALL, "Picture Analysis Applied to Biomedicine," *Computer Graphics and Image Processing,* **2,** 417–432, 1973.

40. K. R. CASTLEMAN and J. H. MELNYK, *An Automated System for Chromosome Analysis—Final Report,* JPL Document 5040-30, July 4, 1976.

41. K. R. CASTLEMAN, J. MELNYK, H. J. FRIEDEN, G. W. PERSINGER, and R. J. WALL, "Computer-Assisted Karyotyping," *Journal of Reproductive Medicine,* **17,** No. 1, 53–57, July 1976.

42. K. R. CASTLEMAN, J. MELNYK, H. J. FRIEDEN, G. W. PERSINGER, and R. J. WALL, "Karyotype Analysis by Computer and Its Application to Mutagenicity Testing of Environmental Chemicals," *Mutation Research,* **41,** 153–162, 1976.

43. R. NATHAN, "Image Processing for Electron Microscopy: I. Enhancement Procedures," *Advances in Optical and Electron Microscopy,* **4,** Academic Press, New York, 85–125, 1971.

44. F. C. BILLINGSLEY, A. F. H. GOETZ, and J. N. LINDLSEY, "Color Differentiation by Computer Image Processing," *Photogr. Sci. Eng.,* **14,** 28, 1970.

45. A. F. H. GOETZ, F. C. BILLINGSLEY, E. YOST, and T. B. McCORD, "Apollo 12 Multispectral Photography Experiment," Proc. of the Second Lunar Science Conference, Suppl. 2, *Geochim. Cosmochim. Acta,* **3,** MIT Press, Cambirdge, Massachusetts, 2301, 1971.

46. F. C. BILLINGSLEY, "Computer-Generated Color Image Display of Lunar Spectral Reflectance Ratios," *Photogr, Sci. Eng.,* **16,** 51, 1972.

47. R. J. BLACKWELL, "Fingerprint Image Enhancement by Computer Methods," 1970 Carnahan Conference on Electronic Crime Countermeasures, Lexington, Kentucky, April 17, 1970.

48. R. J. BLACKWELL and W. A. CRISCI, "Digital Image Processing Technology and Its Application in Forensic Sciences," *Journal of Forensic Sciences,* **20,** No. 2, 288–304, 1975.

49. M. D. LEVINE, D. A. O'HANDLEY, and G. M. YAGI, "Computer Determination of Depth Maps," *Computer Graphics and Image Processing,* **2,** No. 6, 131–150, 1973.

50. D. A. O'HANDLEY, "Scene Analysis in Support of a Mars Rover," *Computer Graphics and Image Processing,* **2,** No. 3/4, 281, December 1973.

51. H. ARP and J. LORRE, "Image Processing of Galaxy Photographs," *The Astrophysical Journal,* **210,** 58–64, November 15, 1976.

52. J. J. LORRE, "Analysis of the Nebulosities Near T Tauri Using Digital Computer Image Processing," *The Astrophysical Journal,* **202,** 696–717, December 15, 1975.

53. J. J. LORRE, D. J. LYNN, and W. D. BENTON, "Recent Developments at JPL in the Application of Digital Image Processing Techniques to Astronomical Images," *Proceedings of the SPIE,* **74,** 234–238, 1976.

54. Special Issue on Remote Sensing, *Aviation Week and Space Technology,* October 17, 1977.

55. F. C. BILLINGSLEY, "Digital Image Processing for Information Extraction," *International Journal of Man-Machine Studies,* **5,** 203–214, 1973.

56. R. J. BLACKWELL and D. H. BOLAND, "The Trophic Classification of Lakes Using ERTS Multispectral Data," *Proceedings of the American Society of Photogrammetry 41st Annual Meeting*, 393–414, March 9–14, 1975.

57. N. A. BRYANT and A. L. ZOBRIST, "IBIS: A Geographic Information System Based on Digital Image Processing and Image Raster Data Type," *IEEE Trans. on Geoscience Electronics*, **GE-15**, No. 3, 152–159, July 1977.

58. A. R. GILLESPIE and A. B. KAHLE, "Construction and Interpretation of a Digital Thermal Intertia Image," *Photogrammetric Engineering and Remote Sensing*, **43**, No. 8, 983–1000, August 1977.

VICAR PROGRAM INDEX

The VICAR software system for digital image processing is described in Chapter 4. It includes a library of FORTRAN language image processing programs. This appendix lists many of the generally applicable VICAR programs by name. A brief functional description is supplied for each. The programs are divided into categories. Within each category, the more general programs precede the more specialized ones.

The VICAR system, and its documentation, may be obtained from

COSMIC
Computer Center
112 Barrow Hall
University of Georgia
Athens, Georgia 30601
(404) 542-3265

The VICAR system is filed under number NPO-13415. As of this writing, all programs followed by asterisks are available from COSMIC. Many of the others may be available there in the future. A set of image processing programs written for the PDP-11 computer is filed there under NPO-14892 and the name Mini-VICAR.

IMAGE GENERATION

The following programs are useful for generating digital images.

GEN* is a program that produces images in which gray level increases lineary, modulo 256, from the upper left-hand corner, at specified horizontal and vertical rates.

GRATE generates digital images with vertical or horizontal stripes of specified gray level on a specified background gray level.

SPOT* produces an image containing a spot. The spot may be Gaussian, conical, reciprocal, reciprocal squared, exponential, or dome-shaped. Spot location, size, and contrast may be specified.

LOCUS* generates images containing 5-by-5-pixel plus signs at specified locations.

COHER generates sinusoidal coherent noise images and can also add sinusoidal patterns to other images.

GAUSNOIS* generates random noise images having a Gaussian histogram of specified mean and standard deviation.

POLYNOIS* generates random noise images having a histogram that is specified by the user.

SYNPIC generates an image in complex format having gray level zero. The user supplies the real and imaginary values of a few specified pixels. This program can be used in conjunction with the inverse Fourier transform to produce images containing sinusoidal components.

POINT OPERATIONS

The following programs implement point operations (gray scale transformations) on digital images.

STRETCH* performs general point operations. The transformation may be specified as linear, piecewise linear, or as a cube root or exponential function. It can also produce contour lines at specified intervals.

ASTRTCH2* first computes the gray level histogram of the input image. The program then analyzes that histogram to determine the point operation required to put the histogram into a specified form. The user may specify either a linear point operation or a uniform or Gaussian output histogram. The linear point operation is designed to produce a specified amount of saturation at each end of the gray scale.

MATCH* performs the point operation required to make one image "match" another. The point operation may be based upon specified areas common to the two pictures. Alternatively, for a single input image, the program will make the mean gray level of specified areas conform to specified values.

ALGEBRAIC OPERATIONS

The following programs perform pixel-by-pixel arithmetic functions.

PICAVE* can average up to 10 input images, which may be linearly displaced with respect to one another.

UNITAVE* divides the input image into a set of contiguous rectangles and averages those rectangles together. In the output image, each rectangle is replaced by the average. This program is useful for noise removal from periodic structures.

DIFFPIC* can add or subtract two images following a linear displacement. A specified linear point operation is also performed on the output image.

F* is a general-purpose pixel arithmetic routine. The function that relates the output image to the two input images has 11 specifiable parameters. These may be chosen to implement addition, subtraction, multiplication, division, exponentiation, and logarithms. The program operates by first generating a 256 × 256 two-dimensional look-up table, which is subsequently used to produce the output image.

F2 performs general arithmetic operations on one or two input images. The arithmetic operation is specified by a FORTRAN-like expression.

PIXC performs complex arithmetic on two complex-valued digital images.

PIXGRAD calculates the magnitude and angle of the gradient vector of an input image. The magnitude is taken as the maximum difference between the current pixel and its eight adjacent neighbors. The angle is specified by an integer from 1 through 8.

PIXH performs addition, subtraction, multiplication, and division on two input images. The output and the two input images may have one or two bytes per pixel. The arithmetic is done in a fractional format.

PIXRMS produces an output image composed of local means or local standard deviations of the input image.

RATIO* is a preprocessor used for comparing two input images. These may be compared on the basis of ratio, log ratio, difference, or log difference. The program generates the parameters required for proper scaling of the output image and fetches program F, which performs the specified operation.

LOCAL OPERATIONS

The following programs perform a variety of specialized local operations on input images. Many of them were designed to remove specific blemishes from images.

SAR* can replace specified rectangular areas with an average of surrounding gray levels. It is also commonly used to copy images from one data set to another.

AUTOSAR locates pixels that deviate from the average of the pixels above and below by more than a specified tolerance. Such pixels are replaced with the average of the pixels above and below.

QSAR* can add or subtract specified values at all pixels within specified rectangular areas of the input image.

PSAR* can add or subtract specified values at all pixels within specified polygonal boundaries.

ERASE* locates and sets to zero the pixels within connected sets having gray level and perimeter below specified thresholds. It is used to remove small objects from an image.

BLEM replaces specified areas in an image by linear interpolation.

ADL* can add or subtract a specified value to all pixels lying along a straight line between any two pixels in the input image. The diagonal line is specified by its end points.

ADESPIKE* replaces the gray level of a pixel if it differs from its four nearest neighbors by more than a specified tolerance. The average of its adjacent neighbors is used for the replacement.

IMAGE MEASUREMENT

The following programs are concerned with extracting and displaying various measurements from an input image.

LIST* can be used to list the gray levels or the histogram of an image on a line printer.

HISTO generates and plots gray level histograms on a line printer.

CLSTR* and **HIST2** compute the two-dimensional histogram of a pair of input images.

LPLOT2* produces a graphic plot of the gray levels along a specified diagonal line in an image.

LAVE* can average all the horizontal or vertical lines in an image. The average values are listed on the line printer and output as a one-line image.

PIXSTAT produces output images which represent the local mean, variance, second moment, or standard deviation of an input image.

LITEXFER* is designed to calculate the light transfer characteristics using input images that are flat fields at various intensity levels. For specified rectangular areas within the set of input images, the program computes and plots the mean and standard deviation of gray level as a function of the input brightness level.

THRESHLD* locates and lists areas of an image containing points above a specified gray level threshold. When such a point is located, a 30 by 50-pixel area surrounding that point is listed on the line printer and set to zero gray level in the image. This program is useful for automatically locating small objects, such as stars, in an image.

GRIDLOCA* and **GRIDLOCB*** are used together to locate the intersections in an image containing a rectilinear grid network. These programs are helpful in the geometric calibration of image digitizers.

DRECK locates line segments in digitized line drawings.

ANNOTATION AND DISPLAY

The following programs are concerned with adding various types of annotation to images or with effecting image display.

ARROW writes arrows into an image at specified locations.

MARK* superimposes rectangular marks centered at specified coordinates in an image.

SCRIBE* places rectangles around specified areas in an image.

GRID* overlays a rectilinear grid network into a digital image.

MAPGRID* overlays an alternating black and white grid (dashed lines) into an image.

OVERLAY* superimposes a latitude-longitude grid onto cartographic projection images produced by program MAP2.

MASK* adds gray scales, pixel coordinate reference marks, label annotation, and a gray level histogram to an image in preparation for display. Most of the digital images in this book were produced using program MASK.

SHADY introduces shading and contour lines into an image. The shaded image is actually a partial derivative image taken in a specified direction. This assists visualization of slowly varying (low-frequency) images.

DNSYMBOL replaces each pixel with a square multi-pixel black and white symbol that represents gray level. When the image is displayed, the user can read the gray level of each pixel by examining the symbol.

DISPLAY* produces images by printing on a line printer. Overprinting of characters is used to achieve up to 64 gray levels.

PRINTPIX also prints images on a line printer. However, it includes facilities for demagnification, variable aspect ratio, and a gray scale transformation.

The following programs are designed for general user specified geometric operations.

GEOM* transforms a specified control grid of contiguous quadrilaterals into a specified rectangular grid. This program is particularly efficient when vertical displacements are small.

LGEOM* performs the same transformation as GEOM but is more efficient when vertical displacements are large. Both programs use bilinear interpolation of displacement and gray levels in the transformation.

GEOMA* performs geometric operations where the transformation is specified as a mapping of quadrilaterals into quadrilaterals rather than into rectangles. This more general format may afford a more convenient specification in some cases. For example, it is possible to degenerate the quadrilaterals into triangles and specify the geometric transformation as a mapping between control grids composed of contiguous triangles. In general, GEOMA runs approximately one-third longer than GEOM on comparable transformations.

POLARECT projects a rectangular image into a sector of a circle, and vice versa.

Rotation and Magnification

The following programs are designed to perform rotation and magnification on digital images.

ROTATE* performs 90° clockwise or counterclockwise rotation.

FLOT* can perform plus or minus 90° rotation or reflection about the horizontal or vertical axis of an image.

ROTATE2* rotates an image through a specified angle about a specified point and places the center of rotation at a specified point in the output image. This program generates the necessary parameters (control grids) and fetches GEOM or LGEOM to perform the rotation.

ANGLE provides the specified rotation and translation necessary to bring two images into registration.

MAG* can generate the parameters necessary to magnify or reduce an image, alter its aspect ratio, or skew the top of the image right or left with respect to the bottom. This program fetches GEOM to perform the operation.

SIZE* can also magnify or reduce an image or change its aspect ratio.

RESAMP* reduces the size of an image by skipping lines and samples.

APAVG* reduces the size of an image by averaging rectangular arrays of pixels to form the new gray level values.

EXPAND* enlarges an image by a specified factor *N* by repeating each pixel in an *N* by *N* array.

INTERP* magnifies an image by a factor of 2 using bilinear interpolation.

CLASP changes the linear aspect ratio of an image by magnification in the horizontal direction. The program can use either linear or cubic spline interpolation.

Image Combination

The following programs may be used to combine images to form larger ones.

CONCAT* can combine up to ten input images of the same size. The images are concatenated side by side and one above the other.

INSECT* can combine two images of unequal size. Specified rectangular areas from each of the two input images are placed at specified locations in an output image of specified size. Where the two areas overlap, the second input image prevails. Areas of the output image where neither input image falls are set to 255.

MOSAIC* is useful for combining multiple overlapping views of an area into one composite output image. The program assumes that the input images are all the same size as the desired output and that the smaller images have already been placed in proper position on a background of zero gray level. Where nonzero portions of images overlap, the program takes values from the input images in a specified order of priority.

HISTLOC generates an output image from a specified two-dimensional look-up table. The gray levels of two input images provide the address into the table. This program is useful for identifying which areas of an image correspond to different clusters in a two-dimensional histogram.

LYNX joins the top half of one image to the bottom half of another. Translation is provided to bring overlapping areas into registration.

Map Projection

The following programs are designed for producing map projections of aerial or spacecraft images.

MAP2* is a general cartographic projection program. The program generates the necessary parameters to project an image from the camera coordinate system into a standard cartographic projection map. The six cartographic projection options are Mercator, Lambert conformal conic, oblique orthographic, polar orthographic, oblique stereographic, and polar stereographic. The program uses LGEOM to perform the actual projection.

MAPTRANS can be used to transform a projected image from one cartographic projection to another. The program uses LGEOM to perform the necessary geometric transformation.

MERCATOR generates LGEOM parameters to transform an image into a mercator projection.

PROJECT generates LGEOM parameters to transform an image from mercator to orthographic projection or vice versa.

GEOTRAN* performs geometric transformation to effect projection from a sphere onto a cylinder or a plane. It can also perform orthographic projections to produce the image that would be obtained if the camera were moved to a specified different position.

CORRELATION AND CONVOLUTION

The following programs perform tasks that are implemented with digital correlation or convolution.

CROSS* compares specified areas within two input images to determine the translation required to register the two images. It computes and prints the sum of squared differences between a stationary rectangle in one image and a moving rectangle in the second image and the relative displacement that results in minimum sum-squared difference. The size of the rectangle and the area of search can be specified by the user. It is useful in many cases requiring cross-correlation or autocorrelation.

REGISTER* generates the GEOM parameters required to bring one picture into registration with another. Specified rectangular areas of the images are compared by first removing low-frequency information, then determining the translation that results in minimum mean square difference, and finally fitting a polynomial surface through the required displacements. A GEOM control grid that spans the image is computed from the polynomial surface.

FILTER* performs general two-dimensional convolution. The input image is convolved with a specified rectangular impulse response. This program assumes that the impulse response has four-quadrant symmetry. It permits either specified or automatic scaling of the output gray scale.

AFILTER* is similar to FILTER except that no assumptions are made about symmetry in the point spread function.

FILTER2* computes the point spread function of a specified two-dimensional transfer function. The program then fetches FILTER to perform the convolution. Thus the user may perform convolution filtering where the filter is specified in the frequency domain.

GFILTER accepts parameters that describe a two-dimensional transfer function having elliptical cross sections and Gaussian profile. It then fetches FILTER2, which computes the corresponding psf and, in turn, fetches FILTER to carry out the convolution. This program makes it convenient to implement lowpass filters of Gaussian profile in the frequency domain.

BOXFLT* convolves an input image with a flat-topped rectangular point spread function. The user specifies the size of the psf. Because of the constrained psf, this program executes much faster than FILTER.

BOXFLT2 is similar to BOXFLT except that it can also produce a highpass filtered output image. This is obtained by subtracting the lowpass filtered image from the input.

FASTFIL2* can produce a highpass and a lowpass filtered version of an input image. The lowpass psf is a two-dimensional rectangular pulse. The highpass filtered image is obtained by subtracting a specified fraction of the lowpass filtered image from the original. Options to reduce ringing at discontinuities are included.

FASTFIL1* is a one-dimensional (horizontal) version of FASTFIL2.

MEDIAN produces an output image in which the gray level represents the median value of surrounding pixels within the input image. The program is limited to one-dimensional (line-by-line) processing.

POLYFILT implements convolution with a spatially variant point spread function. Up to 10 different psfs may be supplied. A second input image specifies the areas in which each of the psfs are to be applied.

SMEAR73* is designed for Wiener deconvolution of linear motion blur. The user specifies the direction and amount of motion blur, and the program calculates the psf of the Wiener deconvolution filter assuming white signal and noise. The program estimates the signal and noise power spectra from the image and fetches AFILTER to implement the convolution.

FOURIER TRANSFORM COMPUTATION

The following programs perform tasks that involve computation of the Fourier transform.

FFT1* computes the forward and inverse one-dimensional complex Fourier transform on a line-by-line basis.

FFT1PIX* can be used to display the complex transforms produced by FFT1.

FFT2* computes the direct or inverse two-dimensional complex Fourier transform of a digital image.

FFTPIC* produces a digital image from the complex data set produced by FFT2. This permits display of the amplitude and phase of the Fourier transform as an image.

POWER* computes the one-dimensional power spectrum of each line in a digital image and displays the root-mean-square power spectrum of all lines on the line printer.

MTF can be used to compute the modulation transfer function of a camera system from digitized images of sine wave test targets.

FREQUENCY DOMAIN FILTERING

The following programs implement linear filtering in the frequency domain.

FREQMOD* multiplies the complex spectrum produced by FFT2 by a user-specified transfer function. The resulting complex spectrum can be inverse transformed by FFT2 to implement linear filtering. The user specifies the profile of an elliptically symmetrical, real, nonnegative transfer function.

FFTFIT can either multiply a complex spectrum by an input picture or make the amplitude of the spectrum proportional to the input picture. In either case, phase is unaltered. This program permits frequency domain filtering with unrestricted real transfer functions.

FFTFIL* performs one-dimensional frequency domain filtering on a line-by-line basis. Each line is transformed, multiplied by a specified transfer function, and inverse transformed. The transfer function may be a bandpass or notch filter with user-specified frequency bands. The program can also modify the spectrum by interpolating across specified frequency bands. This program is useful for removing coherent noise from images. Program POWER may be used to determine the unknown noise frequency.

UNSHADE* implements highpass filtering in the frequency domain. It multiplies a complex spectrum (from FFT2) by a transfer function that is unity everywhere except near the vertical and horizontal frequency axes. The transfer function goes to zero at the frequency axes with a negative Gaussian profile. This removes low-frequency information (shading) from the inverse transformed image.

SPIKMASK multiplies the complex spectrum by a transfer function that is unity everywhere except in specified small rectangular regions. In these regions, it takes on the value zero. This program may be used to remove spikes in the frequency domain produced by periodic noise in the image.

OUTSPIKE* removes spikes from the spectrum of an image containing periodic noise. The program first locates spikes in the amplitude spectrum by searching for local maxima above a specified severity. It then removes the spike by interpolation of the surrounding values. The phase is not altered.

APODIZE* modifies an image near the edges so as to reduce the effects of truncation when the Fourier transform is computed. The program places a quarter cycle of a sine function at each end of each line and column of pixels so that making the image periodic in two dimensions does not produce discontinuities. This prevents the introduction of artifact along the frequency axes in the two-dimensional Fourier transform.

OTF1* can compute the horizontal component of a blurring transfer function given an image containing a degraded vertical edge. It can also compute the horizontal component of a transfer function, given the line spread function. The program differentiates each line through a degraded vertical edge to obtain an estimate of the line spread function. These estimates are averaged together and inverse transformed to determine a component of the transfer function.

OTF2* implements two-dimensional frequency domain image restoration by Wiener deconvolution. The user supplies the complex spectrum of both the input image and the degrading psf. The program assumes white signal and noise spectra, estimates their amplitude, and computes a Wiener deconvolution transfer function. It produces an output spectrum that is the product of the input spectrum and the restoration transfer function. The restored image may be obtained by inverse transforming the output spectrum.

STEREOMETRY PROGRAMS

The following programs are concerned with stereometry and stereoscopic display.

RANGE computes a range image, given a stereo pair of input images.

ELEVMAP generates a topographic map of a surface, given a stereoscopic pair of input images.

VPROFILE* produces a vertical profile plot of a surface in a range image.

STEREO* produces a stereo pair, given a brightness image and a range image. The user supplies the right eye image and the range image, and the program computes the left eye image.

FOURIER TRANSFORMS

■■

This appendix contains graphs of several useful Fourier transform pairs. We have included arbitrary width constants in the graphed functions because it is frequently important to know the relationship between the widths of the functions in both domains. We have not substituted arbitrary multiplicative constants, however, because their use is straightforward under the addition theorem. Neither have we substituted the shift theorem, which unnecessarily complicates the graphics.

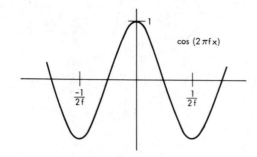

$\cos (2\pi f x)$

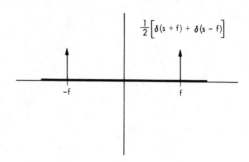

$\frac{1}{2}\left[\delta(s+f) + \delta(s-f)\right]$

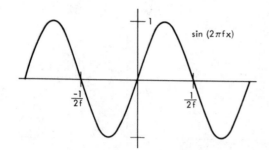

$\sin (2\pi f x)$

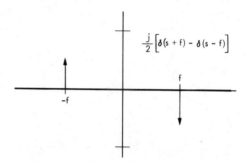

$\frac{j}{2}\left[\delta(s+f) - \delta(s-f)\right]$

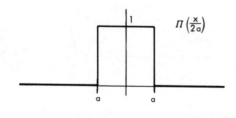

$\Pi\left(\dfrac{x}{2a}\right)$

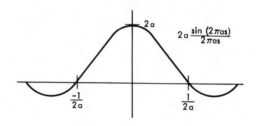

$2a\,\dfrac{\sin (2\pi a s)}{2\pi a s}$

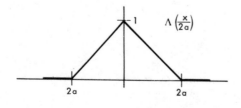

$\Lambda\left(\dfrac{x}{2a}\right)$

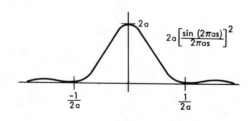

$2a\left[\dfrac{\sin (2\pi a s)}{2\pi a s}\right]^2$

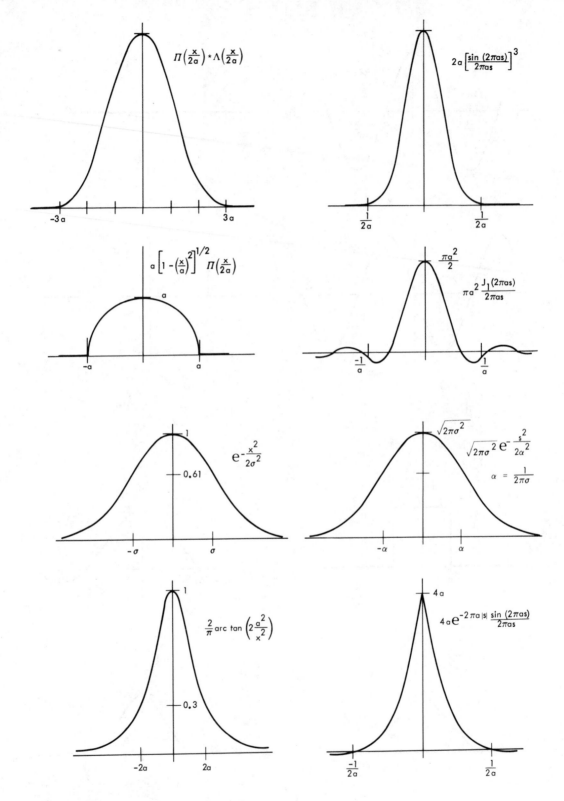

$$\Pi\left(\frac{x}{2a}\right) * \Lambda\left(\frac{x}{2a}\right)$$

$$2a\left[\frac{\sin(2\pi as)}{2\pi as}\right]^3$$

$$a\left[1-\left(\frac{x}{a}\right)^2\right]^{1/2}\Pi\left(\frac{x}{2a}\right)$$

$$\frac{\pi a^2}{2}$$

$$\pi a^2\frac{J_1(2\pi as)}{2\pi as}$$

$$e^{-\frac{x^2}{2\sigma^2}}$$

$$0.61$$

$$\sqrt{2\pi\sigma^2}$$

$$\sqrt{2\pi\sigma^2}\,e^{-\frac{s^2}{2\alpha^2}}$$

$$\alpha = \frac{1}{2\pi\sigma}$$

$$\frac{2}{\pi}\arctan\left(2\frac{a^2}{x^2}\right)$$

$$0.3$$

$$4a$$

$$4a\,e^{-2\pi a|s|}\frac{\sin(2\pi as)}{2\pi as}$$

414

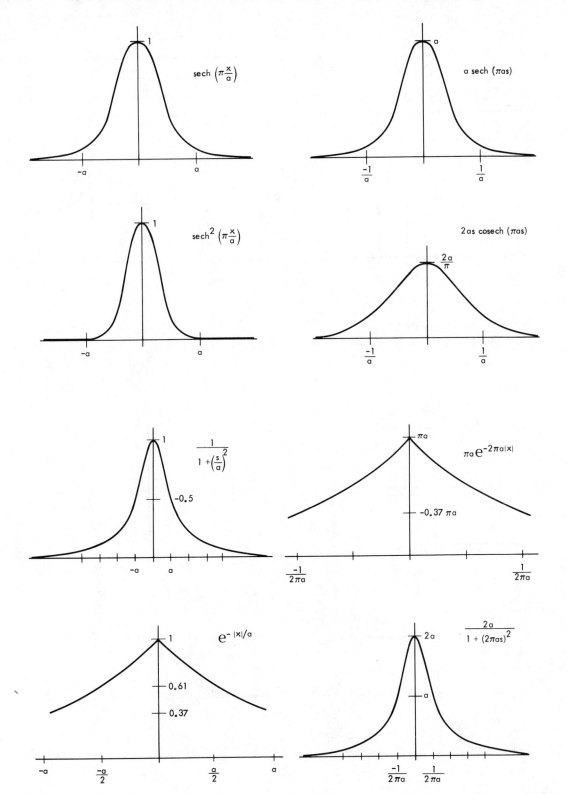

$$\text{sech}\left(\pi\frac{x}{a}\right)$$

$$a\,\text{sech}\,(\pi a s)$$

$$\text{sech}^2\left(\pi\frac{x}{a}\right)$$

$$2as\,\text{cosech}\,(\pi a s)$$

$$\frac{1}{1+\left(\frac{s}{a}\right)^2}$$

$$\pi a\,e^{-2\pi a|x|}$$

$$e^{-|x|/a}$$

$$\frac{2a}{1+(2\pi a s)^2}$$

415

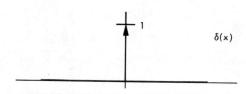

$\delta(x)$

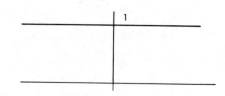

1

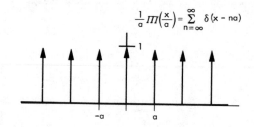

$$\frac{1}{a}\, III\!\left(\frac{x}{a}\right) = \sum_{n=\infty}^{\infty} \delta\,(x - na)$$

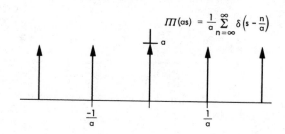

$$III\,(as)\; = \frac{1}{a}\sum_{n=\infty}^{\infty}\delta\!\left(s - \frac{n}{a}\right)$$

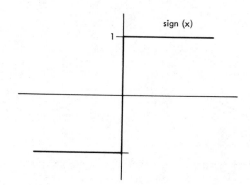

sign (x)

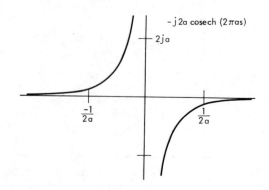

$\dfrac{-j}{\pi s}$

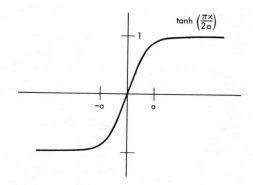

$\tanh\left(\dfrac{\pi x}{2a}\right)$

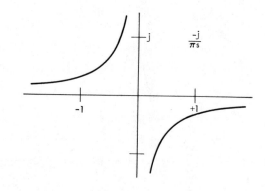

$-j\,2a\;\text{cosech}\,(2\pi as)$

416

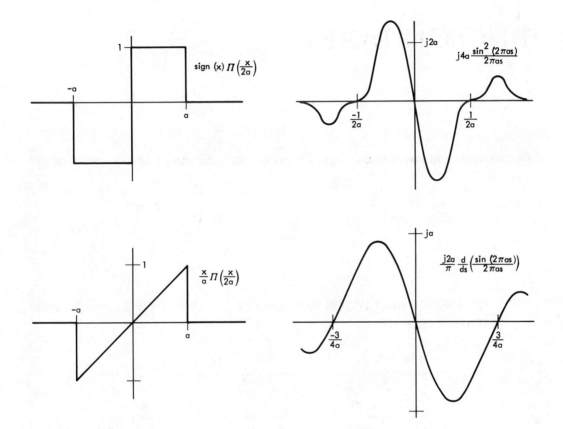

$$\text{sign } (x)\, \Pi\!\left(\frac{x}{2a}\right)$$

$$j4a\,\frac{\sin^2(2\pi as)}{2\pi as}$$

$$\frac{x}{a}\,\Pi\!\left(\frac{x}{2a}\right)$$

$$\frac{j2a}{\pi}\,\frac{d}{ds}\!\left(\frac{\sin(2\pi as)}{2\pi as}\right)$$

FUNCTION TABLES

Tables IV-1 and IV-2 list the values of several functions that are useful in the analysis of linear systems.

Table IV-1

x	$\dfrac{J_1(x)}{x}$	x	$\dfrac{J_1(x)}{x}$	x	$\dfrac{J_1(x)}{x}$
0	0.500	5.0	−0.066	10.0	0.004
0.5	0.485	5.5	−0.062	10.5	−0.008
1.0	0.440	6.0	−0.046	11.0	−0.016
1.5	0.372	6.5	−0.024	11.5	−0.020
2.0	0.288	7.0	−0.001	12.0	−0.019
2.5	0.199	7.5	0.018	12.5	−0.013
3.0	0.113	8.0	0.029	13.0	−0.005
3.5	0.039	8.5	0.032	13.5	0.003
4.0	−0.013	9.0	0.027	14.0	0.010
4.5	−0.051	9.5	0.017	14.5	0.013

Table IV-2

x	$sech\,(\pi x)$	$sech^2\,(\pi x)$	$cosech\,(\pi x)$	$tanh\,(\pi x)$	$\dfrac{sin\,(\pi x)}{\pi x}$	$\left[\dfrac{sin\,(\pi x)}{\pi x}\right]^2$	$e^{-\pi x}$	$e^{-x^2/2}$
0	1	1	∞	0	1	1	1	1
0.1	.95260	.90745	3.13134	.30422	.98363	.96753	.90484	.99501
0.2	.83058	.68987	1.49146	.55689	.93549	.87514	.81873	.98020
0.3	.67659	.45777	.91883	.73636	.85839	.73684	.74082	.95560
0.4	.52656	.27727	.61939	.85013	.75683	.57279	.67032	.92312
0.5	.39853	.15883	.43454	.91715	.63662	.40529	.60653	.88250
0.6	.29683	.08811	.31084	.95493	.50455	.25457	.54881	.83527
0.7	.21911	.04801	.22456	.97570	.36788	.13534	.49659	.78271
0.8	.16095	.02590	.16308	.98696	.23387	.05470	.44933	.72615
0.9	.11792	.01390	.11874	.99302	.10929	.01195	.40657	.66700
1.0	.08627	.00744	.08659	.99627	0	0	.36788	.60653
1.5	.01797	.00032	.01797	.99984	−.21221	.04503	.22313	.32465
2.0	.00373	.00001	.00373	.99999	0	0	.13534	.13534
2.5	.00078	.00000	.00078	1.00000	.12732	.01621	.08209	.04394
3.0	.00016	.00000	.00016	1.00000	0	0	.04979	.01111
3.5	.00003	.00000	.00003	1.00000	−.09095	.00827	.03020	.00219
4.0	.00001	.00000	.00001	1.00000	0	0	.01832	.00034

INDEX

A

Adaptive thresholding, 304-5
Addition Theorem of the Fourier transform,167-68
Airy disk, 264
Algebraic operations, 96-109
  applications of, 101-8
    averaging for noise reduction, 101-2
    image subtraction, 104-6
    multiplication and division, 106, 107
  defined, 96
  and histogram, 97-101
    of difference images, 99
    IOD of noisy image, 99-101
    of sum images, 97-99
  uses for, 97
  VICAR programs for, 403
Aliasing, 231, 243
  bounding of error, 236-38
  unavoidability of, 236
Aliasing error, 231
Analytic continuation, 288
Andrews, H.C., 282, 289

Angiography, 219-20
Annotation and display, VICAR programs for, 405
Apodisation, 261
Area measurement, 323-24
Astigmatism, 262, 282
Atmospheric turbulence, 283, 291
Autocorrelation function, 186
Avalanche photodiode, 21
Average boundary gradient function, 309-10
Axial tomography, computerized, 360-64

B

Back projection, 363-64
Background subtraction, 104-5
Bandpass filter
  general, 193-94
  ideal, 191-92
Bandstop filter, ideal, 192-93
Bayes' theorem, 335-36

421

Bayesian estimation, 338-43
Benton, William, 396
Bilinear interpolation, 113-15, 117
Billingsley, Fred, 384, 387, 388, 396
Blackwell, Richard, 396
Blurring
  spatially variant, 282
  temporally variant, 283
Boundary approach to image segmentation, 303, 311-13
Boundary chain code, 316-17, 324, 328
Boundary peeling, 330
Boundary tracking, 311-12
Boundary tracking bug, 313
Brandt, Roger, 384
Bressler, Stan, 388
Bryant, Nevin, 396

C

Campbell, John, 388
Cannon, T. M., 281
Cartographic projection techniques, 127-29
Castleman, Ken, 396
Cathode-ray tube (CRT), 18, 49
Cathode-ray tube (CRT) film recorder, 42, 50
Center of projection, 128
Central moments, 327
Centromeric index, 340
Cepstrum, 292
Charge-coupled devices (CCDs), 4, 31
Circular symmetry, 184-85
Circularity, 325-27
Classification, 334-44
  classifier design, 322
  classifier performance, 344
  classifier training, 322-23, 337-43
  classifier types, 337
  statistical decision theory, 334-36
Coherent illumination, 253
Coherent point spread function, 257-58
Coherent transfer function, 259
Column digitizing, 28
Coma, 262, 282
Computer graphics, 371, 373
Computerized axial tomography, 360-64
  image reconstruction, 363-64
Concentric circular spot, 306
Constant functions, 164-65
Contour lines, 76, 85
Contrast enhancement, 84
Contrast manipulation, 9
Contrast stretching, 84
Control grid interpolation, 117-19
Control grids, 116-17
Conversion, 8
Convolution, VICAR programs for, 408-9

Convolution filtering, 246-47
  edge enhancement, 157-58
  smoothing, 156
Convolution operation, 145-50
  digital, 149-50
  in one dimension, 146-48
  sampling with a finite spot, 149
  in two dimensions, 148-49
Convolution theorem of Fourier transform, 169-70
Coordinate transformation restoration, 282
Correlation
  and power spectrum, 186-87
  VICAR programs for, 408-9
Corresponding points, 9
Cross-correlation, system identification by, 272-73
Cross-correlation function, 187
Curvature of field, 262, 282

D

Deconvolution, 150, 278-81
  three-dimensional, 354-55
  and Wiener filter, 209-10
Deconvolution filter (Wiener), 280-81
Differential chain code, 328-29
Diffraction-limited optical systems, 254-61
  See also Optics
Digital convolution, 149-50
  See also Convolution
Digital image (*see* Image, digital)
Digital filtering, 246-48
  applications of, 150
  See also Filtering; Filters
Digital image display (*see* Display)
Digital image processing
  defined, 7
  effects of, 240-46
  elements of, 4-5
  functional requirements for, 12-13
  history of at JPL, 383-97
  philosophical considerations, 9-11
  problem solving with, 12
  processing efficiency, 12
  system analysis, 264-73
    See also System analysis
  terminology, 5-9
  three-dimensional, 347-77
    See also Three-dimensional image processing
Digitization, defined, 4
Digitizer, 14-26
  characteristics of, 15-16
  components of, 17-26
    light sensors, 18-22
    light sources, 17-18
    scanning mechanisms, 22-26

Digitizer *(cont.):*
  density, 284
  electronic image tubes, 26-30
  elements of, 15
  other systems, 30-31
  types of, 16-17
Digitizing, defined, 8
Display, digital image, 39-51
  decalibration, 93-94
  defined, 8
  gray scale linearity, 41
  high-frequency response, 45-48
  image size, 40-41
  low-frequency response, 42-45
  noise, 48-49
  permanent, 8, 40, 50-51
  photometric resol. tion, 41
  rectification, 119
  volatile, 8, 40, 49-50
Distortion, 262
Division, 106, 107
Drum feed display, 50
Dual moving mirror-image plane scanner, 30

E

Edge enhancement, 157-58
Edge spread function, 270
Efron, Ed, 388
Electrical filter design, 278, 279
Electron beam scanning, 24-26
Electronic image tubes, 26-30
Electronic log conversion, 36-38
Electronic noise, 293
Elliot, Denis, 394, 395, 396
Emulsion, 32-35
Enhancement
  edge, 157-58
  of features through digital filtering, 150-51
Ergodic random variables, 200-1
Estimation
  Bayesian, 338-43
  maximum-likelihood, 337
Evenness and the Fourier transform, 166-67

F

Feature enhancement through digital filtering, 150-51
Feature extraction, 300
Feature vector, 300
Field curvature, 262, 282
Film grain noise, 284, 293-94
Film scanners, 15, 16

Film scanning, 31-38
  electronic log conversion, 36-38
  photocopying, 35-36
  photographic process, 32-35
  transmittance and density, 31-32
Film-to-film system, analysis of, 265-67
Filter design, 190-225
  electrical, 278, 279
  Gaussian high-frequency enhancement filter, 194-95
  general bandpass filter, 193-94
  high-frequency enhancement filter design, rules for, 195-97
  ideal bandpass filter, 191-92
  ideal bandstop filter, 192-93
  low-frequency response, 197-99
  optimal, 199-224
    matched detector, 210-19
    random variables in, 199-200
    Wiener estimator, 201-10
Filtering, frequency domain, 247-48
  VICAR programs for, 410-11
Filters
  convolution, 246-47
  deconvolution, 278-81
  digital, 246-48
  frequency domain, 247-48
  geometrical mean, 282
  homomorphic, 282
  linear, 246
  linear combination, 286-88
  matched detector, 210-19
  notch, 193
  parametric Wiener, 282
  power spectrum equalization, 281-82
  Wiener deconvolution, 280-81
Flat field response, 42
FORTRAN, 53, 66, 67, 388
Fourier transform, 161-88, 412-17
  computation of, VICAR programs for, 409
  correlation and power spectrum, 186-87
  definition, 162
  existence of, 164-65
  of Gaussian function, 163
  and linear system analysis, 173-80
    See also Linear system analysis
  properties of, 166-73, 187
    addition theorem, 167-68
    convolution theorem, 169-70
    Rayleigh's theorem, 172-73
    shift theorem, 168-69
    similarity theorem, 170-72
    symmetry properties, 166-67
  in two dimensions, 180-86
    circular symmetry, 184-85
    definition, 180-81
    projection property, 184
    rotation property, 183
    separability, 181, 183

Fourier transform *(cont.):*
  similarity, 183
  of useful functions, 165
Frequency, negative, 179-80
Frequency domain filtering, 247-48
  VICAR programs for, 410-11
Frequency sweep target, 271-72
Fresnel approximation, 257
Frieden, Howard, 384, 388, 390
Function tables, 418-19

G

Gamma of emulsion, 33
Gas discharge displays, 49
Gaussian display spot, 42-45
Gaussian function, 151-52
  Fourier transform of, 163, 165
Gaussian high-frequency enhancement filter, 194-95
Geometric decalibration, 119
Geometric mean filters, 282
Geometric operations, 110-35
  applications of, 119-34
    cartographic projection techniques, 127-29
    display rectification, 119
    examples, 131-34
    geometric decalibration, 119
    image format conversion, 123
    image registration, 119, 121
    implementation, 129-31
    map projection, 123, 127
  gray level interpolation, 111, 112-15
  pixel transfer, 111-12
  spatial transformations, 110-11, 115-19
Geometric transformation, VICAR programs for, 406-8
GEOTRAN, 119
Gillespie, Alan, 393, 396
Girard, Mike, 395
Global operation, 9
Global thresholding, 303-4
Goetz, A.F.H., 396
Goodman, J.W., 289
Gradient magnitude, 106
Graininess, 294
Grass-fire technique, 330
Gray level, 5
Gray level histogram (*see* Histogram)
Gray level interpolation, 111, 112-15
  bilinear, 113-15
  higher-order, 115
  nearest neighbor, 112
Gray scale
  linearity of, 41
  manipulation of, 84
  resolution of, 9
Green, Bill, 393, 396

H

Haas, David, 394
Habibi, A., 281
Hankel transform of zero order, 185-86
Hard-copy output, 8
Harmonic signals, 142-45
  response to harmonic input, 142-43
  sinusoids and, 143-44
  transfer function, 144-45
Harris, J.L., 280, 288
Helstrom, C.W., 281
Hermite function, 167
Higgins, G.C., 294
High frequency enhancement filter
  design of, rules for, 195-97
  Gaussian, 194-95
Higher-order interpolation, 115
Histogram, gray level, 68-83
  algebraic operations and, 97-101
    histograms of difference images, 99
    histograms of sum images, 97-99
    IOD of noisy image, 99-101
  defined, 68, 70-71
  and image, 78-83
  and point operations, 86-90
    histogram flattening, 90-91
    histogram matching, 91-92
    output histogram, 87-90
  properties of, 71, 73
  and threshold selection, 305-8
  two-dimensional, 71
  uses of, 73-83
    boundary threshold selection, 76
    digitizing parameters, 73, 74
    integrated optical density, 76, 78
Homomorphic filter, 282
Huang, T.S., 282
Hunt, B.R., 282, 289
Huygens-Fresnel principle, 256-57

I

Illumination, coherent and incoherent, 253
Image, digital, 8
  averaging of, 101-4
  combination of, VICAR programs for, 407
  defined, 7
  generation of, VICAR programs for, 402
  and gray level histogram, 78-83
  measurement of, VICAR programs for, 404-5
  multidigital, 348
  range, 349
  registration of, 119-21
  size of, 40-41
  of spacecraft, 129-31, 134
  spatially three-dimensional, 347-48

Image, digital *(cont.):*
 subtraction of, 104-6
  background subtraction, 104-5
  gradient magnitude, 106
  motion detection, 106, 107, 108
 types of, 6-7
Image digitizer (*see* Digitizer)
Image display (*see* Display)
Image dissector tube, 29-30
Image format conversion, 123
Image restoration, 277-95
 approaches and models, 278-88
  deconvolution, 278-80
  geometrical mean filters, 282
  local stationarity, 284-85
  matrix formulation, 284
  power spectrum equalization, 281-82
  spatially variant blurring, 282
  temporally variant blurring, 283
  Wiener deconvolution, 280-81
 defined, 277
 noise modeling, 293-94
 OTF from degraded image spectrum, 292-93
 superresolution, 288-90
 system identification, 290-92
Image segmentation, 292-93, 299-319
 gradient based methods, 311-13
 optimal threshold selection, 305-10
  analysis of spots, 306-9
  average boundary gradient, 309-10
  histogram techniques, 305-6
  objects of general shape, 310
 process of, 303
 region growing techniques, 313-14
 segmented image structure, 314-19
  boundary chain code, 316-17
  line-segment encoding, 317-19
  object membership map, 314-16
 statistical pattern recognition, 299-303
 by thresholding, 303-5
Imaging systems
 aberrations in, 262-64
 and optics, 251-53
Impulse, 153-54
 Fourier transform of, 165
 properties of, 153
 response of linear system, 153-54
Impulse response, 146
Impulse train, 227
Incandescent lighting, 17
Incoherent illumination, 253
Incoherent transfer function, 259-60
Indirect copy, 62
Integrated optical density (IOD), 76, 78, 97, 323-24
 of noisy image, 99-101
International Business Machines (IBM), 388
Interpolation of a function and sampling, 241-46
Invariant moments, 327-28

Isoplantic optical system, 252
Isoplantic patches, 252

J

JCFILE, 56-57
Jepsen, Paul, 394
Jet Propulsion Laboratory (JPL), history of digital image processing at, 383-97
Jones, Ken, 395

K

Karyotype, 123, 125, 340, 341
King, Tom, 388
Knox, K.T., 283
Kreznar, John, 390, 394

L

Labeyrie, A., 283
Lambert conformal conic projections, 128, 129
Laser
 as light source for digitizer, 17
 display, 49
Latitude of emulsion, 33
Lead screw, 23, 24
Length measurement, 324
Levine, Martin, 396
LIBEXEC, 395
Liebes, Sidney, 395
Light sensors, 18-22
Light sources for digitizers, 17-18
Light-emitting diodes (LEDs), 18
Line pairs, 46
Line segment encoding, 317-19
Line spread function, 269-70
Linear combination restoration, 286-88
Linear filtering, 246
Linear point operations, 86
Linear system analysis and the Fourier transform
 linear system identification, 173-75
 linear system terminology, 173
 negative frequency, 179-80
 sinusoidal decomposition, 175-79
Linear system theory, 139-59
 convolution filtering, 156-58
 convolution operation, 145-50
 digital filtering, applications of, 150
 harmonic signals, 141-45
 linearity, 140-41

Linear system theory *(cont.)*:
  shift invariance, 141
  useful functions, 150-51
Linearity, 140-41
LINK General Precision, 384
Local operations, 9
  VICAR programs for, 403-4
Log conversion, electronic, 36-38
Lorre, Jean, 394, 395, 396
Low-frequency response, 197-99
Lynn, Don, 393, 396

M

McGlammery, B.L., 280
Magnification, 9
  VICAR programs for, 406-7
Map projection, 123, 127
  VICAR programs for, 407-8
Maps, properties of, 27
MAP2 program, 127, 128, 131
Mariner Mars Mission, 387, 389-93
Mariner Venus Mercury Mission, 393-94
Martin, Benn, 396
Matched detector, 210-19
  examples of, 215-17
  optimality criterion for, 211-12
  Schwartz's inequality, 212-14
  transfer function, 214-15
  and Wiener estimator, compared, 217-19
Matrix formulation, 284
Maximum-liklihood estimation, 337
Mean-square error, 202-5, 281
  minimizing of, 203-5
Measurement, 321-34
  of classifier performance, 323
  feature selection, 321-22, 332-34
  of images, VICAR programs for, 404-5
  of psf, 291-92
  of shape, 324-32
  of size, 323-24
Mechanical scanning devices, 23-24
Medial axis transformation, 329-32
Mercator projection, 128, 129
Microprocessors, 4
Microscope digitizing system, analysis of, 267-68
Minimum enclosing rectangle, 324
Monotone spot, 306
Morecroft, John, 384, 387
Mosher, Joel, 394
Motion detection, 106, 107, 108
Moving mirror scanner, 22-23
Mueller, P.F., 280
Multidigital images, 348
Multiplication operation, 106, 107
Multispectral analysis, 348, 350-51

N

Naderi, F., 294
Nathan, Robert, 179, 180, 384, 386-88, 395, 396
Nearest neighbor interpolation, 113
Negative frequency, 179-80
Noise, 16, 48-49, 99-101, 186
  averaging for reduction of, 101-4
  defined, 9
  modeling of electronic, 293
    photoelectronic, 293
    film grain, 293-94
  nonstationary, 283-84
  random, 199
  root-mean-square level of, 41
  white, 215
Noise power ratio, 286-88
Noise removal, 150
Nonparametric classifiers, 337
Notch filter, 193
Numerical aperture, 268

O

Object classification, 300
Object isolation, 300
Object membership map, 314-16
Oddness and the Fourier transform, 166-67
Optical character recognition, 383
Optical density (OD), 31, 32
Optical pattern recognition, 299-301
Optical sectioning, 351-60
  deblurring optical section images, 354-57, 358
  defocus OTF, 357-60
  thick specimen imaging, 351-54
Optical system, defocused, optical transfer
    function of, 357-60
Optical transfer function, 252, 260-61, 262
  of defocused optical system, 357-60
  from degraded image spectrum, 292-93
Optics, 278
  aberrations in an imaging system, 262-64
  diffraction-limited systems, 254-61
    apertures and pupil function, 254
    coherent point spread function, 257-58
    Fresnel approximation, 257
    Huygens-Fresnel principle, 256-57
    incoherent point spread function, 259-60
    incoherent transfer function, 259
    lens shape, 254
    optical transfer function, 260-61
  and imaging systems, 251-53
    coherent and incoherent illumination, 252-53
    image quality factors, 253
    linearity of optical systems, 251-52

Order of moments, 327
Orthographic projection, 128, 129
Output histogram, 87-90

P

Paluzzi, Peter, 396
Parallel processing, 4
Parametric classifiers, 335
Parametric Wiener filter, 282
Pattern recognition system design, 302-3
Perimeter measurement, 324
Periodic functions, 164-65
Permanent displays, 8, 40, 50-51
Phosphor, electroluminescent, 17-18
Photochromic displays, 49
Photoconductive devices, 19, 20
Photocopying, 35-36
Photodiodes, 20-22
Photoelectronic noise, 293
Photoemissive devices, 19
Photographic process, 32-35
Photometric decalibration, 92-93
Photometric resolution, 41
Photomultiplier tubes, 19-20
Phototransistors, 20-22
Phototubes, 19
Photovoltaic cells, 19, 20
Picture, defined, 7
Picture elements, 4, 8
Pixel, 4, 8
Pixel carryover approach, 111
Pixel filling approach, 111, 112, 117
Pixel transfer, 111-112
Plumbicon, 29
Point operations, 9, 84-95
    applications of, 90-94
        display decalibration, 93-94
        histogram flattening, 90-91
        histogram matching, 91-92
        photometric decalibration, 92-93
    and the histogram, 86-90
    linear, 86
    uses for, 85-86
    VICAR programs for, 402-3
Point spread function, 252
Position noise, 48-49
Power spectrum equalization, 281-82
Power spectrum, 186
    parameters, 285-86
Pratt, W.K., 281
Principal axes, 327-28
Projection, 251
    back, 363-64
Projection property, 184
Psf measurement, 291-92

Pupil function, 254-55

Q

Quantum mottle, 364

R

Rabedeau, M.E., 292
Radiography, 360, 361
Radiology, 364
Random functions, 165
Random noise, 199
Random variables, 199-201
    ergodic, 200-1
Range image, 349
Ranger program, 384-87
Raster, 8
Rayleigh criterion, 264
Rayleigh's theorem and the Fourier transform, 172-73
Reciprocity failure, 33
Reciprocity law, 33
Rectangular pulse, 150-51
    Fourier transform of, 165
Rectangular pulse detector, 216-17
Rectangularity, 325
Region approach to image segmentation, 303
    See also Thresholding
Region growing, 313-14
Resolution cells, 264
RETMA scanning convention, 27-28
Reynolds, G.O., 280
Rindfleisch, Tom, 386, 390
Robbins, G.M., 282
Root-mean-square (RMS) noise level, 41
Rotary motion, 282
Rotating drum, 23-24
Rotation, VICAR programs for, 406-7
Rotation property, 183
Ruiz, Reuben, 390, 393

S

Sampling, 227-34
    computing spectra, 234-38
    defined, 8
    digital filtring, 246-48
    with a finite spot, 149
    and interpolating a function, 241-46
    with Shah function, 227-30
    and truncation, 238-40
    undersampling and aliasing, 230-34

Sampling density, 9
Sawchuk, A.A., 282
Scan-in system, 16, 17, 18, 22
Scanning, defined, 8
Scanning mechanisms, 22-26
Scan-out system, 16, 17
Schwartz, Arnold, 393, 394, 395
Schwartz's inequality, 212-14
Seidman, Joel, 390, 395
Self-scanning arrays, 31
Selzer, Robert, 384, 395
Separability, 181, 183
Shaded surface display, 350, 371-77
    imaging geometry, 375-76
    smooth shading, 376-77
    surface description, 374
    surface reflection phenomena, 375
Shah function, 227-30
    behavior under similarity, 227-28
    sampling with, 228-30
Shift invariance, 141
Shift theorem of the Fourier transform, 168-69
Signals
    computing spectra of, 234-38
    nonstationary, 283-84
Silicon diode vidicon, 29
Similarity theorem
    of the Fourier transform, 170-72, 183
Sine wave target, 271
Sinusoidal decomposition, 175-79
Sinusoids and harmonic signals, 143-44
Slepian, D., 281
Smoothing of noisy function, 156
Software, image processing, 52-67
    organization, 53-57
        application programs, 53-54
        control programs, 56-57
        execution parameters, 56
        image format and labels, 55-56
        jobs and tasks, 54-55
    processing sequence, 57-67
        application programs, 66-67
        control statements, 59-66
Soha, James, 393-94, 396
Solar cells, 20
Sondhi, M.M., 281
Spacecraft images, 129-31, 134
Spatial transformation, 110-11, 115-19
    control grid interpolation, 117-19
    control grids, 116-17
Spatially three-dimensional images, 347, 348
Spatially variant blurring, 282
Speckle interferometry, 283
Spectrum, computation, 234-38
Spectrum analyzer, 179
Spherical aberration, 154
Spot analysis and threshold selection, 306-9
Spot position noise, 48-49
Spot profile function, 306

Spuck, Bill, 396
Stars, viewing of, 283, 291
Stationarity, local, 284-85
Statistical decision theory, 334-36
Statistical pattern recognition, 299-303
    optical, 299-301
    system design, 301-3
Step function, 155
Stereographic projection, 128, 129
Stereometric ranging, 364-68
Stereometry, 348, 364
    with angle scanning cameras, 367-68
    VICAR programs for, 411
Stereoscopic image display, 349-50, 368-71
    VICAR programs for, 411
Stereoscopic image pair, 349
Stereoscopic imaging, 365-66
Stretching, 9
Stromberg, William, 396
Stultz, K.F., 294
Superresolution, 288-90
Supervised training, 337-43
Surveyor program, 387-89
System analysis, 264-73
    film-to-film system, 265-67
    microscope digitizing system, 267-68
    system identification, 268-73
        cross-correlation, 272-73
        edge spread function, 270
        frequency sweep targets, 271-72
        line spread function, 269-70
        sine wave targets, 271

T

Temporally variant blurring, 283
Three-dimensional image processing, 347-77
    computerized axial tomography, 360-64
    multispectral analysis, 348, 350-51
    optical sectioning, 351-60
    shaded surface display, 350, 371-77
    spatially three-dimensional images, 347, 348
    stereometric ranging, 364-68
    stereometry, 348-49
    stereoscopic display, 349-50, 368-71
Thresholding, 76, 85-86
    adaptive, 304-5
    global, 303-4
    gradient image, 312-13
    optimal threshold selection, 305-10
        See also Image segmentation, optimal
            threshold selection
Time series analysis, 279
Tomography, 348, 361
    computerized axial, 360-64
Training, 337-43
Transducers, 8

Transduction, 8
Transfer function of a linear system, 144-45
Transient functions, 164
Transmittance, 31-32
Triangular pulse, 151
  Fourier transform of, 165
Truncation, 238-40, 243
Tucker, Robert, 395

U

Unsupervised training, 337

V

VICAR (Video Information Communication
    and Retrieval) system, 52-67, 116, 118,
    338
  program index, 401-11
  See also Software, image processing
Video film converter, 384, 388
Vidicon camera tube, 26-27
Viewpoint, 373
Viking Lander camera, 119, 120
Viking Mars Lander spacecraft, 367
Viking mission, 394-95
VMAST, 56, 57
VMJC, 56, 57
Volatile displays, 8, 40, 49-50
Voyager, 395
VTRAN, 56

W

Wall, R. J., 330
Waveform analyzer, 179
White noise, 215
Width measurement, 324
Wiener, Norbert, 279
Wiener deconvolution filter, 280-81
Wiener estimator, 201-10
  with deconvolution, 209-10
  design of, 205-6
  and matched detector, compared, 217-19
  mean-square error of, 202-5
  optimality criterion for, 201-2
  uncorrelated signal and noise, 206-9
Williams, Don, 396

X

X-ray, 360-362, 364, 395

Y

Yakamovski, Yoram, 396

Z

Zobrist, Albert, 396